Turkey's Water Policy

Aysegul Kibaroglu • Annika Kramer •
Waltina Scheumann
Editors

Turkey's Water Policy

National Frameworks and International Cooperation

Springer

Editors
Aysegul Kibaroglu
International Relations Department
Okan University
Istanbul
Turkey
Aysegul.Kibaroglu@okan.edu.tr

Annika Kramer
Adelphi Research gGmbH
Caspar-Theyß-Str. 14 A
14193 Berlin
Germany
kramer@adelphi.de

Dr. Waltina Scheumann
Deutsches Institut fuer
Entwicklungspolitik
Tulpenfeld 6
53113 Bonn
Germany
waltina.scheumann@die-gdi.de

ISBN 978-3-642-43671-0 ISBN 978-3-642-19636-2 (eBook)
DOI 10.1007/978-3-642-19636-2
Springer Heidelberg Dordrecht London New York

Printed on acid-free paper

Springer is part of Springer Science+Business Media (www.springer.com)

Foreword

This book aims to contribute to understanding the broader picture of Turkish water policy. It is unique in the sense that it comprises a wide range of issues of water policy at national level (Part I) as well as an analysis of Turkey's international water policy beyond the prominent case of the Euphrates and Tigris river basins (Part II).

In the past, Turkey was frequently perceived as a strong upstream riparian that pursues huge water development projects without adequately taking the interests of water-scarce downstream riparians into account. Furthermore, Turkey's attitude towards the development and strengthening of international water law was assessed as being reserved at best, because of the country's reluctance to sign up to international water law. In order to get a full picture of Turkey's position in this regards, one has, however, to consider the national policies and framework conditions that impact water resources management.

Part I of this book, The National Framework, provides detailed analyses of water governance in Turkey, such as national policies and institutional frameworks in the water sector but also in other relevant sectors such as energy, agriculture, and the environment. It further discusses the transformation of Turkish water policy due to pressures and impacts generated at the domestic, regional and international levels. This includes developments ensuing from the European Union Water Framework Directive (EU WFD), from liberalization of national markets, and increasing importance of environmental issues for the Turkish public.

Part II of this book scrutinizes Turkey's international water policy and provides analyses of water management and cooperation in all transboundary river basins that Turkey is riparian to, i.e. the Meric, Coruh, Kura-Aras, Orontes, Euphrates and Tigris basins.

Several chapters of Part II are based on a study by adelphi commissioned by the German Federal Ministry for the Environment, Nature Conservation and Nuclear Safety (BMU) in 2004. Germany has supported various international initiatives and programmes to facilitate and improve transboundary water management at international rivers and lakes, e.g. through the so-called Petersberg Process. Furthermore, Germany is closely cooperating with Turkey within the framework of EU-funded twinning projects. adelphi is continuously assisting German government agencies and international organizations with the analysis of water policies and facilitation of policy dialogues. Against this background, the BMU had commissioned the study in order to improve the body of information on the current state of water cooperation on Turkish transboundary rivers.

The study provided a comprehensive assessment of current management of Turkish transboundary waters and identified existing potentials for cooperation. For this book, the assessment of Turkish transboundary waters has been updated and amended with further analysis of Turkish international water relations. Moreover this perspective has now been enriched with the analysis of national aspects determining water policy by distinguished authors.

This book provides a compendium on Turkish national and international water policy that has up to now been missing on the international book market.

Alexander Carius
Director, adelphi

Acknowledgments

Every book is a curious journey and an adventure. This one was definitely no exception. During the weeks-long editing process in Bonn and Ankara, there were times when we, the editors, were contented with the outcome of our efforts by seeing how they contributed to the individual chapters to reflect what we really had in mind. There were also times when we spent laborious hours and days with a view to clarifying some of the complex issues in the manuscripts for the ease of understanding by our readers. All in all, our lengthy discussions over wide-ranging issues concerning water policy in Turkey were among the most instructive as well as productive times of our academic life. Trying to understand the complexities and peculiarities in the water policy discourse and the practice in Turkey, was a profound and lively learning process for all of us where we could draw from the many years of experience and knowledge we had gained through field studies, consultations and scholarships in Turkey and elsewhere. We realized that these were the moments when we dedicated ourselves truly to the editing process both as insiders and also as outsiders benefiting from our friendship, yet reflecting our objectivity to the work at hand as well.

In realizing this edited volume, we were supported both financially and logistically by our own respective institutions, namely the Middle East Technical University (METU), Ankara, the German Development Institute (DIE, Deutsches Institut fuer Entwicklungspolitik), Bonn, and adelphi, Berlin. Hence, we would like to acknowledge their support as well as the research funding provided by the German Federal Ministry for the Environment, Nature Conservation and Nuclear Safety, which enabled us to draft Part II of this book concerning the analyses particularly of cooperation on transboundary rivers. We would also like to thank those experts and scholars in Turkish water policy circles for sharing their invaluable insights with us throughout the entire working process. Many thanks to Janet Sterritt-Brunner for the language editing as well as to Alina Schellig for preparing maps and up-dating others for Part II of this book. And, last but not least we would like to thank our families for their continuous support and patience all through the research and writing process.

Aysegul Kibaroglu, Waltina Scheumann and Annika Kramer

Contents

Abbreviations

ASAT	Antalya Water and Sewage Administration, Turkey
ASO	Antalya Water and Public Transport Services, Turkey
ALDAS	Antalya Infrastructure Management and Consulting Services Company, Turkey
ANTSU	Antalya Water Management Company, Turkey
APSA	Authority for the Protection of Special Areas, Turkey
BCM	Billion cubic meters
BOD	Biochemical Oxygen Demand
BOO	Build-Own-Operate
BOT	Build-Operate-Transfer
BSEP	Black Sea Environmental Programme
CAMP	Coastal Area Management Programme
CBD	Convention on Biological Diversity
CEE	Chamber of Electrical Engineers, Turkey
COD	Chemical Oxygen Demand
DEFRA	Department of Environment, Food and Rural Affairs, UK
DSD	Dangerous Substances Directive, EU
DSI	Devlet Su Isleri (General Directorate of State Hydraulic Works), Turkey
E_a	Application efficiency
E_c	Conveyance efficiency
EC	European Commission
ECA	Export Credit Agency
EECCA	Eastern Europe and the Caucasus and Central Asia
EFT	Environment Foundation of Turkey
EIA	Environmental Impact Assessment
EIE	Elektrik Isleri Etud Idaresi (Electrical Power Resources Survey and Development Administration), Turkey
EMRA	Energy Market Regulatory Authority, Turkey
ETIC	Euphrates Tigris Initiative for Cooperation
EU	European Union

EUAS	Electricity Production Corporation, Turkey
FAO	Food and Agriculture Organization, UN
GAP	Guneydogu Anadolu Projesi (Southeastern Anatolia Project), Turkey
GAP RDA	GAP Regional Development Administration, Turkey
GATT	General Agreement on Tariffs and Trade
GDAR	General Directorate of Agrarian Reform, Turkey
GDP	Gross Domestic Product
GDRS	General Directorate of Rural Services, Turkey
GEF	Global Environmental Facility
GIS	Geographical Information System
GNP	Gross National Product
GOLD	General Organisation for Land Development, Ministry of Irrigation, Syria
GRID-Europe	One of UNEP's major centres for data and information management
GWh	Gigawatt per hour
HEPP	Hydroelectric power plant
IBA	Important Bird Areas
I&D	Irrigation and drainage
ILA	International Law Association
ILC	International Law Commission
INTERREG	EU-funded programme that helps Europe's regions form partnerships to work together on common projects
INWEB	International Network of Water Environment Centres for the Balkans
IRBM	Integrated River Basin Management
ISKI	Istanbul Water and Sewage Authority, Turkey
IUCN	International Union for Conservation of Nature and Natural Resources/World Conservation Union
JTC	Joint Technical Committee
KfW	Kreditanstalt fuer Wiederaufbau, Germany
kW	Kilowatt
kWh	Kilowatt hour
KOP	Konya Basin Irrigation Project
KOYDES	Village Infrastructure Support Project
LWC	Local Wetland Commission
MAP	Mediterranean Action Plan
MARA	Ministry of Agriculture and Rural Affairs, Turkey
MATRA	Matra Programme for European Co-operation, Netherlands
MCM	Million cubic meters
MEDASSET	Mediterranean Association to Save the Sea Turtles
MED-HYCOS	Mediterranean Hydrological Cycle Observing System
MEDA	Mediterranean Economic Development Area

MedWet	Mediterranean Wetlands Initiative
MoEF	Ministry of Environment and Forestry, Turkey
MoH	Ministry of Health, Turkey
MoU	Memorandum of Understanding
MW	Megawatt
NATO	North Atlantic Treaty Organization
NEAP	National Environment Action Plan of Turkey
NGO	Non-governmental organization
NIMH	National Institute of Meteorology and Hydrology, Bulgaria
O&M	Operation and maintenance
O&M&R	Operation and maintenance and repair
OECD	Organization for Economic Cooperation and Development
OFWAT	The Water Services Regulation Authority, UK
OSCE	Organization for Security and Co-operation in Europe
PHARE	Programme of the EU to assist the applicant countries of Central and Eastern Europe
PHARE-CBC	PHARE Cross-border Cooperation
PIU	Project Implementation Unit
PKK	Partiya Karkeren Kurdistan (Kurdistan Worker's Party)
PPP	Purchasing power parity
RAP	Resettlement Action Plan
REC	Regional Environmental Center
SEA	Strategic Environmental Assessment
SIT	A protection category according to Turkish law
SPO	State Planning Organization (DPT Devlet Planlama Teskilati), Turkey
SPA	Special Provincial Administrations, Turkey
SSD	Suspended Solids
TACIS	The European Union's Technical Assistance to the Commonwealth of Independent States to enhance the transition process in Eastern Europe and Central Asia
TEAS	Turkish Electricity Generation-Transmission Cooperation
TEDAS	Turkish Electricity Distribution Company
TEIAS	Turkish Electricity Transmission Corporation
TEK	Turkish Electricity Authority
TEMA	Turkish Foundation for Combating Soil Erosion, for Reforestation and Protection of Natural Habitats
TETAS	Turkish Electricity Trade and Contract Company
TL	Turkish Lira
TOE	Tons of oil equivalent
TOOR	Transfer of Operating Rights
TOPRAKSU	General Directorate for Soil and Water, Turkey
TMMOB	Union of Chambers of Turkish Engineers and Architects
TSKB	Industrial Development Bank of Turkey

TUIK	Turkish Statistics Institute
TURCEK	Environmental and Woodlands Protection Society, Turkey
UN	United Nations
UNDP	United Nations Development Programme
UNECE	United Nations Economic Commission for Europe
UNEP	United Nations Environment Programme
UNESCO	United Nations Educational, Scientific and Cultural Organisation
USAID	United States Agency for International Development
USSR	Union of Soviet Socialist Republics
UWWTD	Urban Wastewater Treatment Directive, EU
WFD	Water Framework Directive, EU
WWF	World Wide Fund for Nature
WHYCOS	World Hydrological Cycle Observing System
YTL	New Turkish Lira (Yeni Turk Lirasi)

About the Editors

Aysegul Kibaroglu is professor and faculty member in the International Relations Department at Okan University in Istanbul Turkey. Previously she was a faculty member and the Vice Chair in the Department of International Relations at the Middle East Technical University in Ankara. Dr. Kibaroglu spent a post-doctoral fellowship in the International Water Law Research Institute at the University of Dundee, Scotland. Her areas of research include: transboundary water politics; international law; political geography; environmental security and Turkish water policy. Prof. Dr. Kibaroglu has published extensively on the politics of water resources with an emphasis on the Euphrates Tigris river basin including a book volume entitled *Building a Regime for the Waters of the Euphrates-Tigris River Basin* by the Kluwer Law International (2002). She has also worked as Advisor to the President of the Southeastern Anatolia Project Regional Development Administration from 2001 to 2003.

Waltina Scheumann holds her Master in political science and a Ph.D. in engineering. She has been a faculty member at the Chair in Environmental and Land Economics, Technical University Berlin. Dr Scheumann later worked as a senior researcher at the Helmholtz Centre for Environmental Research (UFZ), Leipzig, and is presently working at the Deutsches Institut für Entwicklungspolitik, DIE, Bonn. Dr. Scheumann's work on water-related topics includes cooperation on transboundary waters, water governance issues in irrigated agriculture including drainage and combating salinization as well as the implementation of international standards for sustainable dam development in emerging economies (Brazil, China, India, Turkey) and in developing countries.

Annika Kramer holds a degree in environmental engineering from the Technical University Berlin with a specialisation in water management and international environmental politics. Her work on topics related to water management over the last ten years includes research on cooperation potentials as well as legal and institutional frameworks for transboundary basin management, mainly in the Middle East and southern Africa. Annika is currently working as a Senior Project Manager with adelphi and preparing her Ph.D. on global evolution and diffusion of IWRM norms (University of Osnabrück).

About the Contributors

Argun Baskan holds two Master degrees (Dokuz Eylul University, Turkey and University of Kent, UK) in International Relations and European Union studies, as well as a Ph.D. in the field of International Relations from Middle East Technical University (METU), Turkey. Currently he is working as a research assistant at the Ege University in Izmir, Turkey. Previously he has held the position of a research assistant in the Department of International Relations at METU (2004-2011) and at the Ege University (2000-2004). His research interests cover international security issues and the international political economy of energy and natural resources.

Vera Baumann has held the position of Desk Officer at the Press and Public Relations Division of the German Federal Ministry for Economic Cooperation and Development since June 2009. She holds a Master's degree in Cultural Sciences from the Leuphana University of Lüneburg and specialized in development geography. As a young professional at the German Development Institute, she worked on resettlement regulation and planning in dam projects in Turkey.

Filiz Demirayak, acted as the CEO of World Wildlife Fund for Nature-Turkey (2004-2009). She has nearly 20 years' experience in areas ranging from sustainable development to environmental management, from water issues to climate change. She was an early leader in the fight to protect the habitat of Turkey's endangered sea turtles while balancing the economic concerns of coastal residents; the project received the United Nations Habitat International Award for Best Practices. She served on the board and as a Trustee of WWF-Turkey. She holds her Ph.D. in political science. She continues to work as an environmental management expert, today working as a consultant for the World Bank and Ministry of Environment in preparing the National Watershed Management Strategy.

Buket Bahar Divrak graduated from the City and Regional Planning Department of Middle East Technical University and got her master's degree in Urban Policy Planning and Local Governments from the same university in 2005. She has been working for WWF-Turkey since 2003 on a wide range of projects, gaining broad experience in natural resources management, particularly for water resources management. Her main areas of expertise cover integrated river basin management, water policies, wetlands management, climate change adaptation. For the last two years she has been working as a Deputy Conservation Director in WWF-Turkey.

Axel Klaphake, an economist and political scientist, has a Ph.D. in environmental planning and management. He currently works as a GIZ (Gesellschaft für Internationale Zusammenarbeit) project director for transboundary water resources cooperation in the Nile basin. He joined GIZ in 2006, working as a Senior Water Policy Advisor to the German Federal Ministry for Economic Cooperation and Development (BMZ). Prior to joining GIZ, he worked as a researcher and consultant on international water policy, environmental policy, biodiversity and climate change.

Anna Lena Mueller works as a Project Manager at the KfW Development Bank where she focuses on governance issues in Sub-Saharan Africa. Prior to this, she researched on resettlement regulation and planning in Turkish dam projects within a young professional programme at the German Development Institute. She holds a master's degree in international economics from the Johann Wolfgang Goethe-University in Frankfurt as well as the University Dauphine in Paris.

Cagri B. Muluk holds bachelor's and master's degrees in Biology from the Middle East Technical University. He has notable experience in EU Environmental acquis and accession progress in Turkey and related Turkish institutions. Mr. Muluk has been involved in projects related to the implementation of the EU Water Framework Directive and Groundwater Directive projects in Turkey. Furthermore, Mr. Muluk has participated in various training schemes and courses related to the WFD. During this time he gained knowledge and experience of water management situation in Turkey.

Dennis Mutschler holds his MA in political science and social anthropology. As a young professional at the German Development Institute in Bonn, Dennis Mutschler researched on environmental policies, planning procedures and instruments in Turkey. Since 2009 Dennis Mutschler has been a Project Manager at the Deutsche Gesellschaft für Internationale Zusammenarbeit (GIZ) GmbH within the corporate unit AgenZ (Agency for market-oriented concepts). His focus areas are global governance, political communication and strategic marketing.

Gokhan Orhan is an Associate Professor at the Department of Public Administration, Balikesir University, Turkey. He received his MA in Political Science and Public Administration (Bilkent University) and his Ph.D. from the University of Essex. His academic studies focus on public policy analysis and environmental policy and politics in Turkey, and cover water-related topics.

Ilhan Sagsen holds a BS in International Relations, from Gazi University, Ankara Turkey, since 2001. He obtained his master's degree from the Department of International Relations at the Middle East Technical University in 2006. He is a Ph.D. Candidate and a research assistant at the Middle East Technical University. His main fields of study are water issues; transboundary rivers; environmental problems; climate change and its security dimension; as well as Turkey's accession to the European Union.

Alina Schellig studied geography, social and political science and environmental management in Berlin, Germany and Irvine, USA. She holds a diploma in Geography from the Humboldt University, Berlin. She has gained practical experience in the field of applied environmental research and geo-data management while working for various public and private institutions, including the Humboldt University Berlin, Adelphi Research and the Berlin Senate Department of Urban Development.

Zekai Sen obtained a B.Sc. and M.Sc. from the Technical University of Istanbul, Civil Engineering Faculty, Department of Reinforced Concrete in 1972. His further post-graduate studies were carried out at the University of London, Imperial College of Science and Technology. He was granted Diploma from the Imperial College in 1972, M.Sc. in Engineering Hydrology in 1973 and Ph.D. in stochastic hydrology in 1974. His main interests are hydrology, water resources, hydrogeology, hydrometeorology, hydraulics, science philosophy and history. He has published numerous scientific papers in almost 50 different international journals. He is currently working at the Technical University of Istanbul, Civil Engineering Faculty, and president of the Turkish Water Foundation.

Sylvia Steiner is currently employed at the German Ministry of Economic Cooperation and Development, Dept. Middle East. Before she was assistant to the Director at the Centre for International Migration and Development, a consortium of the Deutsche Gesellschaft für Internationale Zusammenarbeit (GIZ) GmbH and the German Federal Employment Agency. She holds two master degrees in Development Management and Administration (Ruhr University Bochum, Germany and University of the Western Cape, South Africa). As a young professional at the German Development Institute in Bonn, Sylvia Steiner researched on environmental policies, planning procedures and instruments in Turkey.

Vakur Sumer is a Ph.D. Candidate and a researcher at the Middle East Technical University, Ankara, and the Selcuk University, Konya, Turkey. He was a visiting scholar at the University of California, Davis, Deparment of Environmental Science and Policy from 2008-2009. He is currently studying water issues, and Turkey's accession to the European Union in particular.

Ece Tereci holds a B.Sc. in International Relations, with a minor in Public Administration from Middle East Technical University, Turkey. She was a Fulbright Scholar at New York University, USA, between 2005-2007, receiving her

MA in Near Eastern Studies. She worked as an EU Expert in the Turkish Ministry of Energy and Natural Resources, and is presently working for BOTAS Petroleum Pipeline Corporation. Since 2008 she has also worked as a professional reader for Newsweek Turkey.

Sahnaz Tigrek was born in Mardin, a town in Southeastern Anatolia, in 1967. After receiving B.Sc. and M.Sc. degrees from the Middle East Technical University (METU) in 1988 and in 1990, respectively, she went on to the University of Tokyo, Japan to study for her Ph.D. Soon after receiving her doctoral degree in 1996, she returned to METU. Since then, she has been working as a faculty member in the Hydromechanics Laboratory of Civil Engineering Department at METU.

Sevilay Topcu obtained her master in irrigation engineering from the Cukurova University in Adana, Turkey and her Ph.D. in agronomy and crop science from the University of Hohenheim in Germany. She has worked as a faculty member in the Irrigation and Drainage Department of the Agricultural Faculty from 1986 until present. She has worked as a principal investigator as well as senior researcher in numerous national and international projects including topics such as the efficient use of water in irrigated agriculture, environmental impact of irrigated agriculture and modelling agricultural systems and climate variability and change.

Ilter Turan is a professor of Political Science in the Department of International Relations of Istanbul Bilgi University where he has also served as the university president. He held professorships at Koc University and Istanbul University, where he was chair of the International Relations Department. Dr. Turan has been the president of the Turkish Political Science Association and an Executive Committee member, vice president and programme chair of the International Political Science Association. He has published in English and Turkish on Turkish political institutions including the legislature, political parties and political culture; on Turkish foreign policy and on the domestic and international politics of water. He holds a BA from Oberlin College, an MA from Columbia University in Political Science and a Ph.D. in Economics from Istanbul University.

Thomas Walenta holds two Masters degrees in international business administration (awarded by the Viadrina European University, Germany and the Catholic University of Cordoba, Argentina). He currently works for the German promotional bank Kreditanstalt fuer Wiederaufbau (KfW) in the field of financial cooperation with developing and transition countries. As a young professional at the DIE in Bonn, Thomas Walenta was part of the working group Turkey, that focused on the implementation of international standards for the sustainable development of dams. There, he conducted research on the planning process and decision-making setting for hydropower/infrastructure development in the context of Turkey's diversified investment environment.

Paul A. Williams obtained his undergraduate and Masters degrees in Political Science from the University of California, Berkeley and his doctoral degree in the same field from the University of California, Los Angeles. He is Assistant Professor in the Department of International Relations, Bilkent University, Ankara, Turkey and recently completed a visiting scholar position at the Woodrow Wilson International Center for Scholars in Washington DC. In addition to his work on the politics of transboundary water issues, he has researched and published on North-South energy trade relations and the role of Turkey as an energy corridor. His book (co-authored with Ali Tekin) Geo-politics of the Euro-Asian Energy Nexus: The EU, Russia and Turkey was published in November 2010 from Palgrave Macmillan.

Trajectories of Turkey's Water Policy

From state-led water resources development and management to neo-liberal and decentralized approaches

Waltina Scheumann, Aysegul Kibaroglu, and Annika Kramer[1]

Turkey's water policy has undergone continuous reforms since the middle of the past century with significant changes occurring especially throughout the past three decades. The chapters gathered in Part I of this volume cover policy developments in the various fields relevant to Turkey's national water policy. They indicate that significant changes in the water sector have followed the country's overall development paths.

1 State-led water resources development and management

The State was considered the active and dominant agent of Turkey's (economic) development until the early 1980s. Turkey had long followed a Keynesian development model which assigns the State an active interventionist role in the nation's economic life including the provision of public services (Kibaroglu et al. 2009). After World War II, state-led water resources development was fostered by the establishment of a central water bureaucracy, i.e. State Hydraulic Works (DSI), assuming a role similar to that of the US Bureau of Reclamation. One of the main

[1]We appreciate Dr. Asiye Ozturk's comments on a preliminary version of the introductory chapter.

W. Scheumann
Deutsches Institut fuer Entwicklungspolitik, Bonn, Germany
e-mail: waltina.scheumann@die-gdi.de

A. Kibaroglu
International Relations Department, Okan University, Istanbul, Turkey
e-mail: Aysegul.Kibaroglu@okan.edu.tr

A. Kramer
adelphi, Berlin, Germany
e-mail: kramer@adelphi.de

objectives of DSI was to stimulate economic development through water infrastructure financing. It was believed that only the State was able to overcome underinvestment and realize economies of scale. Since its foundation in 1954, DSI has been the major state agency entrusted with water resources development (Kibaroglu and Baskan in this volume). While DSI was not the only agency with this mandate, it has long enjoyed discretion over a significant share of the public budget.

After the military coup in 1960 and the establishment of a civil government, import substituting industrialization became the official development strategy. Accordingly, a central planning body, i.e. the State Planning Organisation (SPO), was established in 1961 to steer economic development by creating a strong industrial base. This also meant that public investments in the water sector were arranged through the national five-year development plans: major water infrastructure such as irrigation systems, storage facilities and multi- and single purpose dams (for e.g. hydroelectricity generation) was state-financed and state-managed. In recognition of the economic and political relevance of the agricultural sector for the country, particular attention was also paid to rural development: along with DSI the early General Directorate of Rural Services (i.e. TOPRAKSU) was mandated to foster development of water and land resources in rural Turkey.

The protectionist development model of import substitution was followed until the end of the 1970s when the Turkish economy ran into a serious crisis with high inflation, growing trade deficit and high unemployment rates (Keyman 2005; Kibaroglu et al. 2009).

2 Early signals of neo-liberal and decentralized approaches

Following the military coup in 1980, the government embarked on a major stabilization and economic liberalization programme, i.e. the beginning of export-oriented industrialization. The architect of this economic transformation was Turgut Ozal, the then Deputy Prime Minister. Ozal was reappointed as Deputy Prime Minister with responsibility for the economy and the implementation of the International Monetary Fund's structural adjustment programme, and became Prime Minister in 1983 and 1989. From 1981 to 1988, the government under Ozal's leadership facilitated the shift to a free market economy, albeit with various negative impacts on social welfare such as the deepening of the sharp disparities between the lowest income groups and the rich segments of the society.

The Constitution of 1982 established the basic principle that water is a public good under the State's trusteeship. The authority to explore and manage water resources is vested in the State, and water resources in the domain of private law and private proprietorship are subject to title deed registration. In this way, the State has maintained control over water resources development while at the same time initiating liberalization in the early 1980s as part of its economic transformation programme:

- With a gradual liberalisation and deregulation of the national energy and electricity sector, the classic investment model where planning, financing, construction and operation is carried out by the public sector only, became obsolete and partly replaced by private investment models. In 1984, models were introduced that promoted private sector investment in the energy and electricity sector. This development was later on followed by the reorganization of the electricity sector with the Electricity Market Law (2001) and the Law on the Utilization of Renewable Energy Resources for the Purpose of Generating Electrical Energy (2005). Recent regulation favours licensing as a means to encourage private companies to develop a river's water resources for generating hydroelectricity based on a water use right agreement (Baskan; Scheumann et al., both in this volume).
- Devolution of large-scale irrigation system management has always been a legal option according to DSI's establishment law (1953), and small-scale irrigation systems using surface- or groundwater were transferred to respective user groups. However, transferring large-scale irrigation system management to a variety of management organizations gained momentum only in the early 1990s (Topcu in this volume). The main motivation behind this shift was to reduce budget allocations to the agricultural sector because management costs of irrigation systems could not be recovered by water charges and was publicly subsidized. This was in line with the cutting off of subsidies for agricultural inputs (e.g. pesticides, fertilizer, credit lines) which had started in the 1980s, the degree of which is still being debated today. However, unlike in the hydropower sector, the public sector is still supposed to invest in irrigation projects and in dams for irrigation and water supply.
- Water supply and sanitation have always been and remain the responsibility of municipalities. Financing of water supply and sanitation was however largely provided by the central government through the Municipalities Fund of the Bank of Provinces. In 1981, a new model was introduced first in the city of Istanbul, i.e. Istanbul Water and Sewage Administration (ISKI). At the beginning, ISKI was independent of the Istanbul Municipality but it was later subordinated to the municipality as a public body with an independent budget. Meanwhile, this model has been extended to 16 water and sewage administrations within metropolitan municipalities. These administrations are encouraged to mobilize their own resources beyond the Bank of Provinces mechanism and to finance large-scale urban infrastructure investments through foreign loans under the Treasury Guarantee Scheme. This has stimulated private sector involvement in the delivery of water and sanitation services (Kibaroglu and Baskan in this volume).

When Suleyman Demirel became president in 1993, after Turgut Ozal's sudden death, the government continued to pay particular attention to dam construction which had already started in the 1950s, when Demirel was the first Director General of DSI. It was his government that decided on the Southeastern Anatolia Project (GAP) and dedicated a huge amount of financial resources towards infrastructure development in the underdeveloped regions of Turkey, among them 22 large dams, 19 hydropower plants and large-scale irrigation infrastructure for irrigating about 1.7 million hectares.

3 The European Union as a trigger for improving environmental policy

Aligning with the European Union (EU) gained momentum when Turkey was officially recognised as a candidate for full membership on 12 December 1999 at the Helsinki summit of the European Council. In December 2004, the European Council decided that official accession negotiations with Turkey would commence in October 2005 with EU membership for Turkey as the possible outcome.

With the so-called Copenhagen Criteria, the EU has developed a set of key conditions that all candidate countries are required to meet in order to ensure successful membership. These three Copenhagen Criteria are: (i) institutions guaranteeing democracy, the rule of law, human rights, and minority rights, (ii) the candidate state needs to prove that it has a functioning market economy and is capable of coping with the competitive pressures within the EU, and (iii) the ability to fully implement all obligations of EU membership. Core to these obligations of membership is the full legal transposition and the practical implementation on the ground of the so called *acquis communautaire*, which is the whole body of EU law in force.

One important area of EU legislation deals with environmental issues of which water management is certainly one of the fields that is already largely shaped by a complex body of European law. The most relevant directive is the EU Water Framework Directive which demands both detailed requirements for national water management and the obligation for EU members to internationally coordinate their activities along river basins in order to achieve the environmental objectives of the directive. Apparently, while Turkey is obliged to develop a national approach to the adoption of the environmental *acquis*, the country's transboundary approach to water management are major topics in the accession negotiations.

Turkey's wish to join the EU has far-reaching consequences for its water policy and has at the same time stimulated the many efforts in changing its national water regime targeting towards compliance with the European environmental and water *acquis* – while reconciling changes with the country's specific conditions and concerns (Sumer and Muluk in this volume).

Triggered by the EU Water Framework's participatory approach, modest changes also took place with regards to the participation of civil society and non-governmental organizations (NGO) in water resources development. Civil society in its relation to the State has seen dynamic and fundamental shifts with the number of NGOs significantly increasing throughout the past decades (Adem 2005, 72). More importantly, their role and relation to the political elite has been changing. This can best be demonstrated by the nature of the NGOs. According to Adem (2005, 73), the founders of the Turkish Association for the Conservation of Nature and Natural Resources established back in the 1950s, were high-level bureaucrats. Many of the founding members of the Environment Foundation of Turkey (EFT) which was created as a non-profit organization in 1978, were lawyers and enjoyed good relations with the state apparatus. They were thus able to push for institutionalizing environmental issues, among them the establishment of the Prime Ministry Under Secretariat for the Environment which had been a key concern of EFT along with creating the legal basis for environmental protection by drafting and promoting the Environmental Act of 1983. Both NGOs still acted on the consensus that the State should play a major role by favouring command and control policies.

During the 1990s, a new generation of environmental organizations emerged which demand the protection of civil rights and freedom, along with environmental issues. Today, several groups are active and demanding a voice in decision-making on water resources development and management, among them experts, academics and intellectuals, civil society organisations and non-governmental organisations, and professional organizations such as the Union of Chambers of Turkish Engineers and Architects (a public umbrella organisation with 23 chambers and about 300,000 members).

The massive river development programmes of the AKP - Justice and Development Party - government are challenged by an emerging discourse which is sceptical towards an infrastructure-centred development path ignoring social and environmental concerns. However, the debate over dams has also become instrumental for political groups in their opposition towards the government (Kaygusuz and Arsel 2005). The well-known campaigns against the Ilisu and Yusufeli dams are but a few examples (Kadirbeyoglu 2005; Scheumann et al. in this volume).

Although protecting the environment / nature has been on the international agenda since the 1970s and especially following the Stockholm Conference in 1972, the 'greening' of Turkey's water policy only started later. It had been shaped with the By-Law for Water Pollution Control enacted in 1988, the By-Law on Environmental Impact Assessment of 1993 and by-laws protecting forests, land, surface water and groundwater, biodiversity including wetlands to mention but a few. The concept of 'Water for Nature' is in the process of being recognized, but is still dominated by developmental over ecological concerns as can be seen in the massive dam development programme which fundamentally alters river ecology. Because a command and control policy approach has not been effective, the state apparatus has yet to consider how to create incentives for resource users to comply with its many regulations (Orhan and Scheumann in this volume). Institutional reform directed towards the strengthening of institutions mandated with environment / nature protection and streamlining ecological concerns into sector policies has yet to be undertaken as well as inclusive forms of decision-making (Divrak and Demirayak in this volume).

4 Turkey's international water policy

The chapters collected in Part II discuss Turkey's current positions in international water law and outline the processes and status of cooperation on those rivers which Turkey shares with its (riparian) neighbours. The European Council's strong attention to transboundary water management within the context of Turkish EU accession was illustrated by the EU-Turkey accession partnership dating from May 2003 (European Council 2003). In this document, the European Council rated Turkish transboundary water management as a priority that needed short-term effort and improvement. More specifically, the Council's decision determined the short-term need for Turkey to "pursue the development of transboundary water cooperation, in line with the water framework directive and international conventions to which the Community is a party" (European Council 2003, 10). The European Commission's annual reports on Turkey's progress towards accession always underline the need to step up cross-border water cooperation with its neighbouring countries (European Commission 2004, 2005, 2006, 2007, 2008, 2009).

Motives for these statements stem from international concerns about unresolved water disputes and potential water conflicts at Turkey's borders, in particular over the Euphrates and Tigris rivers. In the past, Turkey was frequently perceived as a strong upstream riparian that pursues huge water development projects without adequately taking the interests of water-scarce downstream riparians into account. Several commentators have emphasised the important interdependencies between water scarcity and security issues on Turkey's borders turning transboundary river management into an important issue in a region that is largely blighted by tensed political relations anyway (e.g. Beschorner 1992, Lorenz and Erickson 1999, Warner 2008). Furthermore, Turkey's attitude towards the development and strengthening of international water law was assessed as being reserved at best, because of the country's reluctance to sign up to international conventions such as the United Nations (UN) Convention on the Law of Non-navigational Uses of International Watercourses, the United Nations Economic Commission for Europe (UNECE) Convention on the Protection and Use of Transboundary Watercourses and International Lakes and the Environmental Impact Assessment in a Transboundary Context (Cascao and Zeitoun 2010, Scheumann 2003).

However, in-depth knowledge on the current state of water cooperation and unresolved disputes on Turkish transboundary rivers has been lacking so far. Part II of this book comprehensively assesses current use and management of the Turkish transboundary waters including, if available, bilateral and multilateral agreements. The first chapters of Part II cover Turkey's position relating to basic principles and the development of international water law (Kibaroglu and Kramer in this volume) and discuss both the water dimension in Turkey's foreign policy and her water diplomacy (Turan; Williams, both in this volume). The subsequent chapters present our findings on cooperation and conflicts on all Turkish transboundary rivers, such as the Meric River with the riparian states Greece, Bulgaria and Turkey (Kramer

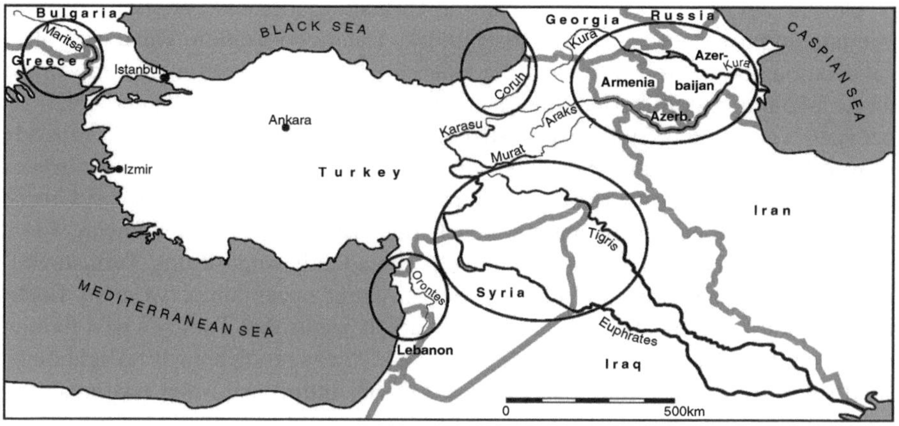

Fig. 1 Turkey's transboundary rivers

and Schellig in this volume); the Kura-Aras river basin with Turkey, Georgia, Iran, Azerbaijan and Armenia (Klaphake and Kramer in this volume); the Coruh River with Turkey and Georgia (Klaphake and Scheumann in this volume); the Euphrates-Tigris rivers system with Turkey, Syria and Iraq as riparian states (Kibaroglu and Scheumann in this volume), and finally the Orontes River with Turkey and Syria (Scheumann, Sagsen and Tereci in this volume) (see Figure 1 and the Annex which includes all bilateral agreements involving transboundary river issues). The last chapter compares the cases with regards to the particular water management challenges and the state and process of cooperation on these transboundary rivers.

Our assessment shows that the situation is often misinterpreted: the somewhat alarming description of Turkish transboundary water disputes as having potential for serious water conflicts appears exaggerated and does not realistically mirror the current situation, even in the most marked water quantity disputes over the Euphrates-Tigris rivers system.

Most disagreements on Turkish transboundary rivers relate to the building of dams which influence water usability downstream. Turkish water policy, with its emphasis on hydropower and irrigation projects by means of infrastructure (e.g. dams), is outlined in Tigrek and Kibaroglu, Sen, Scheumann et al., and Topcu (all in this volume). In principle, we can assess disputes on transboundary waters as the external consequences of the internal economic development strategy putting strong emphasis on the production of agricultural commodities and - even more important - on achieving independency from energy imports. Interestingly, attempts to initiate cooperation on Turkey's transboundary rivers also mainly focus on the development of joint infrastructure. In line with the national discourse, water quality and environmental issues do not play a significant role in transboundary cooperation, with the exception of the recent environmental protection and water quality remediation projects in the Meric and Euphrates and Tigris river basins, respectively.

On the other hand the development of Turkish international relations also has potential to positively influence transboundary water cooperation. The tense political relations between Turkey and the riparian states have seen a shift towards a more favourable political environment recently. For instance, bilateral relations between Turkey and Syria improved considerably after 1998 and have already proved to be favourable for dealing with water-related disputes over the Orontes and the Euphrates-Tigris rivers. Furthermore, political relations between Greece and Turkey are, in the meantime, much more pragmatic, also allowing for a deepening of water cooperation. Similarly, the already intensifying Turkish-Bulgarian water cooperation could benefit from a shared accession perspective. Looking at all initiatives being taken in recent years, it seems that Turkey's new foreign policy outlook which centres on the principle of "zero problems with neighbours" (Perthes 2010; Kramer 2010; Guzeldere 2009) has influenced water relations too.

In realizing this edited volume, we hope to shed light on both Turkey's national and international water policy and politics, and we kindly invite our knowledgeable readers to share their experiences with us.

References

Adem C (2005) Non-state actors and environmentalism. In: Adaman F, Arsel M (eds) Environmentalism in Turkey, between democracy and development? Ashgate, Aldershot, pp 71–86

Aydin Z (2005) The state, civil society, and environmentalism. In: Adaman F, Arsel M (eds) Environmentalism in Turkey, between democracy and development? Ashgate, Aldershot, pp 53–69

Aydin M (2003) Security conceptualization in Turkey. In: Brauch HG, Liotta PH, Marquina A (eds) Security and environment in the Mediterranean, Conceptualizing security and environmental conflicts. Springer, Berlin et al., pp 345–355

Beschorner N 1992: Water in the Middle East. Adelphi paper no 273

Cascao AE, Zeitoun M (2010) Power, hegemony and critical hydropolitics. In: Earle A, Jägerskog A, Ojendal J (eds) Transboundary water management, Principles and practice. Earthscan, London and Washington, DC, pp 27–42

European Commission (2004) 2004 Regular report on Turkey's progress towards accession. Commission of the European Communities, Brussels

European Commission (2005) 2005 Regular report on Turkey's progress towards accession. Commission of the European Communities, Brussels

European Commission (2006) 2006 Regular report on Turkey's progress towards accession. Commission of the European Communities, Brussels

European Commission (2007) 2007 Regular report on Turkey's progress towards accession. Commission of the European Communities, Brussels

European Commission (2008) 2008 Regular report on Turkey's progress towards accession. Commission of the European Communities, Brussels

European Commission (2009) Turkey Progress Report. Commission Staff Working Document, Commission of the European Communities, Brussels

European Council (2003) Council decision of 19 May 2003 on the principles, priorities, intermediate objectives and conditions contained in the Accession Partnership with Turkey. 2003/398/EC

Guzeldere EE (2009) Turkish foreign policy: from "surrounded by enemies" to "zero problems". In: Schäffer S, Tolksdorf D (ed) The EU member states and the Eeastern neighbourhood. CAP Policy Analysis, 1. Center for Applied Policy Research (CAP), Munich, Germany, pp 14–19
Kadirbeyoglu Z (2005) Assessing the efficacy of transnational advocacy networks. In: Adaman F, Arsel M (eds) Environmentalism in Turkey, between democracy and development? Ashgate, Aldershot, pp 101–116
Kaygusuz K, Arsel M (2005) Energy politics and policy. In: Adaman F, Arsel M (eds) Environmentalism in Turkey, between democracy and development? Ashgate, Aldershot, pp 149–165
Keyman FE (2005) Modernity, democracy, and civil society. In: Adaman F, Arsel M (eds) Environmentalism in Turkey, between democracy and development? Ashgate, Aldershot, pp 35–50
Kibaroglu A, Baskan A, Alp S (2009) Neoliberal transitions in hydropower and irrigation water management in Turkey: main actors and opposition groups. In: Huitema D, Meijerink S (eds) Water Policy Entrepreneurs. A research Companion to water Transitions around the Globe. Edgar Elgar, Cheltenham, UK; Northhampton, MA, USA, pp 287–303
Kramer H (2010) Die neue Außenpolitik-Konzeption der Tuerkei. Moegliche Konsequenzen fuer den EU-Beitrittsprozess, SWP-Aktuell 25: 1–4
Lorenz FM, Erickson EJ (1999) The Euphrates Triangle. Security implications of the Southeastern Anatolia Project. National Defense University Press, Washington, DC
Perthes V (2010) Turkey's role in the Middle East: An outsider's perspective. Insight Turkey, vol 12, no 4: 1–8
Scheumann W (2003) The Euphrates issue in Turkish-Syrian relations. In: Brauch HG, Liotta PH, Marquina A (eds) Security and environment in the Mediterranean, Conceptualizing security and environmental conflicts. Springer, Berlin et al., pp 745–760
Warner J (2008) Contested hydrohegemony: hydraulic control and security in Turkey. Water Alternatives 1(2): 271–288

Part I
National Framework

Turkey's Water Policy Framework

Aysegul Kibaroglu and Argun Baskan

1 Introduction

Turkey's water policy and management is a feature of various laws and regulations, and is subject to a range of national ministries and executive administrations. Some of the legislation governing water management dates back to the early years of the Republic.[1] Due to numerous amendments and additions to the existing legislation in the course of time, water management in Turkey ceased to be simple.

The chapter is mainly a descriptive one, which introduces the legal and administrative setting of the water sector in Turkey. However, some analyses will also be presented, particularly when reflecting the critiques on the water policy and management. In the event, criticism of Turkey's water policy originate basically from two different perspectives: *equity* and *environmental protection*. Hence, some local analysts as well as international (e.g. Organization for Economic Cooperation and Development, OECD), supranational (e.g. European Union, EU), and non-governmental (e.g. World Wide Fund for Nature, WWF-Turkey) entities claim that there is a high emphasis on infrastructure development and management (which is no doubt necessary) but protection of water resources and related aquatic eco-systems is weak (see Orhan and Scheumann; Scheumann; Divrak&Demirayak; Sumer&Muluk in this volume). While another group of opponents, the Union of Chambers of Turkish Engineers and Architects (TMMOB[2] in Turkish acronym) in particular, criticize the

[1]For instance, Village Act No. 442, 1924 and the Act No. 831 on Waters, 1926

[2]TMMOB is a corporate body and a professional organization defined in the form of a public institution as stated in Article 135 of the Turkish Constitution. As an umbrella organization, it has 23 chambers and about 300,000 members.

A. Kibaroglu (✉)
International Relations Department, Okan University, Istanbul, Turkey
e-mail: Aysegul.Kibaroglu@okan.edu.tr

A. Baskan
Department of International Relations, Ege University, Izmir, Turkey
e-mail: argun.baskan@gmail.com

A. Kibaroglu et al. (eds.), *Turkey's Water Policy*,
DOI 10.1007/978-3-642-19636-2_1, © Springer-Verlag Berlin Heidelberg 2011

changing water policy discourse and practice and the erosion of the public nature of water resources and resulting negligence of strategic planning and investment. The opponents point to the fact that equitable access to water resources by all has been hindered by neoliberal legal and institutional reforms, which have been introduced since the early 1980s in the name of efficiency and accountability (Kibaroglu et al. 2009; Cinar 2009, 362-63). The chapter refers to these positions when deemed necessary.

2 Legal framework

The Constitution of 1982 (currently in effect) established the basic principles (Article 168) which govern water resources: water is a public good under the State's trusteeship. The authority to explore and manage water resources is vested in the State. The Turkish Civil Code[3] (2001) considers water in two distinct categories: public water resources and water resources in the domain of private law and private proprietorship. This classification is deduced from Article 715 of the Civil Code: the assets under nobody's possession and the commodities at the service of the public shall be under the command and possession of the Government (Coskun 2003, 70). The Civil Code, Article 756 regulates springs as a subject to private ownership. It specifies that "any spring is an integral part of the land, the ownership of a spring may be allowed only together with the ownership of the land" (Coskun 2003, 74). Articles 715 and 756 should be evaluated together, thus, except for privately owned springs, surface and groundwater resources cannot be owned, but are subject to user rights which are granted for beneficial use only, such as domestic and agricultural use, fishing, hydropower generation, industry and mining, transportation and medicinal and thermal uses (Ozbay 2006, 37-8).

Assigned user-rights enjoy the right of prior use, and can neither be sold nor transferred. User-rights to water resources in the domain of private law and ownership are subject to title deed registration. Until 1960 this included groundwater resources which were then transferred from the private to the public domain.

However, legislation on user rights and ownership is not clear for surface waters. Water is allocated, in practice, by a variety of agencies operating independently of each other. These include State Hydraulic Works (Turkish acronym DSI), surface and groundwater water management organizations (irrigation associations, irrigation cooperatives etc.), municipalities and industries. There is no legal framework for surface water rights and only a system of granting access to groundwater, and both are largely open access resources at present. The current national system of recording and harmonizing user rights to water dates back to early years of the Republic and is not well adapted to water short environments. It does not provide security for present users, does not allow for or adequately protect environmental uses of water, and does not provide incentives for an economic use or for orderly transfers among sectors (Harmancioglu et al. 2007, 138, 149, 151).

[3]'Codes' also stand for 'Laws' as well as for 'Acts'.

A prevailing view is that as there is no system for allocating water rights for surface water users, various provisions are present in the different codes, which brings about some complications regarding authority, organizational malfunctions, and legal gaps. These gaps are being filled by means of judiciary decrees (Coskun 2003, 71). However, according to Ozdemir Ozbay, former legal adviser at the DSI, "although there are separate enactments dealing respectively with matters such as rural and urban water supply, groundwater, irrigation and hydropower, Act No. 6200 on the Organization and Duties of the DSI (Establishment Law of DSI, 1953) empowers DSI to coordinate water use at the national level." In this respect, any agency, which embarks on a water development project or is engaged with a water sector-related activity has to cooperate with DSI and must obtain prior approval from DSI concerning the source and volume of water to be used for each purpose (Ozbay 2006, 38-9). Furthermore, when a dispute arises between a user and the supplier (DSI) about public surface and groundwater resources and their utilization, the cases are dealt with by the administrative courts rather than courts of justice. There are numerous administrative courts' and Council of State's (highest administrative court: Danistay, in Turkish) decisions, which confirm that the DSI is the ultimate authority to allocate public water resources.

According to Article 756 of the Civil Code and the Groundwater Law No. 167 (1960) groundwater resources are public waters, and, therefore, shall be under the command and possession of the State. Any spring, in an inherent part of an area under private property, is also considered as private property. Groundwater resources, however, are of public nature and owning a piece of land does not bring along the ownership of water beneath that land. All forms of research, utilization, protection and registration are subject to the provisions of this Law. Later, Law No. 3202, which instituted the General Directorate of Rural Services (GDRS), and the Law No. 2560 which introduced the Istanbul Water and Sewage Administration[4] also began to operate within the domain of Law No. 167. This has resulted in administrative duplications which, in turn, caused improper protection of groundwater resources (Ozbay 2006, 38).

In Turkey, the way for environmental protection has been paved since the 1970s. Environmental awareness rose internationally because negative environmental impacts of industrialization became increasingly apparent. In 1973, following the United Nations Conference on the Human Environment of 1972, environmental concerns were addressed for the first time in the country's third *Five Year Development Plan* (see Orhan and Scheumann in this volume). It took another five years, until in August 1978, the first department, *Prime Ministry Undersecretariat for the Environment*, responsible for the coordination of activities related to environment was established. Moreover, Article 56 of the Constitution grants the right to a

[4]Responsibilities of Water and Sewage Administrations (within the border of all metropolitan municipalities) are to take legal, technical and administrative measures for preventing groundwater pollution and decreasing quantity. Supplying potable water to rural communities by drilling groundwater wells is one of the main duties of the Special Provincial Administrations after the abolition of GDRS (see Vliegenthart et al. 2007, 68-69).

healthy environment and stipulates the prevention of environmental pollution as a duty of the State and its citizens. The law provided the first framework legislation for environmental regulations to come. The first environmental law was the National Parks Law of 1983. It defined a variety of protection areas such as National Parks and Nature Conservation Areas and prohibited any harmful activities within their borders such as the construction of dams (see Scheumann in this volume).

In 1982, the Environmental Law was enacted and came into force in 1983. Its basic principle is that citizens as well as the State bear responsibility for the protection of the environment. It requires that all economic activity takes every measure to minimize pollution. It also incorporates the polluter pays principle (Article 3/e). In 1988, the By-law on the Control of Water Pollution put forward principles for the discharge of waste water into surface and ground waters, defined strategies for the protection of water basins where drinking water is generated, and laid down drinking water standards. The 1988 By-law set ambitious targets but implementation has always been weak.

The By-law on Environmental Impact Assessment (EIA) which was issued in 1993 and revised several times, the latest revision being in 2008, is compulsory for all large-scale projects, including storage facilities (dam reservoirs) with a surface area of more than 15 square kilometers and more than 10 million cubic meters (MCM), and for irrigation and drainage (I&D) projects, water supply and sanitation facilities etc. Development projects on agricultural lands, wetlands, lakes and eco-systems which are rich in biodiversity, as well as lands protected under national law or accord with international agreements which Turkey signed, are subject to EIA studies. These international agreements are the Barcelona Convention and the Ramsar Convention, which concern the protection of the Mediterranean Sea and the protection of wetlands respectively. Turkey has not yet signed the Espoo Convention (1991) which refers to EIA in a transboundary context. However, as EU accession talks progress, Turkey will have to reconsider signing the Espoo, Aarhus and other United Nations Economic Commission for Europe (UNECE) conventions (see Kramer and Kibaroglu in this volume).

The primary sources of Turkish law are the constitution, laws, law amending ordinances (decree-laws), international treaties, regulations and by-laws. With regards to water as a resource, the three most commonly used categories of legislation are laws (acts), decree-laws and by-laws.[5] Water resource development

[5]Article 87 of the Turkish Constitution authorizes the Turkish Grand National Assembly to make, amend, and abrogate laws. Legislative bills are proposed by the Council of Ministers and individual deputies as stated in Article 88. Such bills are debated and adopted by the Parliament in its plenary session in accordance with the Constitution and the Standing Orders of the Grand National Assembly. The Constitution requires a simple majority for the adoption of ordinary laws in Article 96. Legislative bills adopted by the Grand National Assembly are submitted to the President of the Republic in accordance with Article 89 of the Constitution. The President shall either sign the bill within fifteen days or return it partially or entirely to the Grand National Assembly for reconsideration within the same period. However, budget laws are excluded from this power of the President due to time considerations. If the Grand National Assembly readopts the bill without a change the bill shall be promulgated by the President. Laws are only enacted by publication in the Official

and management is regulated by diverse elements of Turkish legislation. One can find provisions related to water use, management and allocation in almost 100 different law (acts), by-laws, decrees etc. (see Box 1) which makes implementation and enforcement of water-related legislation not an easy task (Ozbay 2006, 38).

Box 1 Principal water legislation in chronological order (own compilation based on Carl Bro International (2001) and World Bank (2006))

- Village Act No. 442, 1924 (Articles 1, 6 and 13)
- Act No. 831 on Waters, 1926
- Act No. 1593 on General Hygiene, 1930
- Act No. 4759 on the organization and duties of The Bank of Provinces Law, 1945 (replacing the Bank of Municipalities Law 2301, 1933)
- Act No. 6200 on the Organization and Duties of the General Directorate of the State Hydraulic Works (Establishment Law of DSI), 1953
- Act No. 7478, Village Domestic Water Supply, 1960
- Act No. 167, Groundwater, 1960
- Act No. 1053, Domestic and Industrial Water Supply Law, which authorizes the DSI to provide domestic and industrial water to Ankara, Istanbul and to cities with a population over one hundred thousand, 1968
- Act No. 2872 on Environment, 1983
- Government Decree No. 181 in Force of Law on the Organization and Duties of the Ministry of Health (Article 9/e), 1983
- Agricultural Reform Act No. 3155 (Article 2/c), 1985
- Act No. 3202 on the Organization of the General Directorate of Rural Services (GDRS) (Article 2/d), 1985 (abolished)
- Act No. 3416 Amending Act No. 2872 on Environment, 1988
- Act No. 3621, Coastal Law, 1990
- Government Decree No. 443 in Force of Law on the Establishment and Duties of the Ministry of Environment, 1991 (abolished)
- Gov. Decree No. 441 in Force of Law on the Establishment of Ministry of Agriculture and Rural Affairs, 1991
- Act No. 4950 Amending Act No. 1380 (1971) on Aquaculture, 2003
- Act No. 4856 on the Organization and Duties of Ministry of Environment and Forestry, 2003
- By-law* 19919 on the Control of Water Pollution 1988, which is abolished by By-law 25687 on the Control of Water Pollution, 2004, amendments, 2008

(continued)

Gazette. Law amending ordinances (decree-laws) have the same legal effects as laws. By-laws are regulated in Article 124 of the Constitution. The Prime Ministry, the ministries, and public corporate bodies may issue by-laws in order to ensure the application of laws and regulations relating to their particular fields of operation, provided that they are not contrary to these laws and regulations. The law shall designate which by-laws are to be published in the Official Gazette (see Yazici 2009).

- Act No. 5393 on Municipalities 2005 (Act No. 1580 on Municipalities, 1930 (abolished); Act No. 5215 on Municipalities, 2004 (abolished)
- Act No. 5216 on Metropolitan Municipalities, 2004, which replaces Act No. 3030 (1984)
- Act No. 5177 Amending Act No. 2560 (Act on the Organization and Duties of the Water and Sewage Administration of Istanbul-ISKI, 1981) 2004
- Act No. 5286 on Abolishment of GDRS (1985), 2005
- Act No. 5302 on the Special Provincial Administrations, 2005
- Act No. 5625, Amending Act No. 1053 (1968), which authorizes the DSI to provide water supply to all cities with municipalities without due regard of the population criterion, 2007
- Act No. 5686 on Geothermal and Mineral Waters, 2007

*Recent by-laws concerning water quality and environmental protection are presented thoroughly by Orhan and Scheumann in this volume.

3 Administrative setting

While it is sometimes argued that Turkey's administrative set-up for water resources is highly fragmented with overlapping responsibilities, the organizational structure is actually not so complex: at the strategic decision-making and planning level, the Prime Minister, the State Planning Organization (SPO) and ministries are involved. Governmental organizations under the ministries form the executive level and governmental and private sector organizations implement water policies, and operate and maintain the water infrastructure. The Turkish "water sector" has a distinct pattern:

- Water-related development objectives are part of the government's strategic planning[6] with a central planning organization (i.e. SPO) in place;
- The government has taken the lead in infrastructure development and financing in order to provide water and water-related products (e.g. energy), but there is a trend towards greater private sector involvement in particular for generating hydroelectricity (see Baskan in this volume);
- Water management is organised along sectoral lines with strong central government organizations;
- Managing water and operating water networks is decentralized in urban water supply and sanitation, and in the irrigation sector through irrigation management organizations.

The Turkish administrative system, including the water-related institutions, has three administrative levels: the national, the provincial and the local level

[6]Associated with the strategic role water plays for the Turkish economy, water policy, in terms of funding, puts much more emphasis on water development than on protection.

(i.e. municipalities and villages). Being modelled on the French system, it is highly centralized and linked to strong central government organizations. Administratively, Turkey is divided into 81 provinces with appointed governors, i.e. head of the provincial governments, who are affiliated to the Ministry of Interior. All ministries and their General Directorates have provincial branches. At the intermediate level, the General Directorates are organized in Regional Directorates covering different service areas.[7] Municipalities are elected in all provincial and district centres. Turkey has 3,225 municipalities of which 16 are structured as "Greater City/Metropolitan Municipalities".[8] According to the Turkish Village Law, villages are the lowest administrative units and are self-governing autonomous local administrations.

3.1 Strategic orientation and planning

The State Planning Organization-SPO

At the national level, the SPO which is under the Prime Minister, is the strategic organization established to guide economic and social development through each of the Five Year Development Plans in which experts from all ministries participate. SPO was founded in 1960 in order to accelerate the economic, social and cultural development of Turkey. Each Five Year Plan is a basic planning instrument which defines investment priorities and the allocation of resources for public investment. To illustrate, the Eighth Five Year Development Plan (2001–2005) emphasised the necessity of striking a balance between ecological stability and economic growth, and called for new regulations to increase the efficiency of the EIA process. Under the Eighth Five Year Development Plan, the most important policy has been to increase the ratio of population with access to basic infrastructure facilities. To this end, an integrated planning approach and harmonisation among the organizations involved was strongly emphasised with respect to the construction of municipal water supply, sewerage and wastewater treatment facilities. The Ninth Five Year Development Plan, covering the period 2007-2013, has underlined that necessary legal regulations would be prepared in order to establish a well coordinated and effective national administrative structure, which would be responsible of allocation, use and improvement of water resources as well as protection against pollution (SPO 2007). Given its status, SPO is the organization that is supposed to monitor the completion of the regulation process. Another statement in the same Development Plan is about the coordinating role of SPO, indicating that the effective use of water will be emphasized through "a strong and structural coordination among relevant institutions," an enthusiastic, yet challenging goal.

---

[7]The Regional Directorates of DSI are organized in a way that their respective jurisdictions cover main river basins.

[8]The Union of Municipalities of Turkey, http://www.tbb.gov.tr/istatistik.php Accessed 28 May 2010.

The General Directorate for Electrical Power Resources Survey and Development Administration-EIE

The first agency responsible for stream flow measurement, hydropower planning and design was the Electrical Investigation Administration (Elektrik Isleri Etud Idaresi, EIE in Turkish acronym), founded in 1935. The General Directorate of EIE is affiliated to the Ministry of Energy and Natural Resources, and it is the main agency responsible for hydrometric measurement (part of its responsibility lies with the DSI). The EIE conducts studies and surveys to explore the country's hydropower potential, executes engineering services and designs studies for dams and hydroelectric power plants. It also carries out studies for new and renewable energy sources (wind power, solar energy etc.) and it oversees and makes hydrological plans for Build-Operate-Transfer (BOT) projects (World Bank 2006, 5).

3.2 Major implementing agencies for developing water infrastructure and its management

The General Directorate of State Hydraulic Works-DSI

The General Directorate of DSI is the primary state executive state agency responsible for the planning, implementation and management of hydraulic works. Its principle objective is to develop the water and land resources of Turkey. DSI was established under Act No. 6200 in 1953 as a legal entity. For the most part it acted under the aegis of the Ministry of Energy and Natural Resources. Yet, with a government decree issued (and approved by the President) on August 30, 2007, DSI has been attached to the Ministry of Environment and Forestry (MoEF). It is organised along the 25 major river basins in the country with Regional Directorates being responsible for preparing master plans which set priorities for the development of water resources in the respective basins. Priority projects are submitted to the SPO for their incorporation into each of the Five Year Development Plans and into the annual investment programmes.

DSI plays a leading role in coordinating water sector planning. Any agency and private party is obliged to cooperate with DSI and must obtain prior DSI approval for the source and volume of water to be used by each project and individual. As the licensing authority[9], it approves both the use and the extraction rate of water for different purposes including groundwater. The Law (No. 6200) establishing DSI

[9]Since 2001, with the adoption of the Electricity Market Law No. 4628, an independent public institution, namely the Energy Market Regulatory Authority (EMRA was created for issuing licenses for production activities (including hydropower generation) in the electricity market. For hydropower projects in order for private sector to get licenses a *Water Use Right Agreement* should be signed between the DSI and the private entrepreneur (see Electricity Market Licensing By-Law No. 248364, 2002).

defines it as the main state agency to develop surface and groundwater resources, to make optimal use of them and to develop them in such a way as to achieve optimum benefit. DSI's mandate is to plan, design, construct and operate single and multi-purpose dams and I&D schemes (Bayazit and Avci 1997); the construction of protective facilities for flood and torrent hazards; land reclamation and the drainage of swamps; the construction of hydropower generation facilities[10] and the improvement of the navigability of rivers. It is also entrusted with the provision of water supply for cities with more than 100,000 inhabitants (Act. No. 1053), provided that the government authorises DSI and that the concerned city council also approves.[11]

Greater private sector involvement has been also envisaged in order to construct and operate drinking water plants based on BOT contracts. Under a BOT contract, the private sector finances, builds and operates a new facility in accordance with performance standards set by the government. The government retains ownership, and the facility is reverted to the state after an operation period of typically 10 to 20 years. Despite the Turkish government's guarantee to repay 85 percent of the construction costs, few BOT contracts such as the one for the drinking water plant in the city of Izmit, have been signed (Bennett et al. 1999).

In terms of number of personnel, DSI is the state's most powerful organization with nearly 25,000 employees, of which 4,500 are engineers and technical personnel. The main financial resources of DSI come from the national budget. For a long time, DSI's investment budget had amounted to roughly one third of the state investment budget. However, DSI's budget has been diminished. Examination of past investment figures of public institutions in Turkey reveals that DSI's has decreased from 33.3 percent to 24.5 percent since 2005 (DSI 2009a, 14). Yet, due to the size and capital intensive characteristics of dam projects, the largest share of public investments goes to the water sector. In 2009, DSI had an investment budget of more than US$2 billion, which accounts for around one fourth of Turkey's state investment budget.

Since the mid-1980s, DSI has had an in-house unit dealing with environmental issues, mainly monitoring water quality in rivers and lakes, and issuing EIA studies which became mandatory from 1993 onward (enactment of EIA By-law) for, e.g., dams, I&D projects of a defined size, which have become stricter during the course of time (see Scheumann in this volume).

DSI has the primary task of planning, constructing, managing and operating I&D systems up to the farm level, including the collection of water fees. However, the Establishment Law 6200 entitles DSI to transfer operation and maintenance (O&M)

---

[10]With the Electricity Market Law No. 4628 of 2001 and the Renewable Energy Law No. 5346 of 2005, the Turkish government envisaged the acceleration the development of the underexploited potential by inviting private investors and private financial service institutions (see Scheumann in this volume).

[11]Article 10 of Law No. 1053 has recently been amended. The Amended Law No. 5625 has revoked the city criteria (cities with a population of which is over 100,000) and extended the duties of DSI. Thus, since 2007, DSI has been authorized for the domestic and industrial water supply of 3,225 settlements all over Turkey, which have municipality administrations. The Law stipulates that, if necessary, DSI could give priority to wastewater treatment plants in progress.

of irrigation systems to irrigation management organizations, such as village administrations, municipalities, cooperatives, irrigation associations[12] and other private legal entities (see Topcu in this volume). From the early 1960s DSI had a programme for such transfers relating to secondary and tertiary canals. Until 1993, however, DSI was able to transfer O&M irrigation systems amounting to only approximately 70,000 hectares to various types of irrigation management organizations. The process has gained momentum since 1993, and within the past seventeen years the management of irrigation covering more than 2 million hectares has been handed over to local administrations or to irrigation associations (Kibaroglu et al. 2009). Irrigation management organizations are currently constituted under the Local Government Associations Law No. 5355.

The General Directorate for Rural Services-GDRS and the Special Provincial Administrations-SPAs

In 1960, Law No. 7457 established the General Directorate for Soil and Water (TOPRAKSU, in Turkish acronym), which in 1985 was reconstituted as the GDRS under the Ministry of Agriculture, Forestry and Rural Affairs. However, when TOPRAKSU was closed, some of its major tasks such as surveys and inventory works on soil-water relations were not clearly transferred to the new unit, GDRS. Thus, much essential data and information related to soil surveys on a national scale such as soil type and classification (land capability and fertility), depth, slopes, organic materials, erosion characteristics have not been studied and updated since the late 1970s (State Planning Organisation 2007, 7-13).

From 1993 until recently, the GDRS operated under the Prime Minister's Office. In 2005, as part of the government's decentralization and cost-cutting programme the GDRS was abolished, and its personnel, cadres, equipments and vehicles and other belongings at the headquarters were transferred to the Ministry of Agriculture and Rural Affairs (MARA) while its personnel and belongings at the provincial level were transferred to the metropolitan municipalities in Istanbul and Kocaeli and to the Special Provincial Administrations (SPAs).[13,14] GDRS research stations were also transferred to MARA.

---

[12]A form of transfer considered innovative, where the irrigation scheme covers more than one local administrative unit (for example, a village or municipality).

[13]With the Act No. 5302, adopted in 2005, 81 SPAs are established in the country; one in each province. The SPAs cover areas that fall neither within municipal nor village boundaries. The SPAs are the public entities enjoying administrative and financial autonomy, which are set up to meet the local and common needs of the people dwelling in the province, and whose decision-making branch (general provincial assembly) is elected and made up by electors. They are composed of the general provincial assembly, the provincial council and the governor. The governor (chief executive of the province and principal agent of the central government) is the chief of the Special Provincial Administration and the representative of its legal personality.

[14]Act No. 5286 on the Abolishment of the General Directorate of Rural Affairs, 2005.

In large-scale public irrigation schemes which were constructed and managed by DSI, the GDRS was responsible for all on-farm development activities, i.e. land levelling of agricultural lands; improving small, scattered land parcels owned by the farmers through land consolidation activities; construction of field canals as well as carrying out functions such as construction of rural transport networks and on-farm roads, on-farm irrigation and surface and tile drainage infrastructure (World Bank 2006, 13-14). The agency was concerned with soil conservation; the construction of small dams/reservoirs, and the construction of minor surface and groundwater irrigation schemes (with a capacity of less than 500 litre per second) which were turned over to irrigation management organizations and groundwater cooperatives, respectively. Since 1964, first TOPRAKSU, and then GDRS were also responsible for supplying drinking water to villages and rural households either from surface water or groundwater, regardless of geographic location.

For the time being, it is rather difficult to reach systematic empirical studies[15] on the impact of the abolishment of GDRS, particularly the impacts it had at the provincial level. However, the abolishment of GDRS and its ramifications stimulated a debate among the Chambers, non-governmental organizations (NGOs) and the concerned government offices such as MARA and SPAs officials. The debate culminated in the discussion over the possible negative outcomes of the decentralization of land and water resources management and conservation from the central authority, namely GDRS, to the local entity, namely the SPAs.

TMMOB makes the point that because soil and water resources are strategically significant, and because they are defined as public resources by the Constitution and should be protected and safeguarded, their management and conservation should not be left to local governments. It argues that country-wide strategic planning and investment will always be necessary. Moreover, in their contention, decentralization of these public services will give way to the privatization of land and irrigation water management, which would be against the constitutional order (Articles 44 and 45). Moreover, it will diminish the efforts to remove regional socio-economic disparities between the different regions in Turkey. TMMOB

[15]A World Bank team, conducting a study on key irrigation policy, institutional, and especially investment issues in Turkey in 2006, paid visits to the former Regional Head of GDRS in Izmir who was appointed later on as the head of the technical wing of the SPAs. The following observations are interesting in this respect: "When GDRS was dissolved the SPAs' technical staff were merged with the GDRS staff in the former GDRS building; the investment budget for the unit has reduced considerably since the GDRS period, it is now some 500,000 YTL per annum whereas before it was 60 million YTL for the GDRS region (the region covered 3 provinces, whereas the SPA covers only one province). Moreover, it was reported that the Council of the SPA was not interested in irrigation development, land consolidation or drainage; it is more interested in rural water supply and rural roads, and in urban development. The SPAs have been completing some of the existing irrigation and drainage (I&D) projects but have not planned any further I&D projects" (World Bank 2006, 62).

agrees that "decentralization" of public services provision could be an essential step towards democratization, yet this should not have been started with rural land and water resources protection and management. Furthermore, the Chambers, being the main pressure group, have been observing the poor performance of local governments in terms of protecting agricultural lands particularly those which are located in the proximity of heavily industrialized cities. Thus, TMMOB claims that land and water resources planning, conservation and development should be carried out at the river basin level, hence delegating GDRS tasks to SPAs and decentralizing these main services will hinder these efforts to implement river basin planning and management (Soganci 2005)[16].

The Turkish Foundation for Combating Soil Erosion, for Reforestation and Protection of Natural Habitats (TEMA),[17] a leading NGO in soil protection has, also been vocal in criticizing the government about its decision to abolish GDRS particularly without the consultation of with various stakeholders in the sector. Hence, TEMA argued that while it is possible and useful to devolve the infrastructure-related tasks of the GDRS such as building roads and provision of water supply to the SPAs, which may reduce bureaucracy and increase efficiency and effectiveness, the same does not hold true for protecting soil and water resources, which should be strategically planned and carried out at the basin level.

More time is required to establish more clearly the role of the SPAs. The picture obtained to date is that they are not interested in soil conservation, irrigation development or land consolidation or other on-farm development activities, but focus on rural water supply, sanitation and roads (World Bank 2006, 55-56; Unver 2010; and SPO 2006). In this respect, the Village Infrastructure Support Project (KOYDES) Project[18] was specifically introduced by the government to mobilize substantial financial resources to complete and rehabilitate rural roads and water supply networks. Ozden Bilen, former director general of the DSI, draws attention to the fact that while a consensus has been growing among diverse stakeholders on the issue that 'river basin planning and management' is the most appropriate policy instrument for efficient and equitable water use and management in Turkey, decentralization of the GDRS tasks particularly soil conservation and small-scale water management to the SPAs stands in contradiction to the general consensus on river basin management (Bilen 2009, 300).

---

[16]Soganci is Head of the Board of the TMMOB.

[17]TEMA was founded in 1992 and has been since then one of the leading NGOs in Turkey with specific focus on soil protection, protection of natural habitats and erosion control.

[18]KOYDES (Village Infrastructure Support Project) being underway since 2005, has been implemented by the Directorate General of Local Administrations, Ministry of Interior. The aim of the Project is declared as to solve village problems related to potable water, and roads, with minimum costs and in the shortest possible time under the leadership of governors and district governors, Special Provincial Administration and village service provision units (Kapucu and Palabiyik 2008, 163).

3.3 *Local services: water supply, sewage collection and wastewater treatment*[19]

The Act No. 5393 on Municipalities (2005) assigns numerous powers and duties to municipalities[20] which are, for example, the construction of urban water supply and sewerage systems and wastewater treatment plants. Municipalities usually prefer to combine water and urban transport services as a means of obtaining revenue and cross-subsidizing public services. In the non-metropolitan areas, the primary concern of local government is usually water supply rather than wastewater disposal and treatment. However, separating water supply and sewerage services under different management lines preclude the possibility of an integrated approach (Cinar 2009, 351).

Where the population is less than 10,000, the municipal public works department is responsible for water supply, which is financed from its own budget. In this case, both water supply and sanitation services are grouped with other public services. In municipalities with a population of 10,000–50,000, it is common to have a directorate or 'water office' that is responsible for water supply. These offices do not have a separate legal entity. In municipalities with a population greater than 50,000, water supply is generally combined with other municipal services in a separate operating unit established by the municipal council as a legal entity. These service providers are specific organizations rather than autonomous commercial units, and they have budgets assigned to them (Cinar 2009, 351).

Metropolitan areas have faced serious sewerage problems as a consequence of population increases from the 1980s onward. This has encouraged the establishment of new organizational models which link water and wastewater management. Starting with Istanbul and the establishment of the Istanbul Water and Sewage Administration (ISKI) in 1981, autonomous entities were created with the responsibility for planning, designing, constructing and operating water supply and sewerage services in metropolitan areas. In the beginning, ISKI was independent of the Istanbul Municipality, but after the reorganization of the municipality as a metropolitan administration in 1984, ISKI was subordinated to the Istanbul Metropolitan Municipality as a public entity with an independent budget. This water and sewerage administration model was extended to cover other metropolitan municipalities, such as Ankara in 1987 and Izmir in 1989. Today there are 16 water and sewerage administrations within metropolitan municipalities (Cinar 2009, 351). Metropolitan municipalities and their utilities have been encouraged to mobilize their own resources beyond the Bank of Provinces mechanism and to finance large-scale urban infrastructure investments through foreign loans under the Treasury

[19]Villages, i.e. the lowest administrative units, are self-governing autonomous local administrations. Village mayors and village councils may decide on, for example, the construction of drinking water wells. The concerned agencies, the GDRS until 2005 and then the SPAs, are in charge of providing water to villages.

[20]A municipal administration can be established in settlements having more than 5,000 inhabitants.

Guarantee Scheme. In turn, this has stimulated privatization initiatives in the delivery of local services in Turkey (Cinar 2009, 354).

The General Directorate of the Bank of Provinces (Iller Bankasi, in Turkish) was established in 1933 and restructured in 1945 under the then Ministry of Public Works and Resettlement with a mandate to allocate funds and loans to local governments (municipalities) irrespective of size, in the financing and construction of infrastructure for water supply (drinking water), sewerage and wastewater treatment. Financing is largely provided by the central government through the Municipalities Fund of the Bank. Hence, the Bank was structured to be a financing institution without the responsibility of implementing the projects, yet providing technical support to local governments. However, limitations on access to highly subsidized funding have led to delays or interruptions in the implementation of municipal water supply and sanitation projects. Moreover, rapid population growth and influx of people from rural to urban areas as well as budgetary constraints accumulated further debts in municipalities. As a result, the option of reforming the Bank to enable more efficient transfer of public funds to the municipalities has become a priority item for successive governments (Cinar 2009, 354). Apart from central government's funding constraints, higher investment needs for the provision of local services have forced municipalities to seek alternative financial sources.

A general look at the financial sources at the disposal of the municipalities to undertake water supply and sewerage investments reveals that the Bank of Provinces lost the former role it had played since the mid 1980s (Cinar 2009, 353-54). In the period from 1990s to early 21st century, external financing institutions (e.g. World Bank, German Bank for Reconstruction and Development, Asian Development Bank, Islamic Bank, Council of Europe, Japan Institute for Overseas Investment, European Investment Bank) gradually took the lead in the field. However, it is also clear that the Bank of Provinces has again assumed its former role as the creditor in the 2000s. The reason of this revival was the still ongoing investments of smaller municipalities to realize water services projects, whereas the bigger municipalities had already completed most of their work. Beginning in the year 2000, levels of foreign credit use and municipality budget resources decreased particularly in the case of the Metropolitan Municipalities. This change can be explained by the new policy of the central government to minimize the use of foreign credits, and to prioritize the completion of projects of the Metropolitan Municipalities rather than the construction of new ones. There is a similar pattern for investments in drinking water. The credit agreement signed between the Bank of Provinces and the World Bank under the framework of the 213 million Euros "Municipality Services Project" stipulates that the Bank of Provinces is also likely to begin to rely on foreign credits in the near future just like DSI[21] (Cinar 2006, 230-239).

[21] A comparison of the debt structures of the Bank of Provinces and DSI reveals that 25 to 50 percent of DSI's investments are financed through foreign credits whereas the Bank of Provinces uses its own financial resources (Cinar 2006, 230-39).

Allocations from the national budget still remain the main financial source of the municipalities to provide water services but the municipalities are gradually getting more "soft loans" (state-to-state credits) or loans and/or credits from international financial agencies under the guarantee of the Turkish Treasury. Use of foreign credits obtained from international financial organizations not only results in the debts of the municipalities and rises in the service tariffs imposed on the final consumers, but also had other impacts such as privatization of water services and changes in the institutional structures of the municipalities. In the loan category, the World Bank stands as the leading agency in the debt profiles of the municipalities.[22]

Since the 1999 Helsinki Summit, EU has been supporting water services projects in the municipalities of Samsun, Eskisehir, Mersin, Bursa, Tarsus, Diyarbakir, Adana, Izmit, Ankara, Antalya, and Sanliurfa. EU also supported country wide projects in Turkey through its loans and credits under the Mediterranean Economic Development Area (MEDA) I and MEDA II projects and via credits from the European Development Bank and the German Bank for Reconstruction (KfW Bankengruppe). Moreover, the Social Development Bank of the Council of Europe has provided credits to several Southeast Anatolian municipalities, and official bodies between 1992 and 1999, the highest single credit amount being US$123 million (Topcu 2006, 287-299). EU accession process is expected to increase the investments, use of foreign credits and the participation of the private sector to meet the high EU standards especially for water quality. Furthermore, the Japanese credits provided by the Japanese Official Development Assistance, the Japan Bank for International Cooperation, and the Asian Development Bank as well as the World Bank have financed some drinking water projects in Ankara and Istanbul (Cinar 2006, 244-245 and Cinar 2007).

Antalya was the first case in Turkey where multinational companies were involved in financing urban water infrastructure, management and service delivery. In Antalya, the need for an extensive water supply and sewerage system became very urgent in view of the city's rapid population growth and significant demand for water during the tourist season (Cinar 2009, 355). From the early 1990s, environment and sewerage projects appeared on the policy agenda, and loans were negotiated with the World Bank. A feasibility study was prepared for an alternative model of water and sewerage services in 1992. Discussions were accelerated when Antalya became a metropolitan municipality in 1993. Following the World Bank's recommendations, the Antalya Water and Public Transport Services (ASO) Directorate became the Antalya Water and Wastewater Administration (ASAT), established along the lines of the ISKI model, which is responsible for water supply and sewerage services in Antalya (Cinar 2009, 355). ASAT then launched an international tender for the operation of its water services. One of the loan conditions stipulated that ASAT should establish a separate company for administrative and financial services to oversee the tender process. It thus set up the Antalya Infrastructure Management and Consulting Services Company (ALDAS), which

[22]The World Bank has provided credits worth US$1217.7 million in total for projects in Istanbul, Ankara, Izmir, Bursa, Antalya, and Cesme-Alacati (Cinar 2006, 230-39).

was then contracted for a 12-year period to deliver water and sewerage services, and collection and disposal of all kinds of solid waste within the boundaries of Antalya Metropolitan Municipality.

As agreed with the World Bank, in 1996 ALDAS transferred the water supply operations to Antalya Water Management Company (ANTSU), which is a subsidiary of Ondeo, part of the Suez group, one of the leading multinationals in the water sector. Following a transition period, ANTSU took over the responsibility for providing, managing and operating water and wastewater services in 1997. Under ANTSU the provision of water-supply services was regulated for a ten-year period (Cinar 2009, 356). ASAT was expected to set the drinking water and wastewater tariffs on the basis of a minimum profit rate of 10%, in line with the ISKI law and the operator's margins. In the contract, the operator's payment is based on fixed annual tariffs per cubic meter of water paid by the subscribers. Nevertheless, due to the high inflation rate, charges were actually reviewed each quarter, so inevitably water tariffs rose. During 1996–2002, when the private operator was in charge, the lowest water tariffs for households rose by 26 cents to 84 cents per cubic meter. This trend continued in 2003 after the withdrawal of the private operator from the contract.

Despite the reduction applied to water tariffs in 2004 and 2005, the price of the water was still high, compared with the early 1990s (Cinar 2009, 357). The result was the takeover of the urban water services management by ASAT-ALDAS-ANTSU troika which increased tariffs while service quality in Antalya, a very important tourism city, decreased. Despite the possibility of turning this change into a new source of income for ASAT and the Antalya Metropolitan Municipality, it turned into a loss because of pitfalls in infrastructure, organizational framework and financial flows. There were problematic issues from the beginning like the transfer of "water ownership" of the municipality infrastructure despite the fact that it was only to transfer the "right to operate" (Guler 1999, 150). Before the contract was due to expire in 2007, ANTSU decided to dissolve, in accordance with the provisions of Article 324 of Turkish Commercial Law in 2002, and to seek international arbitration to settle the dispute caused by the unwillingness of ASAT to assume the responsibilities defined in the contract. The reasons for applying international arbitration were stated by the private operator as uncontrollable increases in operating expenses and significant losses due to the misdirected investment, increasing illegal use and inefficiencies in the network. Following the dissolution of ANTSU, delivery of urban water and sewerage services reverted to ASAT, which then introduced new personnel policies resulting in an increase in the number of employees from 50 in 2000 to 350 in 2004 (World Bank 2004).

3.4 *Protection of natural resources including water*

A new legal structure of environmental protection and water management has emerged over the past three decades in Turkey, driven by an increased emphasis in domestic law, the expansion of activity in terms of bilateral and multilateral international agreements and the country's efforts to meet EU criteria toward full membership. The combined result of these three approaches is to raise both the visibility of water and the environmental issues in Turkey and to enhance the ability of the government to act effectively in protecting what its Constitution refers to as its "national wealth" (Republic of Turkey 2003, 93). The leading government body that deals with protection of natural resources including water is the MoEF (see below). In addition, the MARA is responsible for making investigations and preparing projects to protect and improve soil, water, plant, animal and fisheries resources and products, to control wastewater discharges into fish production areas and to monitor nitrates parameters in surface and groundwater in accordance with Decree-Law No. 441. Also, the Ministry of Health is responsible for determining quality standards for drinking and domestic water, monitoring these standards and preparing legislation in these areas (DSI 2009b, 15).

The Ministry of Environment, which was established on 21 August 1991 by the Decree Law (No. decision 443/KHK), replaced the Undersecretary for the Environment, which led to the diversification of the Ministry's responsibilities and to an expansion of its staff. This also led to the administration's empowerment concerning the implementation and enforcement of policies for the protection and conservation of the environment. However, the Ministry of Environment has limited resources and limited competence.

The mandate of the Ministry covers issues such as appropriate land use, protection of natural resources including protection of water and prevention of pollution. Its departments concerned with water resources are the Directorate General of Environmental Management, Directorate General of Prevention and Control of Environmental Pollution (Water Department), the Directorate General of Environmental Protection (Sensitive Eco-Systems Protection Department) and the Directorate General of Environmental Impact Assessment and Planning.

The Authority for the Protection of Special Areas (APSA) was linked to the Ministry of Environment in 1991 and later on to the MoEF after these two Ministries were united on May 1, 2003 (Act No. 4856)[23]. APSA has been carrying its operations as a public institution, having a special budget. It is responsible for protecting and managing the natural and environmental values of 14 Special Protected Areas.

In 2007, DSI has been aligned with the MoEF, which has caused different interpretations by different stakeholders. By government decree approved on August 30, 2007, DSI became part of the MoEF. With this development, water quantity and water quality management is combined under the same Ministry, which seems to be in line with strong recommendations of EU circles (Grontmij Consulting Engineers 2004b and European Commission 2005). However, a significant number of experts assert that this will put DSI into a different position than its entrusted central role in water resources development, possibly eroding its investor,

[23]Act No. 4856 on the Organization and Duties of the Ministry of Environment and Forestry, 2003.

provider and managerial functions by burdening it more with coordinating, supervision and monitoring tasks in the water sector (Yildiz 2009). A different viewpoint emphasizes that environmental protection will be ignored and the clashing discourses and conflict of interests between the DSI and the MoEF will give way to the dominance of the DSI approach in the MoEF. In this context, being a leading NGO in the water sector, WWF-Turkey criticized the government decision to attach DSI, with its large number of personnel and large investment portfolio for water resources development to the MoEF. According to the WWF-Turkey, that merger would damage main functions of the MoEF, which basically encompass protection and monitoring of water resources rather than development (Divrak 2009).

The provincial branches of the Ministry are responsible for taking measures to prevent and minimise pollution, to inspect any activity that might threaten the ecology and cause sea pollution. The Ministry's provincial branches inspect whether discharge of wastewater from industry and domestic sources into rivers comply with legal standards. The Ministry has recently started to publish provincial "state of the environment" reports. Yet, the provincial branches are reported to lack necessary manpower to carry out many of those functions (see Orhan and Scheumann in this volume).

4 The quest for a framework water law and attempts at the reorganization of the water sector

The lack of a comprehensive water law is perceived as the major challenge in the Turkish water sector. A clear need is identified for a framework law which assembles the guiding principles, norms, rules, procedures in water resources management and allocation (Baris and Karadag 2007). It has been emphasized that there is an urgent need to revive the process of formulating, consulting on and passing a modern water law that gives legally enforceable water rights to water users, and which establishes a water resources management and regulatory authority with full legal powers to license and enforce water abstractions and discharges. Conferring this legal authority and establishing individual water rights will force a fundamental change in the manner in which water resources are developed, managed and used. The preparation and enactment of this law will be a fundamental requirement for Turkey's accession to the EU (World Bank 2006, 76).

A framework water law (national water act) has been debated which should address issues such as increasing competition over limited water resources; burgeoning number of actors/stakeholders in the water sector; depletion and degradation of water resources as well as addressing some challenging issues such as 'decentralization,' 'privatization' and 'river basin planning.' New institutions have been established particularly since the early 1980s, under the neoliberal policies of decentralization and privatization. Yet, they were introduced without properly delineating the mandates between these newly introduced institutions and the existing ones (Bilen 2009, 316). To illustrate, when the Metropolitan Municipalities were established in large provinces in 1981, they were entrusted with functions such as flood protection and financing and implementing water supply and sewerage systems, which used to be entirely within the mandates of DSI and the Bank of Provinces. The overlaps and conflicting mandates of these institutions caused not only bureaucratic competition and rivalries but more importantly caused institutional pitfalls in responding to the needs, urgencies and crisis situations. A major water crisis of domestic water shortages occurred during the drought of 2007-08 in central Anatolia, in the capital city of Ankara, in particular. Moreover, devastating floods caused severe damages, even loss of lives in the Thrace and the Marmara region (with 70% urban population) in fall 2009. Hence, much criticism around these incidents has concentrated on the lack of coordination and complementarities among the responsible water authorities (Chamber of Civil Engineers 2009). The debate over water sector reform has culminated in the proposals of a "national framework water law" and a "new ministry of water resources" (Ada Strateji 2010, 18-19).

It is not only that reports and review studies by third parties such as the studies conducted by Carl Bro International or Grontmij Consulting Engineers claim that "current water management structures in Turkey can best be described as complex, confusing and fractionalized," but the government authorities also registered the fact that the reorganization of the institutional set-up and reappraisal of legislation in the water sector is necessary particularly in the period when Turkey undertakes various efforts to harmonize its administrative and legal structure with the EU (Grontmij Consulting Engineers 2004a; Carl Bro International 2001; SPO 2007). In this context, within the framework of the Seventh Five Year Development Plan (1996-2000), the SPO mandated DSI to draft a new comprehensive framework water law. In drafting a framework law the DSI legal experts compiled and analyzed national acts and laws from various countries including the ones which have established legal structures such as France or the ones which have passed through a recent restructuring such as Brazil and South Africa. The draft law was completed by DSI's legal division in 2001. However, the draft law has not yet passed the necessary procedures in the Parliament (Turkish Grand National Assembly, in Turkish acronym), which may take it to parliamentary debate and possible adoption. This delay in the parliament is possibly due to the fact that a water law has not yet been perceived as a priority by the government. Meanwhile, numerous new regulations have been adopted and since 2003 amendments were made to the existing laws to transpose the EU water legislation, regardless of pushing the draft framework law to go through the legislative procedures in the Parliament.

Arising from the review the main principles that need to be covered by the *Draft Water Law* were stated as:

- Equal rights under the law for individuals to access water;
- Prioritization of water allocation: (i) drinking water, (ii) environmental needs, (iii) irrigation, (iv) hydroelectric power, (iv) industry, handicrafts and tourism;

- Registration of all abstractions from the water resource;
- Water should be paid for, it has an economic value;
- Pollution has to be managed and controlled, and the polluter pays principle adopted;
- Transparency in water resources allocation, with wide dissemination of information (World Bank 2006, 38).

Hence the *Draft Water Law* emphasizes that in due consideration of national security, the economic and social development needs of the country, multipurpose development of surface and groundwater resources are carried out to provide adequate and good quality water for citizens. It refers to the utilization, allocation, protection, safeguarding, monitoring of surface and groundwater resources of the country in responding to drinking, domestic water needs of each citizen and natural life as well as providing water for industrial and irrigation uses and hydropower generation.[24]

The *Draft Water Law* also maintains that water resources, including treated and recycled wastewater, are developed and managed at the 'river basin level.' Indeed, within the existing institutional and legal structure in Turkey, there is very limited planning at the basin level with respect to surface water and virtually none for groundwater and wastewater disposal. There is no integrated plan which considers both ground and surface water availability, nor does existing planning consider water quality, wastewater disposal, current and projected land use, anticipated future demand and return flows, or projected future quantity and quality of water resources (Harmancioglu et al. 2007, 149).

There are detailed stipulations, in the *Draft Water Law*, on water resources utilization and allocation, and specific references to adoption of the 'licensing system' in allocating water to various users by the DSI. Yet, still a clear statement needs to be made in the proposed new water law giving rights to water users, and stating the limits of those rights. These rights are particularly important for irrigation management organizations that may be in danger of losing their access to water as a result of agreements made between DSI and private developers of dams for generating hydroelectric power. The granting of water rights to water users will require the strengthening of processes and procedures for the measurement and recording of water abstractions. At present these systems are weak, with insufficient certainty of the quantity of water being abstracted on a daily basis from river and groundwater resources (World Bank 2006).

The *Draft Water Law* involves a section on the establishment of a water information system by the DSI. Provision of water supply for drinking, domestic, industrial, irrigation, hydropower, spring and flood water are all dealt with in detail. A separate part is entirely devoted to the utilization, allocation and preservation of groundwater. There are relevant sections on cost recovery and devolution of the management of the irrigation and domestic water systems.

[24]Su Kanunu Tasarisi Taslagi (Draft Water Law), 2001, on file with the authors.

When it comes to the organizational structure of water management, the *Draft Law* contains only information on the DSI: its organizational structure (headquarters and regional directorates), duties and tasks. There are no references at all to the other organizations' roles, responsibilities, tasks and functions in the water sector. In fact, earlier, there were attempts to get the concerned parties together including NGOs and academia to discuss main problems in water management in Turkey with a specific focus on legislative and institutional structure. The water law review process started with the initiatives of a group of deputies (2001-2002). Intensive consultations with relevant agencies and authorities were made in order to contribute to the water law review process. Some leading government institutions, namely the SPO, the DSI and a number of civil society organizations provided support to this process.

As another push to those efforts, the Ministry of Foreign Affairs coordinated workshops on the reappraisal of water policy in the period from 2004 until 2005. SPO and the Secretariat General for EU Affairs led these workshops related to the institutional and legal review, respectively.[25] Even though these processes were incomplete, it has already been recognized that with the enactment of a comprehensive national law, the roles and functions of existing actors/institutions need to be reviewed and revised. Moreover, within the context of the Social Transformation (MATRA) project,[26] a National Platform, which brought together representatives from concerned ministries as well as representatives from local governments and universities, was established in 2004 as one of the outputs of the project. The meetings among the stakeholders, mainly from the regional offices of the government institutions involved with the Buyuk Menderes river basin, which gathered during the project implementation as well as the ones gathered under the aegis of the National Platform, provided the platform to discuss the institutional and legal issues of water policy and management at national and regional levels.

All in all, these efforts to review water legislation and institutional structure have not ended up in producing a coherent national water act, which could have been taken up through all the inclusive and participatory processes. Efforts to draft the water law has been dominated by the DSI, even though there have been attempts by the concerned government agencies such as the SPO, Ministry of Foreign Affairs and even by the DSI itself to lead the "water law debate" by including the stakeholders concerned. Yet, these discussions remained rather preliminary, not concerted, partially inclusive of non-governmental entities and incomplete. Parallel to these discussions at national level, a series of projects and review processes have been undertaken by international stakeholders, particularly within the context of projects, which have been carried out in the EU harmonization process (see Sumer and Muluk in this volume).

[25]See Sumer and Muluk in this volume for further discussion on Turkey's alignment with the EU water acquis.

[26]The EU WFD Project 2004, "Implementation of European Union Water Framework Directive in Turkey" Project (MAT01/TR/9/3), 2002-2004, Institutional and Legal Strengthening Report.

Meanwhile, Turkish water policy and management have been changing since the early 1980s, under the transformation of the economy towards liberalization (see Tigrek and Kibaroglu in this volume). The reorientation of the economy has resulted in the restructuring of the institutional set-up in the water sector such as the establishment of the irrigation management organizations, water and sewage authorities (ISKI) and the SPAs on the one hand and the abolishment of the GDRS on the other. The legal framework was also subject to change, particularly with the passing of the relevant legislation to include the private sector in water resources development, hydropower generation and water supply management. Major laws and by-laws have also been adopted to introduce or strengthen environmental protection and water quality management. Yet, the picture is not at all clear particularly from the perspective of the majority of the stakeholders, namely the users in rural and urban settings. There is actually a growing consensus that legislative and administrative changes fall short, when it comes to implementation and enforcement. Compliance with the new policies is rather problematic; all the more since most stakeholders do not accept the changes or do not realize that these changes are responsive to their needs.

References

Ada Strateji (2010) Su Kaynaklar Bakanligi Kurulus Kanunu Tasar Taslagi Onerisi. USIAD, Ankara

Baris ME, Karadag AA (2007) Water resources management issues in Turkey and recommendations. Journal of Applied Sciences, vol 7, no 24: 3900-3908

Bayazit M, Avci I (1997) Water resources of Turkey: Potential, planning, development and management. Water Resources Development, vol 13, no 4: 443-452

Bennett E, Grohmann P, Gentry B (1999) Public-Private Partnerships for the urban environment - Options and issues. PPPUE Working Paper Series Volume I. UNDP and University of Yale, New York

Bilen O (2009) Turkiye'nin Su Gundemi (Turkey's water agenda). DSI, Ankara

Carl Bro International (2001) Analysis of environmental legislation for Turkey. Final Report

Chamber of Civil Engineers Working Group on Water (2009) Su Hakki Raporu. Turkiye Muhendislik Haberleri (TMH), Ankara

Cinar T (2006) Turkiye'de Icmesuyu ve Kanalizasyon Hizmetleri: Yonetim ve Finansman. In: Cinar T, Ozdinc HK (eds) Su Yonetimi: Kuresel Politika ve Uygulamalara Elestiri. Memleket Yayinlari, Ankara, pp 227-252

Cinar T (2007) Personal Interview

Cinar T (2009) Privatization of urban water sewerage services in Turkey: Some trends. Development in Practice, vol 19, no 3: 362-63

Coskun AA (2003) Water law: The current state of regulation in Turkey. Water International, vol 28, no 1: 70-78

Divrak BB (2009) Personal Communication, Ankara

DSI (2009a) Water and DSI. DSI, http://www.dsi.gov.tr/english/pdf_files/dsi_in_brief2009.pdf Accessed 23 August 2010

DSI (2009b) Turkey Water Report. DSI, http://www.dsi.gov.tr/english/pdf_files/TurkeyWaterReport.pdf Accessed 23 August 2010

European Commission (2005) 2005 Regular report on Turkey's progress towards accession. Commission of the European Communities, Brussels
Grontmij Consulting Engineers (2004a) Buyuk Menderes River Basin Management Plan (Draft). Houten
Grontmij Consulting Engineers (2004b) Institutional and legal strengthening, Report on the legal and institutional developments required to meet EU legal requirements in the field of water management in Turkey. Houten
Guler AB (1999) Su Hizmetleri Yonetimi: Genel Yapi. Turkiye ve Ortadogu Amme Idaresi Enstitusu, Ankara, Turkey
Harmancioglu NN et al. (2007) Gediz basin management: Problems and possible remedies. In: Proceedings, International Congress River Basin Management, 22-24 March 2007, Antalya, Turkey, Vol. II, pp 138-153
Kapucu N, Palabiyik H (2008) Turkish public administration. International Strategic Research Organization-USAK, Ankara
Kibaroglu A, Baskan A, Alp S (2009) Neo-Liberal transitions in hydropower and irrigation water management in Turkey: main actors and opposition groups. In: Huitema D, Meijerink D (eds) Water policy entrepeneurs. A research companion to water transitions around the globe. Edward Elgar, Cheltenham, UK
Ozbay O (2006) Turkiye Su Mevzuatinin Gecirdigi Evreler. Haber Bulten, Chamber of Geology Engineers, no 3, July-August 2006: 37-40
Republic of Turkey (2003) Turkey Country Report. Prepared for the Third World Water Forum. World Water Council http://www.worldwatercouncil.org/fileadmin/wwc/Library/Publications_and_reports/country_reports/report_Turkey.pdf Accessed 23 August 2010
Soganci M (2005) Koy Hizmetleri Genel Mudurlugu Kapatilmamalidir. Press Release, Ankara
State Planning Organization (2007) 9. Bes Yillik Kalkinma Plani, Toprak ve Su Kaynaklari Ozel Ihtisas Komisyonu Raporu, DPT, Ankara, Turkey
Topcu F (2006) Suda Dis Kredi: Izmit Ornegi. In: Cinar T, Ozdinc HK (eds) Su Yonetimi: Kuresel Politika ve Uygulamalara Elestiri. Memleket Yayinlari, Ankara, pp 287-316
Unver I (2010) Elestirel Gozle 5403 Sayili Toprak Koruma ve Arazi Kullanimi Kanunu. Toprak Su Enerji. http://www.topraksuenerji.org Accessed 10 June 2010
Vliegenthart FJL, Sargin AH, Gorkmen A, Dogdu MS (2007) A new approach for groundwater management in Turkey: A groundwater management plan according to EU Groundwater Directive (80/68/EEC) and Water Framework Directive (2000/60/EC). International Congress on River Basin Management, Antalya, Turkey
World Bank (2004) Implementation completion report (TF-21388 CPL-38930 SCL-38936) on a loan in the amount of US$100.0 million to the Republic of Turkey for Antalya Water Supply and Sanitation Project. Report no 27700, May 28, 2004. The World Bank, http://www-wds.worldbank.org/external/default/WDSContentServer/WDSP/IB/2004/06/03/000112742_20040603123854/Rendered/PDF/277000TR.pdf Accessed 15 March 2007
World Bank (2006) Irrigation and water resources with a focus on irrigation prioritization and management. Economic Sector Work, Working Paper. World Bank, Turkey
Yazici S (2009) Update: A Guide to Turkish Public Law and Legal Research, http://www.nyulawglobal.org/Globalex/Turkey1.htm#_Laws. Accessed 09 June 2010
Yildiz D (2009) Personal communication, Ankara

Strategic Role of Water Resources for Turkey

Sahnaz Tigrek and Aysegul Kibaroglu

1 Introduction

Turkey's water policy can best be characterised by her desire to gain independence from imported energy sources, to increase production levels of agriculture and to achieve food security, to satisfy increasing water demand from industry and urban and rural populations, and to correct regional economic and social imbalances in the country, thus raising the living standard of the population (Kibaroglu et al. 2005). The inclusion of such social aims led to water resources planning and development being carried out by government agencies through public investment (Kibaroglu et al. 2009).

From the 1920s to the 1950s, Turkey was engaged in state consolidation efforts including, inter alia, the investigation, exploitation and management of natural resources, namely water and land resources. The new government institutions, established at the national level, investigated the development potential of water and land. That is, the Ministry of Public Works (1920) and the Electrical Power Resources Survey and Development Administration (1935) were established with missions to explore the country's hydropower potential, and to carry out civil works and land development as well conducting preliminary hydrological surveys.

As the Turkish State further consolidated in the decades between 1950 and 1980, more focused attention has been paid to socio-economic development, based on water and land resources. At the beginning of the 1960s, only 1.2 million hectares out of 8.5 million hectares of the irrigable land were irrigated. Hence, in the 1960s,

S. Tigrek (✉)
Department of Civil Engineering, Middle East Technical University, Ankara, Turkey
e-mail: sahnaz@metu.edu.tr

A. Kibaroglu
International Relations Department, Okan University, Istanbul, Turkey
e-mail: Aysegul.Kibaroglu@okan.edu.tr

A. Kibaroglu et al. (eds.), *Turkey's Water Policy*,

DOI 10.1007/978-3-642-19636-2_2,

the major objective was to irrigate the fertile lands in south-eastern Anatolia, which comprise one fifth of the irrigable land in Turkey, by using the huge water potentials of the Tigris and Euphrates rivers. In this context, the Lower Euphrates Project was implemented to build a series of dams on the Euphrates to increase hydropower generation and to expand irrigated agriculture. Later on, in the late 1970s, the Lower Euphrates Project evolved and expanded into a larger multi-sectoral development project called the South-eastern Anatolia Project (GAP, its Turkish acronym), which includes 22 large dams, 19 hydropower plants and irrigation schemes extending to 1.7 millions of hectares of land in the Euphrates and Tigris rivers system, which accounts for 28.5 percent of the surface water supply in Turkey (Kibaroglu et al. 2005).

Systematic water resource development started in the 1950s with the establishment of the General Directorate of State Hydraulic Works (DSI) (see Kibaroglu and Baskan in this volume). DSI, the central water agency, designated the major river basins, with their recorded potential for water and land resources, for large-scale development projects.

The adoption of the Constitution of the Turkish Republic of 1961[1] paved the way for state-induced economic and social development directed towards overcoming regional imbalances, with the western regions far ahead of the south-eastern and north-eastern provinces. This strategic orientation included the country's water resources being mainly developed from public sources. With the establishment of the State Planning Organization (SPO) in 1960, comprehensive planning activities have been undertaken in Turkey which comprised construction of physical structures to meet energy and food needs for increasing population as well as realizing socioeconomic development goals expected to provide welfare for the citizens. From then on, Turkey has made considerable progress in augmenting its water supply.

The oil crises of the 1970s, in particular, gave additional impetus to developing the country's hydropower potential. As one representative of the DSI stated: "Since the country suffered badly in the oil crises of the 1970s, the government has embarked upon a programme of indigenous resource development, particularly hydropower and lignite schemes to minimise the dependency of the national economy on imported oil."[2] However, population growth, urbanisation and industrialisation have widened even further the supply-demand gap.

The political and economic crises of the 1970s, 1980s and 1990s erupted in the country now and then, hindered these investments from being completed in a timely fashion. Moreover, water needs for drinking, agriculture, industry and energy purposes had increased exponentially in the second half of the 20th century.

[1]The 1961 Constitution of the Turkish Republic was replaced in 1982.

[2]Quoted in International Water Power and Dam Construction, vol 44, no 12, December 1992, 12. Turkey is also developing other renewable energy resources such as geo-thermal power, wind power, biomass energy, but hydropower will provide the greatest share to tackle the energy deficit (see Energy Information Administration 2002).

Furthermore, the physical structures which were built to increase water supply caused the degradation and depletion of natural resources such as water and land and ended up in certain regions of the country being detrimental to the ecology.

Since the early 1980s, however, neoliberal transformation of the Turkish economy has resulted in significant changes in water policy and management. Stemming from decisions made on 24 January 1980, Turkey became one of the first countries in the developing world to make the shift from state-led development strategies to a model of broad market liberalization and opening to the international economy. To illustrate this point, the introduction of a build-operate-transfer (BOT) model to the energy sector in 1984 under Law No. 3096, enabled the private sector to buy the right to generate, transmit and distribute electricity. This system was later modified and applied in other sectors including water, where it was extended to water supply and sanitation services in municipalities, and to the construction, operation and management of infrastructure, such as dams, hydropower plants and irrigation systems.

The new paradigm was reinforced by international frameworks and thus included in the reform packages of multilateral institutions. Hence, the key advocates of the policy change included the World Bank and other international actors such as the Organization for Economic Cooperation and Development (OECD) as well as private national and international corporations. The policy change specifically in the water sector was supported and carried out by bureaucrats from the ministries concerned and their affiliated institutions.

The real push for liberalization, however, came in the 1990s especially with the European Union (EU) process. After the Helsinki decision of the European Council in December 1999 to grant Turkey candidate status, Turkey hastened the process of liberalization in the water sub-sectors. Liberalization of the hydroelectricity sector was introduced in 1984, and reinforced by important legislation adopted in 2001. Commercialization of water services (drinking water and sewerage) has been underway since late 1980s with increasing roles assigned and played by local private business, transnational water companies and international credit agencies. The devolution of irrigation water management in the early 1990s with the guidance and partial financing of the World Bank serves to illustrate this point. Within the framework of an accelerated programme of management transfer, irrigation associations were established to operate and maintain almost all the irrigation systems in the country (see Topcu in this volume).

2 Geography, climate, water and land resources

Turkey lies between 26°-45° eastern longitudes and 36°-42° northern latitudes with land both in Europe and Asia. The Marmara Sea and two water straits form a natural boundary between two continents. Ninety-seven percent of the country is in Asia and called Anatolia, while the remaining three percent is Thrace (Armstrong and Hunkins 1989). Turkey, with a total area of 780,000 square kilometers, is surrounded by the Black Sea, Bulgaria in the north, the Aegean Sea and Greece in the west, the Mediterranean Sea, Syria and Iraq in the south, Iran in the east and Armenia and Georgia in the north-east (DSI 2009).

The average altitude of the country is about 1,100 to 1,200m. A further 80 percent of Anatolia is above 500m, the average height of Europe and Asia being 330m and 1,050m, respectively. Therefore Central and especially Eastern Anatolia, which consists of several high mountain ridges and high plateaus, receive heavy snow in winter.

Turkey has a subtropical, semi-arid climate with extremes in temperature. The average annual precipitation depth in Turkey is around 643mm, which is lower than 800mm, i.e. the average precipitation depth of the world (Usul 2005). Annual precipitation of Turkey is 501 billion cubic meters (BCM) and 274 BCM is assumed to evaporate from surface and transpire through plants.

Diversity of topography brings diversity in climate. As a result, large variations in precipitation occur, ranging from 250 mm in the south-eastern region to over 3,000 mm in the Black Sea coastal area (Republic of Turkey 2003). The construction of dams and reservoirs have been the main means of saving water during the short rainfall seasons to facilitate year round availability.

In hydrological terms, Turkey's territory features 25[3] large river basins (see Figure 1, Table 1) that exhibit a large variation in average annual precipitation, evaporation and surface run-off parameters. In total, average annual run-off is of approximately 186 BCM of which 112 BCM could be exploited at reasonable cost. Surface water contributes 98 BCM and groundwater 14 BCM. At present, Turkey is utilising 46 BCM (41 percent) of its overall capacity (DSI 2009).

The annual flows of many large basins fluctuate and show a high variability throughout the year. Further nearly every 30 years, a drought period occurs. Akkemik et al. (2005) analyzed drought periods in Anatolia over the last 350 years and reported that the duration of the drought is generally one year. Sometimes it lasts two years and occasionally three years. However, the number of drought years shows an increasing trend since 1960. The last drought was in 1994 and lasted five years (Komuscu et al. 2005).

Heavy snow falls in Eastern Anatolia in winter, and then in spring several small streams and brooks join and create two of the world largest rivers, namely the Euphrates and Tigris rivers. Total annual flow rate of these two rivers is about 52 BCM. The total drainage area of Euphrates is 127,304 square kilometers and has a total length of 2,800km. But only 971km of the river is within the Turkey's border. The Tigris has a total length of 1,900km of which 523km lie within Turkey corresponding with 57,614 square kilometers drainage. Both rivers join before discharging into the Gulf. From the face of the Eastern Anatolia Mountains one of the world's fastest rivers - the Coruh River - is born where it flows 150km within

[3]Many sources, including the official ones, had maintained that Turkey had 26 river basins. Yet, the recent documents such as the *Turkey Water Report*, General Directorate of State Hydraulic Works, DSI, 2009 as well as DSI's official website display that Turkey has 25 river basins, where the Euphrates-Tigris is considered as one single basin.

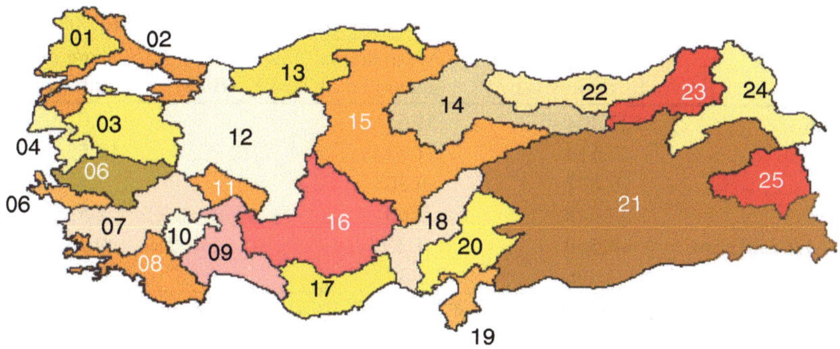

Fig. 1 Turkey's 25 river basins (see Table 1 for legend) (DSI website 2009)

Table 1 Catchment area and discharge of Turkey's 25 river basins (DSI website 2009)

River basins	Catchment area (km^2)	Mean annual discharge (BCM)	Contribution to total (percent)
(21) Euphrates-Tigris	184,918	52.94	28.5
(22) East Black Sea	24,077	14.90	8.0
(17) East Mediterranean	22,048	11.07	6.0
(09) Antalya	19,577	11.06	5.9
(13) West Black Sea	29,598	9.93	5.3
(08) West Mediterranean	20,953	8.93	4.8
(02) Marmara	24,100	8.33	4.5
(18) Seyhan	20,450	8.01	4.3
(20) Ceyhan	21,982	7.18	3.9
(15) Kizilirmak	78,180	6.48	3.5
(12) Sakarya	58,160	6.40	3.4
(23) Çoruh	19,872	6.30	3.4
(14) Yeşilirmak	36,114	5.80	3.1
(03) Susurluk	22,399	5.43	2.9
(24) Kura-Aras	27,548	4.63	2.5
(16) Konya	53,850	4.52	2.4
(07) Buyuk Menderes	24,976	3.03	1.6
(25) Lake Van	19,405	2.39	1.3
(04) North Ege	10,003	2.90	1.1
(05) Gediz	18,000	1.95	1.1
(01) Meriç-Ergene	14,560	1.33	0.7
(06) Kuçuk Menderes	6,907	1.19	0.6
(19) Orontes	7,796	1.17	0.6
(10) Burdur Lakes	6,374	0.50	0.3
(11) Akarçay	7,605	0.49	0.3
Total	779,452	186.86	100

Turkey and after flowing 26km before discharging into the Black Sea in Georgia. The Kura rises in Turkey and Georgia after some 210 km. After 390 km through mountainous terrain the river flows into the Azerbaijan steppes and finally

Table 2 Water potential generated in Turkey's transboundary river basins (DSI 2009)

Transboundary river basin	Catchment area in Turkey (km^2)	Mean annual flow generated in Turkey (BCM)	Share of total usable potential*)
Euphrates	127,304	31.61	17.0
Tigris	57,614	21.33	11.5
Coruh	19,872	6.30	3.4
Kura-Aras	27,548	4.63	2.5
Meric/ Maritsa/Evros	14,560	1.33	0.7
Asi/Orontes	7,796	1.17	0.6

*) The total usable potential is 112 BCM of which 98 BCM is surface water and 14 BCM is groundwater.

discharges into the Caspian Sea. The Aras also rises in Turkey and after 300 km the river forms several borders: between Armenia and Turkey, for a very short distance between Azerbaijan and Turkey, between Azerbaijan and Iran, between Armenia and Iran, and finally again between Azerbaijan and Iran. Eighty kilometers after crossing the border with Azerbaijan the Aras joins the Kura. Other transboundary rivers are the Meric/Maritsa/Evros River shared between Turkey Bulgaria and Greece, and the Asi/Orontes River which traverses Lebanon, Syria and Turkey. Table 2 shows the contribution of transboundary rivers to Turkey's available water resources; the Tigris and Euphrates rivers system alone account for 28.5 percent.

Almost one third of the total area of the country (78 million hectares) is classified as agricultural land which is about 28.05 million hectares. One third of 28.05 million hectares agricultural land can be classified as irrigable land. Hence, an estimated 8.5 million hectares (7.9 million hectares with surface water and 0.6 million hectares with groundwater resources) is economically irrigable under available technology. In other words, the agricultural area is 36 percent of total land, and 30 percent of the agricultural area can be irrigated technically and economically. Presently 5.28 million hectares of land are being irrigated.

3 Infrastructure development

Due to the high population and urban growth rates (4 percent), many regions of the country (south-east, Marmara, Aegean and Mediterranean) are already facing seasonal or even chronic water shortages therefore necessitating infrastructural development in the water sector. Today, an extensive network of dams and reservoirs is maintained of which the larger dams serve multiple purposes (e.g. flood control, irrigation, domestic water supply, and hydropower) (see Table 3).

Table 3 Multi-purpose water infrastructure in Turkey (in operation and planned as of March 2009) (DSI 2009)

	In operation	Under construction or in program
Dams	673 Large scale projects: 260 Small scale projects: 413	146 Large scale projects: 63 Small scale projects: 83
Hydropower plants	172	Under construction: 258
Capacity	13,700 MW	10,846 MW
Annual production	48,000 GWh	39,404 GWh
Irrigation (million hectares)	5.28	0.23
Domestic water (BCM)	3.16	0.50
Flood control (million hectares)	1.0	0.4

Box 1 History of water works in Anatolia

The need for storage facilities in Anatolia is not confined to the present. In many parts of Anatolia, the remains of water structures dating from early civilizations can be easily found. The Urartus and the Hittites used to build water structures including conveyance channels and dams. The Hittites (from 1,800 to 850 BC) practiced irrigation in Central Anatolia. The remains of at least six dams from Hittites were found in Central Anatolia. Among them Koyluotu Dam which has a height of 25-30m and a crest of 900m can be classed as a large dam according to present standards. It is believed that dams served for water supply, some for irrigation some for domestic usage (Ozis 1999). In the 13th century BC, the Urartus irrigated land around the city of Van. They built a 56km long Samran channel which has a capacity of 2-3 cubic meters per second in order to carry fresh water to the city (Belli 1997). Today, this channel is used to irrigate 2,000 hectares of land.

In the Byzantine and Roman era, water storage structures, aqueducts and different conveyance systems were built (Ozis 1999). Among the conveyance systems with a length of 240km the Pinarhisar-Istanbul conveyance system is the largest. Another remarkable system is Magradag-Pergamon which has a length of 195km and also included a lead-pipe inverted siphon under the largest pressure during the Hellenistic period. There are several remains of dams from Roman times. Especially those dams that include Roman concrete are classified as Roman dams. Among them the Cavlik Dam in Southern Anatolia can be distinguished from others by its function. It diverts a creek into a tunnel in order to prevent siltation of the antique harbour of Seleucecia Pieria.

The Seljukides built several impressive bridges in Anatolia, some of which are still in use. Among them the Malabadi Bridges in Batman has a 39m single span, and is believed to be one of largest masonry bridges in the world (Ilter 1978). Although Seljukides built several dams in Iran and in Central Asia, in Anatolia only two dams are found in the Konya region. The reign of

(continued)

Seljukides in Anatolia corresponds to conflicts and wars. That can be taken as the reason why they built many bridges and local water structures, but not big storage facilities. This trend continued during the Ottoman period as well. The Ottomans also built aqueducts, conveyance systems, cisterns, public fountains, but not large storage structures, with only some exceptions in Istanbul. Water supply for the capital city of Istanbul was a crucial question for the Ottomans, as it had been for the previous empires. During the Ottoman Empire, groundwater supplies were not sufficient and transmissions lines were built to divert water from springs to public fountains. Later facilities for water transfer from Lake Terkos in 1869 and Elmali-I Dam were built in 1888. Some of them are still in use (Altinbilek 2006). The first treatment plant was built in 1926 in Istanbul. However, there were no systematic irrigation activities during the Ottoman era. In 1875, a severe drought affected the Konya region. Consequently, the Ottoman government decided to build a diversion weir and irrigation channel. The project, which was named as the Cumra Irrigation and Drainage Project, is the first modern irrigation project in Anatolia being carried out between 1908 and 1914 in Konya (Demir 2001).

Hydropower generation has a much shorter history in Anatolia. In 1902, a small hydroelectric power plant with 60 KW for lighting was established in Tarsus. In 1914, the production and distribution of electricity in Istanbul was started by companies which were founded with the special permission of the Ottoman Sultan. In 1923, at the time of the foundation of the Republic of Turkey, electricity was available only in Istanbul, Izmir and Tarsus with a total installed capacity of 29.7 MW. Usage of electrical energy for purposes other than lighting began after 1930 when industrial development started (Altinbilek 2002).

4 Water use per capita and by sectors

Turkey still has a growing population despite the fact that the pace of population increase has slowed down considerably in the last several decades. While the population of Turkey has steadily grown, the rate of population growth has decreased from 2.52 percent to an annual rate of 1.31 percent.[4] The rise of the population has a two-fold impact:[5] population growth not only increases the demand for food, and thus of water, but also causes the decline of per capita

[4]As of December 31, 2008, http://www.tuik.gov.tr. Accessed 17 November 2009.

[5]A "water stress index" based on the approximate minimum amount of water per person necessary to maintain an adequate quality of life in a moderately arid zone, was developed by M. Falkenmark: "The massive water shortage in Africa: why isn't it being addressed?" Ambio. 18(2):112-18, 1989. The Falkenmark water stress index measures per capita water availability and considers that a per capita water availability of between 1,000 and 1,600 cubic meters indicates water stress, 500–1,000 cubic meters indicates chronic water scarcity, while a per capita

water resources. Due to population growth and urbanisation, water and energy demand is expected to increase as well. According to the DSI statistics, annual per capita water availability is 1,600 cubic meters with population of about 73 million in 2010. By the year 2023 this amount will decline to 1,125 m^3 per capita/year with an expected population of 100 million (DSI 2009).

As of 2009, water use, related to sectors, was as follows: the irrigation sector used 34 BCM/year (74 percent), domestic water 7 BCM/year (15 percent), and industry 5 BCM/year (11 percent). In total, 46 BCM/year (41 percent) of the usable water potential is utilised.

Although *agriculture*'s contribution to the Turkish economy is declining (from 35 percent in 1970 to 8.8 percent in 2008), agriculture is still vital to the national economy employing 30 percent of the country's work force. Crop production on the 5.28 million hectares of irrigated land creates the basis of agricultural exports to European countries and to Near East and North African regions. Export of agricultural and agro-industrial commodities were valued at US$ 4.4 billion and accounted for 16 percent of Turkey's total export value in 2001. According to DSI estimates, 8.5 million hectares of land are technically and economically irrigable and subject to further development. It is expected that the high share of water consumption in agriculture will decline from 74 percent at present to 64 percent in 2023 through the use of modern irrigation techniques. On the other hand, domestic and industrial use would increase to 16 percent and 20 percent in this period, respectively.

Domestic water use accounts for 15 percent of the water resources developed, showing high variations throughout the country. Domestic water use is highest in the Marmara Region, and far below the national average in north-eastern and eastern Anatolia. With more than half of Turkey's population living in urban areas, construction of water supply, sewerage and wastewater treatment plants has received high political attention. Population growth together with high internal migration from rural to urban areas over the last 30 years has caused domestic demand to increase. In urban areas, access to a drinking water supply was 83 percent in 1990 and 81 percent in 2000; in rural areas, it was 72 percent in 1990, and 86 percent in 2000 (Republic of Turkey 2003). In general, 83 percent of the population (urban 94 percent, rural 62 percent) of Turkey has access to improved sanitation. 40.2 percent of the total population is served by wastewater treatment plants. 78.4 percent of the total municipal population is connected to a sewage system. The other municipalities lack a treatment system or have only primary (physical) treatment, or they lack the capacity to operate the established sewage treatment plants.

---

water availability below 500 cubic meters indicates a country or region beyond the 'water barrier' of manageable capability. See Falkenmark and Widstrand (1992).

Box 2 Inter-basin water transfer projects for "thirsty" Istanbul

Istanbul has been challenged to meet water needs throughout its history. In its early years, water was supplied by groundwater, small springs and cisterns. In the Roman period four long gravity transmission lines were built, and during the Byzantine Empire one BCM storage capacity had been reached. During the reign of Sultan Mehmet the Conquerer (1453 AD), the city's population was estimated at around 100,000. After conquest five new transmission lines and eight dams with a capacity of 1.7 MCM were built. In 1869, a French company constructed necessary structures to bring water from the Terkos Lake to the European side, and another company was founded in 1888 for the Asian part. In 1933 private companies were abolished and water supply was transferred to the Istanbul Water Administration. Later in 1971, DSI prepared the Water Supply Master Plan for Istanbul (DAMOC 1971). In 1981 water and wastewater services were combined under the Istanbul Water Sewerage Administration with the responsibility of planning, construction and operation of water supply, sewage collection and treatment.

Today the population of Istanbul is around 11.5 million of which two thirds live on the European side. However, water sources on the European side dramatically decreased. In the interim, therefore 126 MCM per year is transported from the Asian to the European side by two under-marine pipelines. The 1971 Master Plan for Istanbul's water supply suggested developing eleven water sources, six of which have already been put into service. However, the remaining five could not be implemented due to geological problems, unsuitable water quality and because of low capacities. DSI has started two large water supply projects, namely the Yesilcay and the Greater Melen projects (Altinbilek 2006). The Yesilcay project has been designed to convey water from the Goksu and Canak rivers to Istanbul from Omerli Dam over a distance of 60 km, to meet the medium-term water need of Istanbul. 335 MCM will be supplied annually in order to meet the demand of 1.5 million inhabitants. Total investment of the Yesilcay system is estimated at US$ 271 million. The Greater Melen system which is divided into four stages, will supply 1.18 BCM, and is expected to secure Istanbul's water needs until 2040. Its source is the Melen River which is located 170km east to Istanbul. Raw water will be pumped to a treatment plant via a transmission line, and treated water will be conveyed to a service reservoir via a tunnel and a pipeline under the Bosporus. The first stage of the Melen system (25km tunnel, 180 km transmission line) will provide an additional 268 MCM/year of water to meet the potable and domestic water needs of an additional 2.75 million people. Its total cost is US$ 1,180 million. However, after 2040, no more inland water resources can be developed, and sea water will have to be desalinized.

Box 3 Domestic water crisis in Ankara

Ankara was chosen as the capital city of the new Turkish Republic gaining administrative importance and weight in the region. Since then it has continuously expanded in size and population. Geographically Ankara is situated between two large river basins, i.e. Sakarya to the west and Kizilirmak to the east. Long ago, Ankara itself was the birthplace of several small streams. However during the early planning stage of the city these water systems were not protected. In early 1920s Ankara had 20,000 to 30,000 inhabitants; and when the number climbed up to 300,000, the city faced serious water shortage problems between 1940 and 1950. As a consequence, the Ankara Water Administration was established.

The Ankara Water Administration searched for new water resources from both groundwater and surface water, and three more dams were built. In 1969 the Ankara Water and Wastewater Master Plan was prepared, which provided for the construction of the Camlidere Dam and the Ivedik treatment plant. During the rearrangement of the governance mechanism of large municipalities in the early 1980s Ankara Metropolitan Municipality and Ankara Water and Wastewater Works Administration were established. Two master plans which were commissioned by DSI, i.e. the 1969 Master Plan and the 1995 Master Plan, proposed to implement a system on the Gerede River North West of Ankara to meet future demand. The Kizilirmak River east of Ankara, was also evaluated as a long-term solution. However, the 1969 Master Plan did not consider the Kizilirmak because it contains high levels of salts. By 1995, the required reverse osmosis and other desalinization methods were marketable technologies, so Kizilirmak was fully assessed. The 1995 Master Plan advised Ankara to implement the Gerede system by 2003, and the Kizilirmak system in 2027 (Master Plan 1995).

Nevertheless, the municipality of Ankara did not follow the plan. Between 2006 and 2008 Ankara experienced a serious drought and water levels in the dams were lower than usual. In August 2007, the city had to cut off the water supply. The Municipality started with the Kizilirmak Emergency Plan which would transfer water from the Kesikkopru Dam on the Kizilirmak River to Ankara. The structure was constructed in haste from April 2007 to March 2008, and cannot be considered as a thoroughly planned project, let alone its financing. Water is pumped from the intermediary reservoir of the Kizilirmak system, Kesikkopru, and consists of three adjacent pipelines crossing a distance of 128km. The water travels 100km mainly northwest from Kesikkopru, and then due north 28km where it reaches the Ivedik treatment plant north of Ankara. According to the Master Plan another treatment plant should be constructed south or east of Ankara rather than north (where Ivedik is located). The total length of the transmission will then extend from 85km to 128km in order to reach the treatment plant. Costs will increase the overall project costs are estimated in the order of 572 million new Turkish Liras.

The percentage of water use in *industry* has not changed considerably over the past few years, being slightly over 11 percent (52 percent from surface water, 48 percent from groundwater). The major water consuming industries are steel, chemical, paper manufacturing, petroleum refining and agro-industry. In 2000, the greatest industrial demand came from the highly industrialised Marmara Region. Other industrial centres developing in the context of GAP will not change the overall percentage of industrial water use, and will only change the regional distribution.

The *annual per capita energy consumption*, which is at present far below the world average, is expected to increase from 1,840 kWh (1999) to 6,794 kWh (2020). To achieve this growth rate and reach energy consumption levels of the OECD countries, huge investments are envisaged (Altinbilek, no year). Currently, Turkey has 172 *hydroelectric power plants* in operation with a total installed capacity of 13,700 MW generating an average of 48,000 GWh/year, which is 35 percent of the economically viable hydroelectric potential. Annual energy consumption per capita in Turkey has reached 2,900 kWh which is above the world average of 2,500 kWh. The average energy consumption for the developed countries is 8,900 kWh, but it varies from 12,322 kWh in the USA to 827 kWh in China. Annual increase in energy consumption is 8-10 percent in Turkey except for the recession years (DSI 2009). In the 1970s Turkey was seriously hit by the energy (oil) crises and after 1997 became an importer of electricity. At present, hydropower provides about 40 percent of the total power generated, but there is more additional potential. The hydropower share is expected to increase in particular through the construction of power plants on the Euphrates and Tigris (see Sen in this volume).

Based on these overall water use and energy projections, Turkey considers herself not to be a water rich country. With 1,600 m^3 per capita per year (2008) and an expected decline to 1,125 m^3 in 2023, Turkey is moving from a relatively water-endowed country to one where water availability will reach critical levels. This projection explains why DSI argues in response to the World Commission on Dams' Final Report that: "dam construction is a vital and unavoidable programme for the country. [...] while the countries being in the leading positions of the [WCD] process have developed their water resources with about the level of 100 percent, the prejudiced findings of the report may probably prevent the water resources development projects planned by the developing countries, such as India, Turkey, with the development level of 30 percent, and China." (DSI 2001) Turkey, having developed only about 30 percent of her water potential would be in dire need of producing and providing cheap energy, and improving the living standard of her citizens by providing adequate water (DSI 2001).

While Turkey's major focus is on continuing water resource development because of their economic and social potential, protection of water-based ecosystems in rivers, lakes and deltas, and water pollution control is increasingly acknowledged, but has yet to reach satisfactory levels (Ministry of Environment 1998; Republic of Turkey 2003). However, both Turkey's National Environmental Action Plan and the Ninth Five Year Development Plan give top priority to these issues (State Planning Organisation 2007).

5 Strategic role of the Euphrates and Tigris rivers system: South-eastern Anatolia Project (GAP)

While Turkey intends to develop water resources all over the country, GAP is of particular importance for generating hydropower and producing agricultural commodities. Futhermore, it is the government's aim to stabilise this under-developed region politically by significantly raising the population's standard of living.

GAP is Turkey's largest integrated development project and perceived as being vital to the Turkish economy. It has the potential to meet the rising demand for hydropower caused by population growth along with urbanisation and the country's impetus in industrialisation. Upon the completion of the GAP project 1.7 million hectares of land will be brought under irrigation (1.08 million hectares on the Euphrates, 600,000 hectares on Tigris), accounting for nearly one-fifth of Turkey's irrigable land; energy production in the region will reach 27 billion kWh, per capita income will rise by 209 percent; and about 3.8 million people will be provided employment opportunities. This would be accomplished through the construction of 22 dams, 19 hydropower plants, and extensive irrigation and drainage networks (Kibaroglu 2002, see also Box 4).

Importantly, GAP was conceived and implemented as an integrated regional development project in one of the most backward and under-developed regions of Turkey. The basic development objectives of GAP are phrased as: to raise the income levels in the GAP region by improving the economic structure in order to narrow the regional income disparities; to increase the productivity and employment opportunities in rural areas; to enhance the assimilative capacity of larger cities in the region; to contribute to the national objective of sustained economic growth, export promotion, and social stability by the efficient utilisation of the region's resources. To these ends, GAP has shifted from a pure infrastructure development project into a project in support of sustainable development with additional investments made in urban and rural infrastructure, agriculture, transport, industry, education, health, housing and tourism.

Although there are visible economic and social achievements,[6] the GAP project and in particular the construction of large dams has come in for sharp criticism. The objections refer particularly to the resettlement issue, environmental and cultural aspects, and the implications of sharing water with Syria and Iraq (the latter issue will be discussed in Part II).

Box 4 The South-eastern Anatolia Project (GAP) in brief
The GAP project area lies in south-eastern Turkey, covering nine provinces, corresponding to approximately 10 percent of Turkey's total population. The project area includes the watersheds of the lower Euphrates and Tigris rivers

(continued)

[6]See the evaluation of the actual impacts of the Ataturk Dam by Tortajada (2000).

and the upper Mesopotamian plains. Its centrepiece is the Ataturk Dam, which was completed at the beginning of the 1990s, with a total storage capacity of 48.7 BCM, and an installed electricity-generating capacity of 2,400 MW. There are 13 large sub-projects altogether, seven of which are on the Euphrates River[7] and six on the Tigris.[8] Major works are the Sanliurfa Tunnels, the Birecik and Karkamis dams on the Euphrates and the Ilisu and Cizre dams on the Tigris.

GAP's aim is to increase the irrigated land from 2.9 percent to 22.8 percent of the total area of the region, which subsequently would lead to a decrease of rain-fed agriculture from 34.3 to 10.7 percent. With the irrigation systems envisaged, Turkey is determined to develop agriculture and agro-industrial production for export and to raise the standard of living in the region, in that way also stopping migration from the region to metropolitan cities. To achieve these ambitious goals would require putting 100,000 hectares into production in the Euphrates basin each year beginning in 1993, and another 60,000 hectares per year in the Tigris basin (Unver et al. 2003).

Due to high investment cost, GAP is considered to be a very costly project: US$ 32 billion is the estimated total cost of which US$ 16 billion have been spent so far. Due to the transboundary flows involved, the Turkish Government was not able to secure international finance, an exception being German and Swiss credits which could be obtained for purchasing equipment. The severe economic and budgetary crisis in Turkey along with, for example, the slow pace of land redistribution caused a considerable delay in implementing the projects. Despite these drawbacks, Turkey is persistently pursuing its plans to harness the Euphrates and Tigris rivers. As of June 2008, some 272,972 hectares have been brought under irrigation. As of 2008, out of 22 dams 13 are operating, and out of 19 hydropower plants 9 are completed and in operation.

6 Conclusion

This chapter has aimed at introducing the political geography of water in Turkey, by highlighting the physical (climate, water, land resources) conditions and the political economic framework. Turkey is situated in a semi-arid climatic zone, with a limited amount of freshwater resources. Water demanding sectors have increased and diversified, such as environment, tourism, industry along with the traditional ones like agriculture and domestic uses. Nearly one third of the (surface) freshwater

[7]The Lower Euphrates Project includes the Ataturk Dam, the Sanliurfa Tunnels and five more sub-projects, i.e. Karakaya, Euphrates Border, Suruç-Baziki, Kahta-Adiyaman, Gaziantep, Gaziantep-Araban.

[8]Tigris, Kralkizi, Batman, Batman-Silvan, Garzan, Ilisu, and Cizre.

resources in the country are transboundary, which plays a decisive role in water policy and management. Water and land resources development had played a key role in consolidating the Turkish State during the early decades of the Republic. State-led economic planning and investment culminated in massive infrastructure (dams and irrigation systems) development since the late 1960s. Results and impacts of water-based regional socio-economic development manifest themselves in the GAP, which aims at harnessing the waters of the Euphrates-Tigris rivers system for "sustainable human development." Liberalization of the Turkish economy since the early 1980s enabled the involvement of the private sector in water resources development and management. Hence, water policy discourse and practice in modern Turkey can only be properly understood with due consideration of the macroeconomic context.

References

Akkemik U, Kose N, Aras A, Dalfes N (2005) Anadolu'nun son 350 yilinda yasanan onemli kurak ve yagisli yillar. Avrasya Yer Bilimleri Enstitusu, Turkiye Kuvaterner Sempozyumu, TURQUA-V http://www.eies.itu.edu.tr/TURQUA/pdf/turqua5/TurquaV_129-135_Akkemik_vd.pdf. Accessed 20 August 2010

Altinbilek HD (2002) Hydropower development in Turkey. Paper presented at HYDRO 2002: Development, Management, Performance. 4-7 November 2002, Kiris, Antalya

Altinbilek HD (2006) Water management in Istanbul. International Journal of Water Resources Development, 22(2):241- 253

Armstrong DG, Hunkins FP (1989) Merrill World Geography. Merrill Publishing Company, Colombus, Ohio

Belli O (1997) Dogu Anadolu'da Urartu Sulama Kanallari. Arkeoloji Sanat Yayinlari, Istanbul

DAMOC Consortium (1971) Master plan and feasibility reports for water and sewerage for the Istanbul region. DAMCOC, Istanbul

Demir A (2001) Su ve DSI Tarihi. Devlet Su Isleri Vakfi, Ankara

DSI (2001) Response to the Final Report of the World Commission on Dams. World Commission on Dams. http://www.dams.org/report/reaction/icold_turkey.htm. Accessed 15 May 2009

DSI (2009) Turkey Water Report. DSI, Ankara

DSI website (2009) Toprak Ve Su Kaynaklari [Soil and water rersources]. DSI. http://www.dsi.gov.tr/topraksu.htm. Accessed 17 November 2009

Energy Information Administration (2002) Turkey: Environmental Issues. Energy Information Administration, U.S. Department of Energy, Washington DC http://www.eia.doe.gov/emeu/cabs/turkenv.html. Accessed 23 March 2004

Falkenmark M and Widstrand C (1992) Population and Water Resources: A delicate balance. Population Bulletin. Population Reference Bureau, Washington DC

Ilter F (1978) Osmanlilara Kadar Anadolu Kopruleri, Karayollari Genel Mudurlugu Matbaasi, Ankara

Kibaroglu A (2002) Building a regime for the waters of the Euphrates-Tigris river system. Kluwer Law International, London, The Hague, New York

Kibaroglu A, Baskan A, Alp S (2009) Neo-Liberal transitions in hydropower and irrigation water management in Turkey: main actors and opposition groups. In: Huitema D, Meijerink S (eds) Water policy entrepeneurs. A research companion to water transitions around the globe. Edward Elgar, Cheltenham, UK

Kibaroglu A, Klaphake A, Kramer A, Scheumann W, Carius A (2005) Cooperation on Turkey's Transboundary Waters. Research report, German Federal Ministry for Environment, Nature Conservation and Nuclear Safety, Berlin

Komuscu AU, Erkan A, Turgu E (2005) Normallestirilmis Yagis Indeksi Metodu (SPI) ile Turkiye'de Kuraklik Olusumunun Cografik Analizi. Turkish State Meteorological Service. http://www.meteor.gov.tr/2005/arastirma/dogalafetler/kuraklik.htm. Accessed 30 September 2009

Master Plan Report on Ankara Water Supply Project (1995), Bayindirlik ve Iskan Bakanligi (4 Volumes)

Ministry of Environment (1998) National Report prepared for the 7th Meeting of the Conference of the Contracting Parties to the Convention on Wetlands (Ramsar, Iran, 1971). Ankara: General Directorate of Environmental Protection, Ministry of Environment, Republic of Turkey. http://www.ramsar.org/cop7/cop7_nr_turkey.htm. Accessed 3 June 2005

Ozis U (1999) Historical dams in Turkey. Nurol Printing and Packing Industry Inc., Ankara

Republic of Turkey (2003) Turkey Country Report. Prepared for the Third World Water Forum. World Water Council http://www.worldwatercouncil.org/fileadmin/wwc/Library/Publications_and_reports/country_reports/report_Turkey.pdf. Accessed 23 August 2010

State Planning Organization (2007) 9th Development plan, Land and Water Special Committee report. State Planning Organization (DPT), Ankara

Tortajada C (2000) Evaluation of Actual Impacts of the Ataturk Dam. Water Resources Development, vol 16, no 4:453-464

Usul N (2005) Engineering Hydrology. METU Press, Ankara

Unver O, Gupta R, Kibaroglu A (2003) Water Development and Poverty Reduction. Kluwer, The Hague, Boston

Challenges for Turkey to Implement the EU Water Framework Directive

Vakur Sumer and Cagri Muluk

1 Introduction

The European Union's (EU) Water Framework Directive (WFD) has been designed to be the centrepiece of legislation for the management of European waters. It was finally adopted in 2000 following a series of lengthy negotiations among the parties involved. It is accepted as the "constitution" of water-related legislation in the European Union (Cicek 2010). According to Estevan and Naredo (2004), the WFD demands a fundamental change in the way water resources planning and management is understood. It is even regarded as "the most ambitious and complex piece of legislation on environment ever enacted in the EU" (Garrido and Llamas 2009, 175). It aims to harmonize and streamline existing water legislation throughout the EU in order to achieve a "good water status" of all EU waters by 2015, at the latest.

In its march towards EU membership, Turkey will face the challenge of transposing and implementing all elements set out in the WFD. As the WFD has become part of the *acquis* (the total body of EU law accumulated thus far), Turkey is obliged to comply with it by date of accession. Yet, for this to be realized, tremendous efforts as well as numerous investments are needed. Further, despite giving ample room for flexible solutions which will enable Member States to determine their own solutions when addressing their country-specific problems, the WFD has a challenging schedule for implementation. The main issues to be addressed by Turkey could be summarized as follows: creating a reliable inventory of water data; establishing a proper monitoring system; setting up pricing systems for all sectors

V. Sumer (✉)
Department of International Relations, Middle East Technical University, Ankara, and Selcuk University, Konya
e-mail: vsumer@gmail.com

C. Muluk
GIZ Office Ankara, Turkey
e-mail: cmuluk@gmail.com

A. Kibaroglu et al. (eds.), *Turkey's Water Policy*,
DOI 10.1007/978-3-642-19636-2_3, © Springer-Verlag Berlin Heidelberg 2011

taking into account the "full cost recovery" principle; institutionalizing participation of all interested parties in water management; delineating river basins, designing respective management plans and river basin organizations mandated to implement measures to reach the WFD's environmental objectives for all water bodies.

This chapter first summarizes the evolution of water policy in the EU, the origin of the WFD and describes its basic characteristics. Following this, both accomplishments and shortcomings of Turkey's endeavours to comply with the WFD will be examined. It is argued in this chapter that harmonization with the WFD is a complicated undertaking for Turkey due to different types of challenges including, but not limited to, the need for organizational/institutional reforms, improving institutionalized public consultation processes and for stakeholder involvement, setting up new pricing schemes, and mobilizing sufficient financial means for investing in sectors critical to achieving the WFD's objectives.

2 The EU Water Framework Directive: an integrated approach towards European water management

The issue of water has generally been taken as an environmental issue within the EU. Water quality matters had always the upper hand when compared with quantity issues because of the relative abundance of water in most of Europe. All in all, water legislation was developed within the broader sphere of environmental legislation. In the literature, the EU's water policies have been analyzed in terms of three waves.[1] The first wave (1975–1980) was characterized by setting "environmental quality standards" on the one hand and "emission limit values" on the other. Following this so called "dual approach", binding quality targets for drinking water, fish waters, shellfish waters, bathing waters and groundwater were set. In other words, the main aim of the first wave was to define water quality objectives for particular water types. The second wave (1981-96) comprised the review and improvement of existing legislation and the setting up of e.g. the Urban Wastewater Treatment Directive (1991), the Nitrates Directive (1991), and the Directive for Integrated Pollution Prevention Control (1996). The third wave started in 1995 and is still on-going. Instead of a "patchwork", a more coherent legislation has been envisaged by the EU Member States where two preceding approaches were combined to target at developing a common framework for European water policy. It is considered as an "integrated approach" paradigm by replacing the water directives that are currently operational (Kaika 2003). The work of the WFD lasted from mid-1995 until 2000. Finally, the WFD was adopted in 2000.

[1]See http://ec.europa.eu/environment/water/water-framework/info/intro_en.htm. Accessed 5 June 2010; http://www.freedrinkingwater.com/water_quality/quality1/1-europe-water-quality-issues.htm. Accessed 2 December 2009.

The WFD introduces a new, combined approach to EU water policy aimed at protecting and improving all types of European waters. For the first time in the development of European water policy it constitutes a single piece of framework legislation that aims at coordinating environmental objectives and measures instead of addressing improvements in different water-related sectors separately. Within this framework, the term "integration" is central (Borja 2005), whereas integration refers to cross-sectoral integration, considering quantitative and qualitative aspects, and the water cycle as a whole. This becomes clear in the wording of the WFD Directive which stipulates that "control of quantity is an ancillary element in securing good water quality and therefore measures on quantity, serving the objective of ensuring good quality, should also be established" (Recital 19). Further, the WFD considers that "the quantitative status of a body of groundwater may have an impact on the ecological quality of surface waters and terrestrial ecosystems associated with that groundwater body" (Recital 20). In sum, an ecological and holistic approach is introduced together with river basin planning, a strategy for eliminating pollution from dangerous substances, public information and consultation as well as the application of economic incentives.

The Directive proposes a three-stage process for implementation: In the first one, characteristics of every river basin district will be assessed (Art 5) (by the end of 2007). In the second stage, programmes of measures aiming at the protection and sustainable use of water resources for each river basin district will be identified, and river basin management plans have to be set up (Art 11 and 13) (by the end of 2009). In the last stage, programmes of measures will be operational (Art 11) (by the end of 2012). Each of river basin management plans is a detailed account of how the environmental objectives – such as ecological status, quantitative status, chemical status and protected area objectives – are to be reached within the timescale required. According to the WFD, "River basin management plans shall be reviewed and updated at the latest 15 years after the date of entry into force of the Directive and every six years thereafter" (Art 13.7).

3 Challenges for Turkey to harmonize its water policies with the EU water *acquis*

As assessed by the European Commission (EC), the level of Turkey's alignment with water-related EU legislation has hitherto been "low" (European Commission 2009, 81) due to reasons which lie inside and outside the water sector. As Turkey does not see a clear membership perspective, it remains reasonably reluctant to pressure for implementing the WFD, particularly with regards to transboundary rivers (Saner 2008).

To get a better understanding of the challenges and the reasons behind slow implementation, one has to take a closer look at Turkey's water sector policies and management which can be characterized by supply-side orientation with very few demand management tools; lack of coordination among responsible institutions; a centralized approach to river basin planning in which utilization of water resources remains the core theme; low level of public participation in decision-making; insufficient pollution control; and weak enforcement of regulations. Furthermore, Turkey shares a number of transboundary river basins with its neighbours and one of the problematic issues for Turkey will be the implementation of WFD Article 3 which requires close cooperation among riparians in river basin planning and management (see Part II in this volume).

Environmental objectives

According to WFD Article 4, its objectives are to prevent the deterioration in status of surface water; to "protect and enhance all artificial and heavily modified bodies of water", with the aim of achieving a "good surface water status" in the year 2015, at the latest; and to achieve the objectives applicable to Protected Areas established under EU legislation. The WFD classification scheme for natural water bodies encompasses five classes: high, good, moderate, poor and bad. According to the Commission, 'high status' is defined by biological, physical, chemical and hydromorphological parameters associated with no or very low human pressure and serves as the 'reference condition' or benchmark because it is the best status achievable. The ecological status of water includes reference to biological and morphological characteristics in addition to the more commonly used chemical classifications. Chemical status includes reference to the achievement of water quality standards (Chave 2001, 19). These environmental objectives are separately laid down for surface and ground water and protected areas. Assessment of the water bodies is based on the extent of deviation from these reference conditions. 'Good status' means 'slight' deviation.[2] Recognizing country-specific circumstances, the 2015 deadline may be extended (Art 4.4) and the environmental objectives may be less stringent for specific water bodies if they are either so affected by human activity or if their natural conditions would render the environmental objectives infeasible or disproportionately expensive (Art 4.5).

Achieving environmental objectives set out by the WFD will be a serious task for Turkey. According to initial evaluations, only 22-30 percent of surface waters are estimated to be of a "good status" (Ministry of Environment and Forestry 2010b, 1).[3]

The General Directorate of State Hydraulic Works (DSI) has established a classification of river waters following the Directive (By-law) for Water Pollution Control where Class I waters refer to high quality waters, Class II to minimal pollution, Class III to polluted water and Class IV to very polluted water. The classification which is based on (i) physical and inorganic chemical parameters,

[2]Http://ec.europa.eu/environment/water/waterframework/objectives/status_en.htm#_Assessment_of_water. Accessed 3 December 2009.

[3]Assessments for ground waters and coastal waters are yet not completed.

(ii) organic parameters, (iii) inorganic pollution parameters and (iv) bacteriological parameters, serves as a starting point for assessing water bodies.

Combating water pollution

Reaching environmental objectives implies an effective pollution control policy which is difficult to achieve for many reasons. For instance, dealing with non-point pollution sources, particularly from agriculture, usage of fertilizer and pesticides may be restricted, and some of the latter may even be forbidden. However, farmers will have to adjust farm operations and apply best agricultural practices including biological pest control. To our knowledge, the Ministry of Agriculture has yet not set up any programme to implement the e.g. Directive on Protection of Waters against Nitrate Pollution from Agriculture (see OECD 2008, 73-74). Reducing pollution from point sources, i.e. households and industry, has progressed with investments being made in sewage systems and wastewater treatment plants. In spite of this, the Organization for Economic Cooperation and Development (OECD) estimates that about "3,000 new plants remain to be built in towns with a population over 2,000" to follow the EU Urban Wastewater Treatment Directive for which EUR 18 billion is needed for investment in and rehabilitation of treatment facilities (OECD 2008, 66). The EU funds are expected to support 40 percent while local administrations will have to co-finance these projects through credits from the Bank of Provinces.

Development and protection of water resources

In Turkey, priority has been given to full utilization and development of the country's water resources and hydropower potential. For instance, according to the Electricity Market and Demand Safety Strategy Document (2009), the technically and economically feasible hydroelectricity potential of Turkey is planned to be fully utilized by 2023 (Ministry of Energy and Natural Resources 2009). Although, it seems contradictory to the requirements of the WFD, the intensive development of water resources seems compatible with the WFD requirements. The WFD gives room for developing new water infrastructure, even if it would result in the failure of reaching a good status and of further modifying natural water bodies (WFD, Art 32).

However, this provision comes with a number of strict conditions: all measures must be taken to mitigate adverse effects. Conditions may be imposed on e.g. dam projects such as the construction of fish ladders and the maintenance of a minimum in-stream flow in order to protect a river and maintain its ecological integrity. The Environmental Impact Assessment (EIA) is one instrument to assess potential environmental (and social impacts), and thus the EU EIA Directive ranks high among the other components of the environmental *acquis* due to its horizontal nature. The current Turkish EIA by-law (2008) is largely in line with the EU EIA

Directive. However, its provisional Article 3 allows exemption of many dam projects from issuing an EIA, and subsequently the development of environmental management plans (see Scheumann et al. in this volume).

Other directives are linked with the WFD's objective of achieving "good water status" such as the Flora Fauna Habitat Directive (92/43/EEC) which requires from the Member States that they prevent the deterioration of natural habitats and the disturbance of species in designated areas (the Natura 2000 sites), which implies the protection of those sites from damaging effects of water infrastructure development. While Turkey is still in the process of designating Natura 2000 sites, membership to e.g. the Ramsar Convention and national laws protect environmental goods and resources such as Natural Parks, Nature Conservation Areas, Wild Life Conservation and Wild Animals Settlement Areas, Cultural and Natural Heritage, Archaeological Protected Areas and Special Environment Protection Areas. However, WFD (Art 4, 1(c)) addressing protected areas calls for compliance "with any standards and objectives specified in Community legislation under which individual protected areas have been established". This, in turn, implies that national legislation should be aligned, in particular for protecting wetlands, and that new areas will be designated for being protected.

Incomplete data

Incomplete data not only prevents correct estimates of the costs of implementing the WFD, but slows down the whole implementation process which rests upon reliable data. According to the Draft National Implementation Plan for the WFD in Turkey, which is the product of the Twinning Project (TR06-IB-EN-01) "Capacity Building Support to the Water Sector in Turkey", "almost no data is available to perform economic analyses properly" (Ministry of Environment and Forestry 2010b, 2). Further, the data available is dispersed among many institutions and locations, such as ministries, provinces, municipalities, banks, industries etc. (Ministry of Environment and Forestry 2010b, 2). The most developed set of water data is being collected by DSI. DSI has over 1,000 water monitoring sites for surface waters, 20 to 22 laboratories throughout Turkey and a large laboratory in Ankara which can perform all necessary chemical and biological analyses. However, DSI collects and processes data relevant to its mandate which aims at developing the country's water resources and supplying water of a quality that suits usage (e.g. drinking water, irrigation).

Since the WFD puts great emphasis on water quality and ecological aspects, the network of monitoring[4] has to be expanded geographically, covering more parameters than used currently, which would require a great number of trained personnel as well as funding to build and sustain such an extensive network (Cicek 2010).

[4]During the course of the Buyuk Menderes River Basin Project, 40 surface water monitoring stations are determined (Ministry of Environment and Forestry 2010a, 73). It would be sensible to expect that around 1,000 or more stations will be needed throughout Turkey.

River basin approach: planning, management, competent authority

Article 3 of the WFD asks for the identification of river basins lying within the territories of the member states, and designates them as individual river basin districts. Such districts are natural geographical and hydrological units with no regard to administrative or political boundaries. River basin districts covering the territory of more than one member state will be assigned to an international river basin district. By the end of 2003, member states needed to ensure appropriate administrative arrangements including the designation of a competent authority for each of the river basin districts that would be responsible for the application of the WFD.

In this respect, one of the main steps that would be necessary for a move towards compliance with the WFD in Turkey is the determination of river basin districts and the delegation of tasks and responsibilities to respective units. In this regard, basic steps have been taken by e.g. the Twinning Project which determined 25 river basin districts, which unlike the Social Transformation (MATRA in Dutch) project had advocated only six (see below). The Twinning Project has taken advantage of the fact that these river basins already serve as planning units for the DSI with its 25 Regional Directorates, the mandated organization for planning overall water resources development and authorizing all water use.

However, discussion is ongoing on who will be designated as the competent authority (see below the Twinning Project). Despite the fact that DSI has been the 'competent' authority for water resources planning and management, and the fact that it is the only institution in Turkey organized on the basis of "basins" (Bilen 2009b, 306), DSI does not focus on river basin management as a whole, but rather implements water development projects which generally downgrade environmental objectives. Furthermore, the numerous public institutions and non-governmental actors at all levels, i.e. central, regional, provincial and municipal, have rightful concerns in water resources management and may wish to be included in decision-making.

Turkish officials assert that "making a formal document listing River Basin Districts and the Competent Authorities for implementing the WFD" is not yet a formal requirement for Turkey (Ministry of Environment and Forestry 2010b, 9). This is simply because Turkey is not an EU Member State and thus, is not bound by the formal requirements. Turkish authorities might have thought that formally designating a competent authority may, in the future, constrain consideration of alternative institutions as competent authorities. For this reason, Turkey chose to follow a different path and implement further steps in the WFD schedule (such as creating river basin management plans) without designating the competent authorities for river basins.

Nevertheless, Turkey will eventually become bound by this requirement, i.e. identification of a competent authority (Article 3.3). In this context, according to Cicek, an official from the Ministry of Environment and Forestry, who is in charge of WFD activities, the Ministry will most likely be declared as the competent authority in implementing the WFD rules (Cicek 2010).

Public information and consultation

WFD, Article 14 provides new opportunities for all interested parties to get involved in water planning and management: "Member States shall encourage the active involvement of all interested parties in the implementation of this Directive, in particular in the production, review and updating of the river basin management plans" (Art 14). Article 14 further stipulates that "On request, access shall be given to background documents and information used for the development of the draft river basin management plan". Yet, who are "all interested parties" and how should they be "actively" involved is subject to national definitions and policies. The Guidance Document on participation outlines (European Commission 2003b) some tools and procedures for "best practices".

In water resources planning, public participation in Turkey has yet to be institutionalized beyond pilot projects. Planning is subject to central bureaucracy (Alpaslan et al. 2007b) including the State Planning Organization (SPO) and the relevant ministries. Only the EIA process provides a regular framework for public consultations related to specific projects, but it takes place at a stage where decisions have almost been made. However, water management has been decentralized providing opportunities for participation in the agricultural sector with irrigation associations managing irrigation systems (see Topcu in this volume) and the water supply and sanitation sector which operate under the mandate of elected municipalities (see Kibaroglu and Baskan in this volume).

There are further attempts to open decision-making to the public by the Law on the Right of Access to Information (2003) and the 2006 amendment to the Law on Environment (1983) beyond improvements in the EIA context. This too will affect environmental non-governmental organizations (NGOs) operating at the national and regional level, the participation of which is yet not institutionalized and relies on the good will of public authorities. There are promising local initiatives and pilot projects (for instance, the Konya Closed Basin Project, see Divrak and Demirayak in this volume) which could contribute in preparing the ground for a participatory framework. All in all, developing a participatory framework in water management will be a long-term objective for Turkey, because, given the history of water management in Turkey, changes in legislation in this particular field have not been rapidly translated into changes in practice.

Recovery of costs for water services

Article 9 of the WFD obliges Member States to take into account the principle of full cost recovery for water services based on the Polluter Pays Principle[5] and include environmental and resource costs. Heinz (2005) summarizes these costs as

[5]The Polluter Pays Principle is part of international environmental law where the polluting party pays for the damage done to the natural environment. It is regarded as a regional custom because of the strong support it has received in most OECD and European Community countries.

follows; financial costs confine to the costs of the equipment needed to ensure the provision of the services; environmental costs are related to environmental damage caused, for instance, by overuse of waters or excessive water pollution; and resource costs can be interpreted as those costs that are caused by a failure to use water resources optimally (WATECO 2003).

The Article further specifies that Member States have to ensure two objectives in this respect: (i) water pricing policies should provide an *adequate* incentive for the efficient water use and thereby contribute to the achievement of environmental objectives; (ii) an adequate contribution from different water users to recover the cost of water services (Article 9). From a strict environmental perspective, the wording "provide an adequate incentive" seems problematic as it gives ample room for interpretation (EEB 2001). The second obligation specifies that water uses are at least disaggregated into industry, households and agriculture, for which pricing scales can be tailored.

Domestic and industrial water tariffs are set by the municipalities in Turkey. Operation and management, amortization, rehabilitation and expanding costs are generally considered when setting the drinking water and wastewater tariffs. Environmental and resource costs are hardly covered (Bilen 2009a). Yet, municipalities try to cover certain costs determined by the relevant legislation: for instance, Istanbul Water and Sewage Authority (ISKI) which operates under the Istanbul Metropolitan Municipality, collects a "maintenance cost" (in accordance with the Act No. 2464 on the "Incomes of Municipalities") and a "tax for cleaning the environment" (in accordance with the Act No. 5035 on "Amendments of Several Acts") from all users, proportionate to their use of water.[6] Apart from these measures, municipalities seek to provide incentives for water savings by applying increasing block tariff structures (Republic of Turkey 2003).

In Turkey, the agricultural sector is by far the largest user of freshwater, accounting for around 74 percent of total water use (DSI website 2009). Irrigation water charges are one factor which causes excessive water use leading to environmental problems such as waterlogging and salinisation of soils (Cakmak et al. 2007, Unver and Gupta 2003). Irrigation water charges in Turkey are generally based on the area irrigated and the type of crop irrigated, not on actual water consumption (volume), and set to recover operation, maintenance and repair costs for providing irrigation water. They are not designed to function as an incentive to stimulate investments in water-saving technology (e.g. drip irrigation) and water savings *per se* (see Topcu in this volume), but to internalize negative externalities of water use, i.e. social costs.

Considering the principles laid down by the WFD, and given the lack of consideration of environmental and resource costs, it is reasonable to expect an increase in both urban and agricultural water tariffs in Turkey. Yet, it has been argued that the average income of Spanish citizens, for instance, is sufficient to pay higher prices for urban water supply, eventually including environmental and

[6]Personal correspondence with concerned officials at the ISKI, June 2010.

resource costs (Cabrera et al. 2009, 81-82). The same might hold true for the urban centres in Turkey as well, but not for those segments of the Turkish agricultural sectors where small holdings prevail and farmers are poor (Bilen 2009a), while commercially oriented farm holdings might not be enthusiastic if confronted with increased production costs. In order to influence farmers to use water more efficiently, a combination of raising water tariffs and providing subsidies for irrigation technologies and techniques is perhaps a more promising solution.

Transboundary cooperation

The WFD encourages transboundary cooperation and demands that its rules are implemented even beyond the territory of the EU, in cases where river basins extend beyond the Community's borders. Article 3 requires Member States to assign transboundary river basin districts for basins lying within the EU's jurisdiction. Where international river basins extend beyond the territories of the Union, the WFD requires from Member States only an "endeavour" to establish appropriate coordination with the relevant non-Member States, with the aim of achieving the objectives of the WFD throughout the river basin district.

With regards to transboundary water relations, Turkey continues to develop its dialogue with its neighbours. Recent processes between Turkey and Syria, and Turkey and Iraq exemplify these efforts. There are long-lasting talks with Greece and Bulgaria aimed at facilitating cooperation in managing waters of the Meric. However, the institutionalization of such dialogues whose goal is to establish comprehensive river basin management plans, applicable in all riparian countries, is apparently lacking. This problem is constantly mentioned as one of the themes in the Progress Reports prepared by the Commission (see European Commission's regular reports 2001ff.).

4 In the process of aligning with the Water Framework Directive

This section discusses the status of aligning with the WFD in Turkey. First the annual progress reports of the EC are summarized because they highlight the points to which the Commission attributes importance. Then, concrete steps such as projects, studies and initiatives that Turkey has taken so far are presented.

4.1 European Commission's perspective on Turkey and the Water Acquis: "moderate steps towards the right direction"

Following Turkey's candidacy for the EU membership in December 1999, EU-Turkey relations gained momentum. Both the EU and Turkey began to handle more systematically those issues stemming from candidacy of Turkey and the idea of an eventual Turkish membership in the Union.

Since 2001, the *water acquis* has always been a significant part of the EC's progress reports. In the 2001 regular report, the Commission's comment was very brief as well as negative. The Commission simply stated that "Turkey's water legislation is not in line with the *acquis*" (European Commission 2001).

The 2002 regular progress report stated that "A By-law on the Conservation of Wetlands was adopted in January 2002. This By-law partially complies with the provisions of the Birds, Water Framework and Habitat *acquis*" (European Commission 2002). Despite noting this specific positive development, it was concluded in the same document that there was no progress concerning the water quality *acquis*. Besides, the 2002 progress report underlined the fact that Turkey itself had already realized the urgent need for a new water framework law: "As regards water quality, the 7th and 8th Five Years Development Plans stress the need for a new framework law on water resources and for bringing drinking water standards and wastewater discharge into line with the *acquis*" (European Commission 2002).

The 2003 regular report mentions only a few developments concerning the water *acquis*. The Commission asserted the need for further efforts to transpose and implement the *acquis*, including a new framework law on water resources and to bring drinking water and wastewater discharge standards in line with the *acquis* (European Commission 2003a).

Emphasis on the transboundary dimension of the WFD is probably the most significant point in the 2004 regular report. The Commission argued that "cross-border cooperation needs to be stepped up with the neighbouring countries" (European Commission 2004).

The 2005 progress report differs from earlier ones, because this time, the Commission provided a more detailed analysis on Turkey's progress regarding the water related *acquis*. Several reasons could be listed for this change, but two of them come to the forefront. First, this time there was some progress that the Commission could evaluate. Second, and perhaps more importantly, Turkey started preliminary membership negotiations in that year, i.e. the Commission began to pay more attention to Turkish efforts regarding the *acquis*. On the whole, by the end of 2005 it is evident that despite some noteworthy developments on several occasions, the Commission seemed to be dissatisfied with both the level and pace of progress in Turkey (European Commission 2005).

The 2006 progress report mentions positive developments such as the transposition of both directives covering urban wastewater treatment and the quality of bathing water. Nevertheless, the Commission was clear enough to underline the necessary steps to be taken: "financing plans remain to be established; to make new investments complying with the *acquis* no steps were undertaken to align with the WFD; Turkey did not initiate steps to develop trans-boundary water cooperation, in particular with Member States; no progress has been made as regards alignment, or further alignment, concerning other directives in the water quality field, the level of transposition remains low, particularly as regards nitrates, ground water and drinking water" (European Commission 2006, 65).

From the perspective of the EU Commission, 2007 was not a fruitful year for Turkey's progress concerning water quality. The Commission emphasized the intimate connection between the low level of overall alignment concerning the water quality *acquis* with the non-transposition of the WFD (European Commission 2007).

For the 2008 progress report, the Commission notes the amendment made to the legislation on prevention of water pollution, but concludes that the overall level of alignment remained low. Like earlier reports, the Commission argues that the institutional framework for water management is not organized on river basin basis and that trans-boundary consultations on water issues are at an early stage (European Commission 2008).

The 2009 progress report did not mention any positive development with regards to Turkey's alignment with the water quality acquis. The report states that "there has been no progress in the area of water quality. The overall level of alignment remains low. The institutional framework for water management is not organized on a river basin basis. Transboundary consultations on water issues are at a very early stage" (European Commission 2009).

There are several themes in the progress reports which the Commission mentions repeatedly. One such theme is the current status of the institutional framework for water management in Turkey which does not comply with the requirement of river basin management. Besides, the Commission continuously stresses the weak institutional capacity due to "unclear division of responsibilities". And thirdly, the Commission keeps reporting that transboundary water cooperation is in its infancy. A clear and basin-based institutional framework and the achievement of a cooperative dialogue with neighbours seem to be crucial benchmarks from the EC's point of view.

4.2 Steps towards implementing the Water Framework Directive

Several initiatives, studies and projects have been hitherto undertaken to support the adoption of the EU water-related *acquis* in Turkey – with varying degrees of success. The most important ones are summarized below together with their main outcomes. There are also a number of local projects which are directly linked to the WFD, such as wastewater treatment projects[7] and drinking water facilities[8] in a number of urban centres. It seems likely that the number of projects will increase in the coming years because negotiations on the Environment Chapter was opened on 21 December 2009.

[7]Examples of such projects include wastewater treatment plants for Erzurum, Bartin, Ceyhan, Adiyaman, Polatli, Siverek, Seydisehir, Carsamba, Diyarbakir, Erdemli, Aksehir, Aksaray, Merzifon, Luleburgaz and Soma.

[8]Examples of such projects include Akcaabat, Bulancak, Dogubayazit, Ercis, Erzincan, Manavgat, Nizip and Silvan.

4.2.1 National Programme for the Adoption of the *Acquis* (2001, 2003 and 2008)

The National Programme is an official document prepared with contributions from relevant ministries of Turkey, setting out the details, timetables and costs for fulfilling each priority area as defined in the Accession Partnerships. The First National Programme, which basically listed the EU legislation to be adopted by Turkey, was published in 2001. A revised National Programme for the Adoption of the *Acquis* was adopted on 24 July 2003[9]. With respect to the water sector, the EU legislation that was prioritized under the heading 22.1 "Improvement of the Water Quality" was studied by relevant ministries. Subsequently, by-laws have been prepared and entered into force in Turkey: the By-Law on the Waters from which Drinking Water is Obtained or Planned to be Obtained (2005), the By-Law on Urban Wastewater Treatment (2006), By-Law on Control of Pollution by Dangerous Substances in Water and its Environment (2005), By-Law on Protection of Waters Against Nitrate Pollution from Agricultural Sources (2004), By-Law on Water Intended to Human Consumption (2005), By-Law on Bathing Water Quality (2006) and the By-Law on EIA (2003, amended in 2008).[10]

The National Programme, again revised in 2008[11], proposed a harmonization schedule for "water management". According to the 2008 National Programme, harmonization with the WFD will be realized when a clear perspective of EU membership appears. Apart from WFD and other water-related directives (Bathing Water Quality Directive, 2006/7/EC, Assessment and Management of Flood Risks Directive, 2007/60/EC and Directive 91/676/*EEC* on Nitrates from Agricultural Sources), transposition of most of the remaining water-related directives are said to be finished in 2009. Table 1 provides an updated list what has been transposed so far (as of June 2010).

A new National Programme will be published in 2011 to set out further work required for aligning with the WFD. The new National Programme is likely to mention the year that Turkey would comply with the WFD (Cicek 2010) which will be either 2027 or 2033[12]. This shows a change in Turkey's attitude, because formerly Turkey was refraining from setting up an exact date for alignment with the WFD, and previous official documents clearly stated that full alignment with the WFD would be considered when Turkey becomes an EU Member State.

---

[9]Secretariat General for EU Affairs (2003) National Programme of Turkey for the Adoption of the EU Acquis, Official Gazette Dated 24 July 2003, No. 25178.

[10]See http://www.cevreorman.gov.tr. Accessed 16 December 2009.

[11]Secretariat General for EU Affairs (2008) National Program of Turkey for the Adoption of the EU Acquis. Official Gazette Dated 31 December 2008, No. 27097. http://www.abgs.gov.tr/index.php?p=42260&l=2. Accessed 6 January 2010.

[12]In the Draft National Implementation Plan for the WFD, which is an output of the Twinning Project (TR06-IB-EN-01) "Capacity Building Support to the Water Sector in Turkey", 2027 and 2033 are envisaged as the possible dates for achieving the "good water status" objectives (Ministry of Environment and Forestry 2010b).

Table 1 Water *Acquis* and Turkey's Progress (own compilation based on Ministry of Environment and Forestry, Draft National Implementation Plan, 2010, and Bilen 2009b)

European Directive	Transposition Status & Date	Leading Ministry
Water Framework Directive (2000/60/EC)	Deadline for transposition tentatively set for 2011	MoEF
Dangerous Substances Directive (76/464/EEC)	Accession of Turkey is not envisaged before the repeal of this Directive, no transposition is required	MoEF
Daughter Directive on Priority Substances (2008/7/EC)	Full transposition after 2015	MoEF
Bathing Waters Directive (New) (2006/7/EC)	Date for full implementation will be determined through the proposed project titled "Harmonization of the New Bathing Water Quality Directive and Strengthening the Monitoring System of the MoH" submitted to 2010 IPA Program	MoEF and MoH
Bathing Waters Directive 76/160/EEC	Transposed on 09.01.2006	MoEF
Directive on the Quality of Water intended for Human Consumption (98/83/EC) (Drinking Water Abstraction Directive, 75/440/EEC is repealed in 2007)	Transposed on 17.02.2005	MoEF
Urban Wastewater Treatment Directive (91/271/EEC)	Transposed on 08.01.2006, with a period of implementation until 2023	MoEF
Nitrate Directive (91/676/EEC)	Partially transposed in 2004, full transposition not sooner than 2013	MARA and MoEF
Integrated Water Pollution Prevention and Control Directive (96/61/EC)	Transposition by the end of 2012	MoEF
Major Accidents (Seveso) Directive (96/82/EC)	Transposed in 2009, implementation until 2014	MoEF
Sewage Sludge Directive (86/278/EEC)	Transposed on 31.05.2005	MoEF
Plant Protection Products Directive (91/414/EEC)	Transposition will be completed by the end of 2010	MARA

Water for Freshwater Fish Directive (78/659/EEC-consolidated version 2006/44/EC))	Accession of Turkey is not envisaged before the repeal of this Directive, no transposition is required	MARA and MoEF
Flood Risks Assessment and Management Directive (2007/60/EC)		MoEF
Marine Strategy Framework Directive (2008/56/EC)		MoEF
Environmental Impact Assessment Directiv (85/337/EEC)	Partially transposed including clauses on public participation	MoEF
Habitat (92/43/EEC) and Birds (79/409/EEC) Directives	Partially transposed, full transposition after 2011	MoEF
Daughter Directive on Groundwater (2006/118/EC)	Deadline for transposition tentatively set for 2011	MoEF
Water for Shellfish Directive (2006/44/EC)	Transposed through a circular in 2008	MoEF
Sampling and analysis of surface water intended for the abstraction of drinking water Directive (79/859/EEC)	Transposed on 20.11.2005	MoEF
Groundwater Directive Dangerous Substances (80/68/EEC)	Accession of Turkey is not envisaged before the repeal of this Directive, no transposition is required (will be repealed in 2013)	MoEF
Shellfish Directive (79/923/EEC)	Accession of Turkey is not envisaged before the repeal of this Directive, no transposition is required (will be repealed in 2013)	MoEF

4.2.2 Working groups under the coordination of the Ministry of Foreign Affairs (2004)

In 2004, to foster the efforts for transposing the WFD and to coordinate different institutions, two separate Working Groups were established under the coordination of the Ministry of Foreign Affairs with the participation of all institutions having a role in water management in Turkey. The aim of Working Group one, which was chaired by SPO, was to screen the existing task and responsibilities of the organizations playing a role in water management. The second Working Group chaired by the Secretariat General of EU Affairs focused on the transposition of EU regulations into Turkish legislation. Upon their establishment these two groups had two meetings during the first period of six months. During the meetings of these two groups it was agreed that there was a need for a framework water legislation in accordance with the requirements of the WFD.

According to the first working group the obstacles in Turkey regarding harmonization of the Turkish water legislation with the EU legislation can be summarized as: the incoherence of legislations resulting in institutional conflicts and overlaps; fragmentation of water quality and quantity management; incapacity of the regional level administrations to promote sustainable water use and problems associated with it (planning, financing, permissions and sanctions); lack of effective pollution control; lack of enforcement despite the existence of legislation; lack of updating the establishment laws of relevant public organizations, and lack of effective coordination and cooperation among institutions (Sorman 2006).

The reports prepared by these two working groups were later sent to the Prime Ministerial Office through the Ministry of Foreign Affairs. However, the Prime Ministerial Office has not yet responded to these reports because, according to Cubukcu (2007), their outcome was not fully conform with their institutional priorities. This project could rather be seen as an exploratory study through which the scope of further work was determined upon the definition of problems by official institutions involved in different aspects of water planning and management.

4.2.3 MATRA project "Implementation of the Water Framework Directive in Turkey" (2002-2004)

Regarding the implementation of the WFD, the project "Implementation of European Union Water Framework Directive in Turkey" was carried out between 1 January 2002 and 30 April 2004. It was the first project for the harmonization of the WFD in Turkey. The main aim of the project was to support Turkey with the implementation of the WFD on national and regional levels. The Project was supported by the Dutch Government under the Dutch MATRA pre-accession programme (MAT01/TR/9/3)[13]. It was proposed that one demonstration area be selected to serve as a pilot area for the entire project. The Buyuk Menderes river basin was selected, where a river basin management was developed on a river basin scale. Key issues were the enhancement of cooperation between different decision-making bodies involved in water management as well as enabling the public to participate in the process.

The results of the Project were listed as follows:

- Delineation of Turkish river basins in accordance with the WFD requirements
- Greater knowledge of the WFD and other relevant European legislation among the organizations involved in integrated water management
- A manual of established methodologies for the implementation of the WFD in Turkey
- A management plan for the Buyuk Menderes basin which will act as an example for other river basins
- Information to the public and policy-makers about the project results achieved, and about the consequences of implementing the WFD in Turkey[14]

Within the framework of this project, aimed at facilitating discussions and decision-making on the national level, a *National Platform on Water Management* was established with the participation of the MoEF, the DSI, the Ministry of Health (MoH), the Ministry of Agriculture and Rural Affairs (MARA), the Ministry of Tourism, the SPO, the Secretariat General for EU Affairs and other relevant official organizations. The National Platform acted as a consultative and discussion group for all government stakeholders in the field of water management, and thus contributed to better coordination and cooperation between institutions involved. However, the Platform did not continue after project completion, despite the widespread expectations expressed by the Platform participants to do so (Bruini et al. 2003).

Assessing the legal and institutional set-up necessary to meet the WFD's requirements was an important part of the project. The Project concluded that in order to facilitate sound water management practices, by taking due account of the views of all stakeholders in the river basin management plans, river basin commissions should be established. It further recommended that the competent authority could be the DSI Regional Directorates (Grontmij Consulting Engineers 2004, 27). The Project laid out the main tasks to be done in order to meet the WFD requirements:

---

[13]The MATRA project aimed at assisting pre-accession countries in adopting the *acquis communautaire*, and establishing relations between Dutch government institutions and recipient countries.

[14]Http://www.evd.nl/business/zoeken/showbouwsteen.asp?bstnum=113198&location=&highlight=, (Website of the Ministry of Economic Affairs, the Netherlands). Accessed 08 July 2010.

- Coordination and cooperation among state institutions
- Transfer of powers and responsibilities to the regional level (River Basin Districts)
- Integrated water management approach focusing on water bodies and water users
- Information sharing and dissemination
- Public consultation and stakeholder participation
- Economic incentives and measures (Grontmij Consulting Engineers 2003, 11)

4.2.4 "Environmental Heavy-Cost Investment Planning in Turkey" (2002-2005)

In order to achieve the objectives of the WFD, particularly "good water status", heavy cost investments are required in Turkey. European directives that call for the highest investments, include the Urban Wastewater Treatment Directive, the Drinking Water Directive, the Quality of Surface Water Intended for the Abstraction of Drinking Water Directive, the WFD, the Directive on Dangerous Substances Discharged into Water, the Nitrate Directive and the Bathing Water Directive.

The "Environmental Heavy-Cost Investment Planning in Turkey" project (EuropeAid/114715/D/SV/TR) was carried out from 2002 to 2005, and estimated the amount of investment required for the implementation of directives. According to estimates of the ENVEST Planners (i.e. international consortium), Turkey needs to invest 65 to 70 billion Euros in the environmental sector in order to meet the environmental *acquis*. The Project also discussed economic instruments in order to provide incentives for implementing water quality requirements (Ozdinc 2006, 173).

4.2.5 "Restructuring of the Turkish Water Sector for the Implementation of EU Water Directives" (2005)

One of the studies facilitating the WFD's implementation in Turkey was a report prepared by United Kingdom (UK) Department of Environment, Food, and Rural Affairs (DEFRA) in 2005. DEFRA's Report, "Restructuring of the Turkish Water Sector for the Implementation of EU Water Directives", was prepared for MoEF. The Report proposed that for an efficient implementation of EU water directives in Turkey, "environment agencies" should be established, and powers and responsibilities of DSI should be transferred to those agencies. According to the report, DSI should only be responsible for large construction works, upon request by municipalities. All the relevant work for the implementation of the WFD (such as preparation of River Basin Management Plans, monitoring of water quality, operation of programme of measures) will be conducted by environment agencies. The Report also recommends the establishment of a council, similar to "Ofwat"[15] in the UK and Wales, acting as a regulatory body for supervising water tariffs and investment proposals of environment agencies (Bilen 2009b, 308).

4.2.6 "Strengthening the Capacity of Sustainable Groundwater Management in Turkey" (2006-2008)

With regard to the Directive on the Protection of Groundwater against Pollution Caused by Certain Dangerous Substances (80/68/EEC), the project "Capacity Building with Regard to Sustainable Groundwater Management in Turkey", which is supported by Government of Netherlands under the MATRA pre-accession programme (MATRA-PPA05/TR/7/8), was carried out between January 2006 and January 2008. The project was to provide support for the implementation of the provisions of the Groundwater Directive (80/68/EEC), its revision (2006/118/EC), and the WFD. It focused on three aspects: assisting Turkish government with transposition of the Groundwater Directive and the WFD; assisting the counterpart organizations in developing an institutional structure for groundwater management in Turkey, with a clear overview of tasks, responsibilities and instruments to be used; assisting the counterpart organizations with hands-on training to co-develop a groundwater management plan, including skills and tools needed. This also involves economic tools as specified in the WFD (Vliegenthart et al. 2007).

Where institutional issues are concerned, the question of how to divide major tasks and instruments between Turkish water management authorities, notably between DSI and MoEF was discussed (note that DSI was not under the MoEF for most part the Project period). The starting point of the discussion was the nomination of the competent authority(-ies) for the directives. However, due to the belated ambiguity regarding the status of DSI (which was then brought under the MoEF as a separate legal entity), the discussion on who is the competent authority could not be finalized within the project timeframe.

As a pilot study for the project, Kucuk Menderes river basin was selected, and a Draft Groundwater Management Plan according to the Groundwater Directive and related parts of the WFD was prepared. Experts from counterpart organizations (DSI and MoEF) at national and local level contributed to the preparation of this plan. During the preparation of this plan various trainings regarding the drafting of groundwater management plan according to the directive were given to the members of the Pilot Project Working Group.

As a result of this project, an advisory action plan regarding the harmonization and transposition of the EU groundwater legislation was prepared, the transposition

[15]Ofwat, i.e. Water Services Regulation Authority, is the regulator of the water and sewerage sectors in England and Wales.

of which was completed in 2008 (Bilen 2009b). Further, a draft by-law on groundwater was prepared to be transposed in 2011.

4.2.7 The EU Twinning Project "Capacity Building Support to the Water Sector in Turkey" (2008-2009)

The Twinning Project started in 2008, aiming to assist Turkey in improving water management in line with EU water legislation, in particular the WFD, the Urban Wastewater Treatment Directive (UWWTD), and the Dangerous Substances Directive (DSD). The project had three components; (1) Legal analysis of the WFD, the UWWTD and the DSD; (2) Developing implementation plans for the WFD and DSD and (3) Pilot implementation of the principles of the WFD, UWWTD and DSD in the Buyuk Menderes River Basin. The project also aimed to strengthen management practices in Turkey with appropriate, advanced decision-support tools to be used in real-time operational river basin information management system. For the pilot basin the aim was to provide an integrated operational turn-key system. The system would be compatible with the WFD and provide an online water quality and quantity monitoring system, a basin information system, basin simulation models as well as a Geographical Information System (GIS) and expert system. The project was designed to help provide basic tools for sustainable management of water basins in Turkey as well as to establish economic efficiency, environmental integrity and social development. The project has taken into account the results of the already finished projects such as the MATRA projects.

The Project was officially completed on 18 February 2010. In the ceremony organized for completion of the Project, Turkish authorities stated that the Project significantly contributed to capacity development in Turkey concerning the water quality sector which constitutes one of the most costly and critical parts of the Environment Chapter.[16]

The most important outcome of this project is the preparation of a "Draft National Implementation Plan" for the WFD. The Plan mentions three categories of actions which aim to ensure that the river basin management plans are completed by the end of 2017. The first category involves the transposition of water-related directives and actions necessary for enabling the start of river basin planning at regional level; the second action focused on the timely production of river basin management plans; and the third includes "practical support in the form of guidance material and training as well as the steps that ensure the coherence between all river basin management plans, such as review and standardization" (Ministry of Environment and Forestry 2010b, 5).

The project concluded that, in order to achieve an integrated approach to water management in Turkey, there is an apparent need for a "Framework Water Protection Law". This need mainly emanates from the fact that authorities and

[16]For details, see http://www.dsi.gov.tr/basinbul/detay.cfm?BultenID=202. Accessed 25 February 2010.

responsibilities of institutions involved in water planning and management in Turkey are guided by their respective establishment law resulting in duplications and inefficiency. Preparations for this law are conducted by the General Directorate of Environment Management of the MoEF and a Draft Law was released as early as 2000 (see Kibaroglu and Baskan in this volume). The proposed Framework Water Protection Law will possibly come into force in 2011 (Sarikaya and Cicek 2010, 11).

In order to coordinate all WFD activities (along with other EU Directives), a Division of Environmental Administration (Cevre Idaresi Baskanligi) will be established within the Ministry of Environment.[17]

Turkey has already made progress towards the requirements of the Directive in some areas (see Table 2).

The Twinning Project has defined "a set of 25 River Basin Protection Actions Plans" being a precursor to the River Basin Management Plans (Ministry of Environment and Forestry 2010b). These plans aim to improve water quality in river basins in line with the environmental objectives of the WFD, to carry out necessary studies and to realize sustainable and efficient planning. They are based on active involvement of stakeholders and ensure the implementation of priority measures to achieve good water status. The core contents of the River Basin Protection Actions Plans include the characterization of the existing situation, description of important pressures, listing of required measures, prevention of pollution, calculation of environmental flows and carrying out studies with regards to implementation of measures with participation of stakeholders. The main difference between River Basin Management Plans and the River Basin Protection Actions Plans is that, the former is broader in considering biological, hydromorphological and chemical issues (Ministry of Environment and Forestry 2010b, 16). Out of 25 river basins, four basins now have River Basin Protection Actions Plans, preparations are underway for 13 basins and for eight river basins preparation for River Basin Protection Actions Plans will start at the end of 2010. It is planned that by late 2014, all River Basin Protection Actions Plans will be transformed into River Basin Management Plan (Cicek 2010).

Apart from these initiatives and projects, a number of new steps have been taken recently by Turkey, aimed at supporting the WFD implementation. For instance, "A Plan for Setting up Necessary Administrative Capacities at National, Regional and Local Level and Required Financial Resources for Implementing the Environmental Acquis" was published in September 2009. This plan set out a strategy for achieving "good water status" by 2027. Yet, interim evaluations indicate the need to extend the period to achieve good water status by another six years (i.e. 2033) (Ministry of Environment and Forestry 2010b, 4). Further, a department was established within the Environment Reference Laboratory (Cevre Referans Laboratuvari) for improving monitoring, and a new twinning project on capacity building on water quality monitoring will start in late 2010 which will come under its responsibility.

---

[17]See http://www.cevreorman.gov.tr. Accessed 11 February 2010.

Table 2 Summary of progress as of 2010 for Water Framework Directive (Draft National Implementation Plan 2010, 11-12)

Item	Little or no progress	Partial progress	Sufficient progress
River basin district designation		Competent authority designation	Boundaries
Method of typology		Rivers Lakes Transitional Coastal	
Delineation of water bodies at the national level	Surface water (national) Groundwater (national)	Groundwater (5 river basin districts)	Surface water (5 river basin districts)
Pressures and impacts	Analysis of pressures and impacts (national)	Analysis of pressures and impacts (5 river basin districts)	
Economic analysis	Cost-benefit analysis	Water use Cost recovery Cost-effectiveness analysis mechanism	
Monitoring	Biological	Chemical Physico-chemical	
Environmental objectives	Draft environmental objectives Reference conditions Final environmental objectives		
Protected areas	Objectives and data		
Public participation	Active involvement	Information Consultation	
Water Framework Directive outputs	External communications strategy Significant water management issues Draft and first river basin management plans	Characterisation	

5 Concluding remarks

This chapter has shown that Turkey already made progress towards the requirements of the WFD in some areas but they are far from being sufficient to enable Turkey to fully align with the WFD. The bulk of the work is still pending. In this respect the Annual Progress Reports of the EC are not very enthusiastic about Turkey's efforts and emphasize the topics not yet addressed more than praise its achievements. The opening of negotiations on the Environment Chapter will be instrumental in promoting the implementation of the WFD and other directives in releasing more financial means. Meanwhile the Ministry of Environment and Forestry has taken leadership in implementing the WFD which has been improved and reinforced by recent projects and activities.

The success of aligning with the WFD will largely depend on the continuous political will and efforts of the Turkish government and the public administrations concerned. This process of alignment could possibly move faster and easier, if and when Turkey gets a clearer membership perspective from the EU.

Depending on this perspective, and of course on internal forces, Turkey stands at a cross-roads where two ways are open: one is the minimum implementation of the WFD without changing the essence of the water management in the country. A second is to transform the way water is managed, and to try to implement the WFD in order to reach "good status" for its waters. Turkey's success in the implementation of the WFD depends on which path it will choose to follow.

References

Alpaslan AH, Atac A, Kimence T (2007) Water resources management in Turkey and European Union. Proceedings International Congress on river Basin Management, Antalya, Turkey, pp 230–244

Bilen O (2009a) Personal communication, Ankara

Bilen O (2009b) Turkiye'nin Su Gundemi: Su Yonetimi ve AB Su Politikalari. DSI, Ankara

Borja A (2005) The European Water Framework Directive: A challenge for nearshore, coastal and continental shelf research. Continental Shelf Research 25:1768–1783

Bruini FLM de, Jaspers FGW, Gupta J (2005) The EU Water Framework Directive: Challenges for institutional implementation. In: Vermaat JE, Bouwer L, Turner K, Salomons W (eds) Managing European coasts: Past, present, and future. Springer, Berlin, Heidelberg, pp 153–172

Cabrera E, Cabrera Rochera E, Cobacho R (2009) Water Supply in Urban Areas. In: Garrido A, Llamas MR (eds) Water Policy in Spain. CRC Press

Cakmak B, Ucar Y, Akuzum T (2007) Water resources management, problems and solutions for Turkey. Proceedings International Congress on river Basin Management, Antalya, Turkey, pp 868–880

Chave P (2001) The EU Water Framework Directive: An Introduction. IWA Publishing

Cicek N (2010) Personal communication, Ministry of Environment and Forestry, Ankara.

Cubukcu Y (2007) Personal communication, Turkish Embassy, Brussels

DSI website (2009) Land and water resources. DSI (General Directorate of State Hydraulic Works) http://www.dsi.gov.tr/english/topraksue.htm. Accessed 22 April 2009

EEB (2001) EEB Handbook on EU Water Policy under the Water Framework Directive. European Environmental Bureau, Brussels
Estevan A, Naredo JM (2004) Ideas y propuestas para una nueva política del agua en España (Ideas and Propositions for a new water policy in Spain). Bakeaz
European Commission (2001) 2001 Regular report on Turkey's progress towards accession. Commission of the European Communities, Brussels
European Commission (2002) 2002 Regular report on Turkey's progress towards accession. Commission of the European Communities, Brussels
European Commission (2003a) 2003 Regular report on Turkey's progress towards accession. Commission of the European Communities, Brussels
European Commission (2003b), Guidance Document No.8, Brussels.
European Commission (2004) 2004 Regular report on Turkey's progress towards accession. Commission of the European Communities, Brussels
European Commission (2005) 2005 Regular report on Turkey's progress towards accession. Commission of the European Communities, Brussels
European Commission (2006) 2006 Regular report on Turkey's progress towards accession. Commission of the European Communities, Brussels
European Commission (2007) 2007 Regular report on Turkey's progress towards accession. Commission of the European Communities, Brussels
European Commission (2008) 2008 Regular report on Turkey's progress towards accession. Commission of the European Communities, Brussels
European Commission (2009) Turkey Progress Report. Commission Staff Working Document, Commission of the European Communities, Brussels
Garrido A, Llamas MR (eds) (2009) Water Policy in Spain. CRC Press
Grontmij Consulting Engineers (2003) Uygulama El Kitabi: Su Cerceve Direktifi'nin Turkiye'de Uygulanmasi. Houten
Grontmij Consulting Engineers (2004) Institutional and Legal Strengthening, Report on the Legal and institutional developments required to meet EU legal requirements in the field of water management in Turkey. Houten
Heinz I (2005) The economic value of water and the EU Water Framework Directive: how managed in practice? IWA International Conference on Water Economics, Statistics, and Finance, Rethymno, Greece, 8-10 July 2005
Kaika M (2003) The Water Framework Directive: A new directive for a changing social, political and economic European framework. European Planning Studies, vol 11, no 3:299–316
Ministry of Energy and Natural Resources (2009) Electricity Market and Demand Safety Strategy Document. http://www.enerji.gov.tr/yayinlar_raporlar/Arz_Guvenligi_Strateji_Belgesi.pdf. Accessed 05 March 2010
Ministry of Environment and Forestry (2010a) Buyuk Menderes River Basin Management Plan, Final Draft. Ministry of Environment and Forestry, Ankara
Ministry of Environment and Forestry (2010b) Draft National Implementation Plan for Water Framework Directive. Ministry of Environment and Forestry, Ankara
OECD (2008) OECD Environmental Performance Review of Turkey (Paris: OECD)
Ozdinc HK (2006) Avrupa Birligi'nde Su Politikalari. In: Cinar T, Ozdinc HK (eds) Su Yonetimi: Kuresel Politika ve Uygulamalara Elestiri. Memleket Yayinlari, Ankara
Republic of Turkey (2003) Turkey Country Report. Prepared for the Third World Water Forum. World Water Council http://www.worldwatercouncil.org/fileadmin/wwc/Library/Publications_and_reports/country_reports/report_Turkey.pdf. Accessed 23 August 2010
Saner E (2008) Personal communication, Secretariat General for EU Affairs
Sarikaya HZ, Cicek N (2010) Su Kaynaklarinin Yonetimi, AB Sureci ve Cevre ve Orman Bakanligi Uygulamalari. Gunce, TUBA Sureli Yayinlari, Nisan 2010, vol 40:5–13
Secretariat General for EU Affairs (2003) National program of Turkey for the adoption of the EU Acquis. Official Gazette dated 24 July 2003, no 25178. Available online at http://www.abgs.gov.tr/index.php?p=196&l=1. Accessed 5 January 2010

Secretariat General for EU Affairs (2008) National program of Turkey for the adoption of the EU Acquis. Official Gazette dated 31 December 2008, no 27097. Available online at http://www.abgs.gov.tr/index.php?p=42260&l=2. Accessed 6 January 2010

Sorman U (2006) AB Su Cerceve Direktifi ve Turkiye Uygulamasi Hakkinda Gorusler, TMMOB Su Politikalari Kongresi Bildiriler Kitabi, 2. Cilt, 205–209

Unver O, Gupta RK (2003) Water pricing: Issues and options in Turkey. International Journal of Water Resources Development, vol 19, no 2:311–330

Vliegenthart FJL, Sargin AH, Gorkmen A, Dogdu MS (2007) A new approach for groundwater management in Turkey: A groundwater management plan according to EU Groundwater Directive (80/68/EEC) and Water Framework Directive (2000/60/EC). International Congress on River Basin Management, Antalya, Turkey

WATECO (2003) Economics and the environment - The implementation challenge of the Water Framework Directive. Common Implementation Strategy for the Water Framework Directive (2000/60/EC), Guidance document no 1. European Communities. http://circa.europa.eu/Public/irc/env/wfd/library?l=/framework_directive/guidance_documents/guidancesnos1seconomicss/_EN_1.0_'a=d. Accessed 6 January 2010

Water for Energy: Hydropower is Vital for Turkey

Zekai Sen

1 Introduction

The Turkish economy recorded a significant growth in its gross national product (GNP), with the annual growth rate reaching 8.2 percent in 2008. Parallel with economic development, energy demand in Turkey has increased significantly over the past decades. Today already, Turkey's domestic energy resources are not sufficient to meet increasing demand and the country is highly dependent on energy imports to such an extent that in 2008 the imported energy rate reached 76 percent (MENR 2009). Society faces possible energy shortages and even power cuts if necessary measures and investments are not considered and implemented in due time. Since the increase in energy consumption is parallel to the increase in the GNP, this further emphasizes the key role of energy for Turkey's economic development. The socio-economic development of Turkey, therefore, urgently requires the exploitation of additional energy resources.

Turkey ranks among those countries with a high potential for hydroelectric power generation. Hydroelectric power plants (HEPP) are currently supplying about 40 percent of the total electricity generated in the country. Nevertheless, the hydroelectric power potential has not exploited to its full extent and the Turkish Ministry of Energy and Natural Resources (MENR) seeks to increase hydropower generation as well as to develop other natural resources (MENR 2009). Among the domestic energy alternatives, hydroelectric energy generation is given priority due to its sustainable supply depending on the local and regional precipitation patterns. However, its initial investment costs are very high, and financing is not secured. Furthermore, hydropower dams encounter costs in terms of negative environmental and social impacts as well as cultural and archaeological losses.

The increase of renewable sources including hydropower in the composition of the national energy production profile is expected to increase steadily in the next

Z. Sen (✉)
Turkish Water Foundation, Istanbul, Turkey
e-mail: suvakfi@suvakfi.org.tr

A. Kibaroglu et al. (eds.), *Turkey's Water Policy*,
DOI 10.1007/978-3-642-19636-2_4,

25 years. Unfortunately, even if all potential renewable energy sources are developed, Turkey cannot cover its energy demand by domestic supply alternatives. However, the development of the hydropower capacity with the recent addition of privately operated small scale hydroelectric power stations would make a positive contribution to the national energy production profile.

This paper will discuss the existing hydroelectricity potential and its limitations in Turkey in addition to possible climate change effects on water resources.

2 Energy demand and supply

According to Electrical Power Resources Survey and Development Administration (EIE, Turkish acronym) (2006) reports, Turkish total energy demand has increased at an annual rate of 3.8 to 8 percent over the past 15 years. The Environmental Indicators of the Ministry of Environment and Forestry (2008) reported that in 1990 energy consumption was 15,800 tons of oil equivalent (toe) for households, 14,960 toe for industry, 8,300 toe for transportation and almost 2,000 toe for the agricultural sector, whereas for 2007 these amounts had risen to 32,371 toe, 25,471 toe 17,282 toe and 3,945 toe, respectively.

The overall energy intensity of the country, in terms of the amount of energy it takes to produce US$ 1,000 of Gross Domestic Product (GDP), increased from 0.377 toe/$1,000 (2000 prices) of GDP in 1990 to 0.382 toe/$1,000 (2000 prices) of GDP in 2004. This trend reflects the shift towards heavy industries such as steel production. The energy intensity is almost equal to the world average of 0.32 toe/ $1,000, but it is higher than the average of Organization for Economic Cooperation and Development (OECD) countries of 0.20 toe/$1,000. When energy intensity is measured against GDP based on purchasing power parity (PPP), Turkey's was 0.12 toe/$1,000 in 2003, while the world average was 0.21 toe, and the OECD average 0.19 toe/$1,000 in 2003 (MENR 2006). Higher energy intensity can increase a country's climate change impact and contribute to resource depletion. .

Major electricity demands come from industry, household, service and agricultural sectors. Use of electrical energy in Turkey rose from 57,543 GWh in 1990 to 175,690 GWh in 2006 indicating an increase of almost 305 percent in 16 years as shown in Figure 1. It is obvious that there is a steady increase in general with a slight exception in 2001.

Electricity consumption per capita per year in Turkey is 1,800 kWh - which is below the world average of 2,200 kWh and far less than the average consumption of 5,000 kWh for developed countries (MENR 2006). On the average, the consumption rate is on a steady increase from 8 to 10 percent per year.

Among the primary energy supply of the country the following percentage shares are effective in 2006; hydroelectric power 32.3 percent, natural gas 31.43 percent, coal 25.17 percent, liquid fuel 6.12 percent, wind 0.14 percent, geothermal 0.06 percent and others 4.85 percent (Rumeli 2006). See Table 1 which is modified

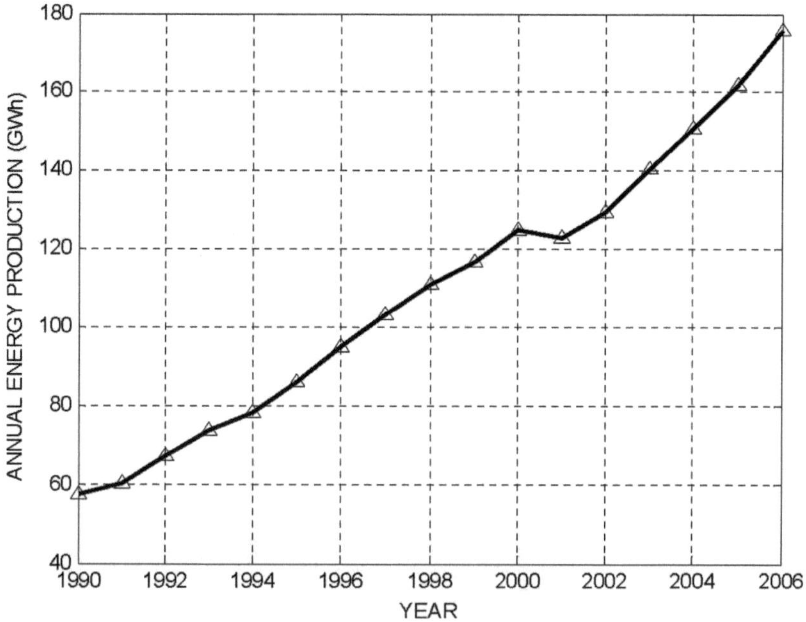

Fig. 1 Annual energy production of Turkey (Basmaci 2006)

from the MENR (2009) website indicating primary energy production by sources from 2004 to 2006 in GWh.

In the 1970s, when sources such as wood, biomass (plant and animal surplus), petroleum and coal stood out as basic energy sources, in 1976 natural gas became utilized, in 1987 solar energy sources and wind energy came into use, and since 1998 have taken their place among basic energy sources.

In the year 2000, a decrease was observed in energy produced from biomass, while a major increase was observed in the utilization of natural gas. Total solid fuel usage still nominally continues (Ministry of Environment and Forestry, Environmental Indicators 2008). A special emphasis has recently been placed on alternative energy sources such as wind and geothermal power and there have been some steps taken towards introducing the use of nuclear power as well (DSI 2010).

3 Gaps and challenges

The major domestic primary energy sources in Turkey are hydropower and lignite, whereas almost all oil and natural gas as well as high quality coal are imported. About 35 percent of the 76.6 million toe of total primary energy demand were supplied by local sources in 1999. This ratio is expected to decrease to 28 percent in the year 2010 and 24 percent in the year 2020 (MENR 2009). In other words,

Table 1 Primary energy sources of Turkey (MENR 2009)

Years		2004		2005		2006		2007		2008		2009	
Sources		Production (GWh)	Contribution (percent)	Production (GWh)	Contribution (percent)	Production (GWh)	Contribution (percent)	Production (GWh)	Contribution (percent)	Production (GWh)	Contribution (percent)	Production (GWh)	Contribution (percent)
THERMAL	Natural gas	62,241	41.30	73,444	45.36	77,239	43.96	95,024	49.6	98,685	49.7	94,396	48.6
	Coal import	9,520	6.32	10,281	6.35	11,055	6.29	11,847	6.2	12,566	6.3	12,798	6.6
	Domestic coal	24,927	16.54	32,912	20.33	35,031	19.94	41,584	21.7	45,148	22.8	42,186	21.7
	Other	7,867	5.22	5,667	3.50	8,013	4.56	6,896	3.6	7,900	4.0	7,315	3.8
Total thermal		*104,557*	*69.38*	*122,305*	*75.53*	*131,339*	*74.76*	*155,352*	*81.1*	*164,301*	*82.8*	*156,969*	*80.7*
Hydraulic		46,083	30.58	39,560	24.43	44,221	25.17	35,850	18.7	33,269	16.8	35,879	18.5
Wind		57	0.04	59	0.04	129	0.07	355	0.2	846	0.4	1,483	0.8
TOTAL (Turkey)		**150,698**	**100**	**161,924**	**100**	**175,690**	**100**	**191,558**	**100**	**198,417**	**100**	**194,059**	**100**

Turkey will be able to produce only about one fourth of her primary energy demand by exploiting her own sources in the near future.

In addition to increasing energy demand and decreasing energy autonomy, the Turkish energy policy has to face the fact that many rural and remote areas are still undersupplied. Other challenges are the big losses during distribution and transmission of energy, which result mainly from the long distances between the locations of supply and demand. There are significant energy losses due to inefficient transmission lines: about 15 to 25 percent of energy is lost in the transmission lines (Rumeli 2006).

Unfortunately, a very significant amount of energy production is lost due to old electric transmission lines, illegal usage and inefficient administration. In Turkey the electric energy losses reached a maximum of 25.2 percent in 2002. Since then losses have been decreased to 16.5 percent in 2006. While it is not possible to avoid all the losses completely, at least a target of a 7 percent loss can be achieved in Turkey (EIE 2006). In order to achieve this, it will be necessary to make the population more aware and provide education in this field. On the other hand, existing rules and regulations that have been adopted in recent years need to be enforced. This can only be achieved in a harmonious manner, if contributions of end users are taken into consideration. Furthermore, super structural and infrastructural improvements must be adopted in addition to the electricity generation, transmission and consumption rules and regulations. If experts are not consulted in the planning, construction, operation and maintenance stages of energy production projects, then the consequent losses are expected to be proportionally high.

A related problem is that illegal usage of electricity causes local fires in settlement areas, which may cause human and property losses. About 80 to 85 percent of illegal electricity uses are in households and about 10 to 5 percent in the small-scale production industries. It is estimated that all the electric energy losses in the form of leakages or illegal usages, together with their consequences as interruptions in the industrial production processes and fires, added up to about US$ 2 billion losses in 2006 all over Turkey (Rumeli 2006). This figure alone shows how important energy conservation and efficiency is in Turkey as it is in many other countries.

The following issues may be addressed so as to optimize energy consumption:

- Anticipated long term energy needs of a local town, country or region must be taken into consideration in order to ensure better development.
- Public and private energy sectors should know their responsibilities and limitations in the energy management and development.
- Various impact assessments should be taken into consideration on a local, country or regional basis. For instance, long term assessments of climate change impacts expose significant effects (Sen 2008).
- Clean, efficient, flexible and dependable energy consumption must be carried out with an integrated system of credits and incentives.
- In cases of important issues, duly found common decisions must be taken without delay after mutual negotiations towards the solution of the problem at hand, especially where important structures are concerned.

- Any decision made at lower stages must be transferred without any delay to higher decision-makers, who should make right and timely decisions with subsequent applications.

Another problem is the optimum operation and management of the interconnected energy sources throughout the whole country at all times in an efficient manner. The management aspect of different sources presents in itself a major problem that should be solved by the introduction of new alternatives such as nuclear and solar-hydrogen energy alternatives into the system.

4 Turkish energy policy puts focus on hydropower

The basic target of Turkish national energy policy is the provision of cheap electrical energy in sufficient amounts and on time, within a reliable and competitive energy market. Reliable energy requires responsibility for all energy procurement aspects on the demand and supply side including the development and management of end users' strategies, daily cash market evaluations, energy market analysis and delivery scheduling. On the other hand, a competitive energy market requires individual electricity generators who can make independent commitment decisions instead of, or in addition to, a centralized commitment. Their management purposes should seek to maximize profits against the predicted market clearing price. Hence, their strategy can be expressed in bids, so that they shutdown or start-up when the market price implies such activity.

In this regard, the energy policy as determined in the Five Year Development Plans includes various aspects such as:

- provision of qualified, reliable and cheap energy for sustainable socio-economic development,
- guaranteeing security in supplying energy,
- encouragement of private sector investment, expediting privatization activities in the power sector,
- addition of new and renewable sources as soon as possible to the energy supply cycle.

Increasing the share of renewable sources including hydroelectric power in the composition of the national energy production profile is expected to increase steadily over the next 25 years. Unfortunately, even if the total potential of renewable energy sources is developed, Turkey cannot cover all its demand by national supply alternatives. However, developing the hydroelectric power capacity with the involvement of private power stations would positively contribute to the national energy production profile.

In order to prevent an energy deficit in the near future, the steadily increasing demand trend for energy should be offset primarily by hydroelectric power generation and other renewable energy sources within the country. Hydroelectric power generation has top priority in Turkey because only one-third of the total potential is being used currently. According to Basmaci (2006), several arguments are in favour of hydroelectric power generation:

- From an environmental point of view, hydropower is a renewable resource and hydroelectricity production does not involve significant emissions of CO_2, SO_2, NO_x and ash.
- Exploiting the hydropower potential contributes towards Turkey's energy security. Further development of the resource will therefore not increase the dependency on energy imports and will not be effected by fluctuations in oil prices.
- Hydropower plants further provide flexibility in supply, as electricity from hydropower plants can be fed into the electricity network simultaneously and adaptively, whereas in thermal plants the boilers must be heated prior to any energy production. The adaptation capability of hydroelectric power is very swift and can replace the production almost simultaneously should another energy generation unit break down. In addition, hydroelectric generation regulates frequency and voltage as well as providing a balance at peak energy hours during the day by reducing the energy deficits.
- Hydropower dams with reservoirs may have further benefits such as regulating the natural flows of rivers, thus mitigating the impacts of floods and droughts; supplying water for domestic and industrial usage, for irrigation and for recreational sites with sport activities. They also store electric energy in water reservoirs and produce energy at the time of demand.
- HEPPs can respond to peak power demand in the cheapest way: in the long term, their production cost per kWh is about 0.92 US cent whereas it is 4,362 and 4,184 US cent for natural gas and lignite plants, respectively.
- The initial investments of hydropower plants are comparatively very high, but production and management as well as operation and maintenance costs are relatively low. It repays the initial investment within a short time after going into operation. Furthermore, the economic lives of dams and any elements such as spillways, derivation tunnels etc. are very long from 100 to 200 years; the active life of turbines is almost 40 to 50 years.
- Almost 80 percent of the investment costs are for construction and the remaining 20 percent are for the electro-mechanical costs. On the contrary, in thermal power plants 20 percent are for the construction, whereas 80 percent are for the electro-mechanical costs.
- The construction of any hydropower plant provides employment for engineers and workers.

Accordingly, HEPPs should be preferred because of their environment-friendly technologies with the lowest risk potential. These plants are able to respond to unexpected demand fluctuations. As explained before they are operated as peak power plants in Turkey as well as in other countries. "Hydroelectric power is environment-friendly, clean, renewable, able to meet peak demands, highly efficient (over 90 percent), involves no fuel cost, is a balancer of energy prices, has a long life-span (200 years), its cost recovery is short-run (5-10 years), its operational costs are low, (approximately 0.2 US cent/kWh), and it is an indigenous source of energy which is national and natural" (DSI 2010).

5 Development of hydroelectric power: existing capacity and future potential

Although small dam construction for water supply to Istanbul were started during the Ottoman period, almost simultaneously with such developments in Europe, the construction of large dams for hydroelectric energy generation in Turkey only started in the 1950s. While total energy generation in Turkey in the 1950s was a mere 800 GWh, this figure has increased about 190 times, reaching 151,000 GWh/year today. The current installed capacity in Turkey is 37,500 MW, which could generate an average of 220,000 GWh/year. However, total generation remains at 151,000 GWh for reasons such as failures, maintenance and repair activities, operation policy, economic recession, low demand, drought, efficiency, etc. In other words, average capacity utilization remains at 69 percent. Capacity utilization has been 59 percent in thermal plants and 105 percent in HEPPs.

As of the beginning of 2009, 673 big and small scale hydropower plants are in operation, and a further 146 hydropower plants are under construction. The theoretical gross hydroelectric potential is 433,000 GWh/year. However, for the time being only 216,000 GWh/year are technically feasible and of this number again only 127,381 GWh/year are economically feasible (DSI 2010). As of 2009, hydroelectric power generation was 47,871 GWh/year. This figure indicates that so far only 32 percent of the exploitable potential has been developed leaving plenty of scope for further development.

The final goal of installing 45,000 MW capacities has been planned for the fully fledged production through 1,738 hydroelectric plants to produce 140,000 GWh per annum (DSI 2010). For the time being there are bureaucratic difficulties such as non-cooperation between some governmental departments regarding the share of water available for their activities. The majority of these problems have been eased recently due to the cooperation and involvement of the private sector, in addition to efforts for further privatization, especially with regards to the construction of small HEPPs designed to augment hydropower generation potential of Turkey. In spite of restricted budgets, Turkey strives to invest more in energy projects especially in domestic hydroelectricity generation. The maximum possible amount from the national budget is being allocated for the completion of ongoing dam projects; with respect to new projects emphasis is given to dams producing hydropower. Besides that other financing models for the implementation of large-scale hydropower projects are being promoted to overcome the energy bottleneck in the next century.

6 Impact of climate change on water resources and on hydropower generation

Climate change scenarios show that Turkey together with other European countries is located in the worst impact location with regards to the change of the annual mean precipitation (IPCC 2007). The major climate change impacts on Turkish water resources and hence on the hydroelectric power production is analysed (Sen 2003): in the long run, Turkey is bound to face a decrease in rainfall and increase in temperature, which are expected to enhance the intensity and areal extend of regional droughts. Table 2 indicates variations in the next 50 years for the various drainage basins in Turkey.

These climatic effects will impact on the overall water resources availability in particular in the Southern Anatolian region of Turkey where most of the large

Table 2 Water resources development estimates for the next 50 years (Sen 2003)

Basin	Area (km^2)	Annual average precipitation (cm) (2000)	Annual average runoff (cm) (2000)	Annual average precipitation (cm) (2050)	Annual average runoff (cm) (2050)
1. Firat	127,304	582	248.2	261	100.2
2. Tigris	56,616	814	437.4	366	176.7
3. East Black Sea	24,077	1,291	581.2	580	234.8
4. Antalya	19,577	910	574.1	409	231.9
5. West Black Sea	29,598	803	339.2	361	137
6. West Mediterranean	20,953	865	370.3	389	149.6
7. Marmara	24,199	766	316.2	344	127.7
8. Seyhan	20,450	629	345.2	283	139.4
9. Ceyhan	21,982	758	328	340	132.1
10. Kizilirmak	78,180	459	80.6	206	32.5
11. Sakarya	58,180	534	103.7	240	41.8
12. Coruh	19,872	540	327.1	243	132.1
13. Yesilirmak	36,114	556	153.4	250	61.9
14. Susurluk	22,399	730	238.8	328	96.4
15. Aras	27,548	462	201.1	208	81.2
16. Konya Plain	53,850	437	62.4	197	25.2
17. B. Menderes	24,976	656	118.1	295	47.7
18. East Mediterranean	22,048	669	556.5	301	224
19. Van Golu	19,405	507	133.5	228	53.9
20. North Aegean	10,003	730	220	328	88.9
21. Gediz	18,000	639	100.6	288	40.6
22. Meric-Ergene	14,560	640	85.9	288	34.7
23. K. Menderes	6,907	740	162.2	333	65.5
24. Orontes	7,796	837	153.9	377	62.2
25. Burdur Golleri	6,374	436	48.6	196	19.6
26. Akarcay	7,605	472	59.2	212	23.9

dam projects are located. Accordingly, climate effects will result in a significant decrease of the overall hydroelectricity capacity which may be as much as 5 or 10 percent. Variability in climate already causes fluctuations in hydroelectric generation. Climate changes that reduce overall water availability will reduce the productivity of hydroelectric facilities. Reliable increases in average flows would increase hydropower production. Changes in the timing of hydroelectric generation can affect the value of the energy produced. Specific regional impacts are not well established (Sen 2008).

Due to droughts the overall sedimentation into the dams is expected to increase, which will cause a reduction in the active life of the dams and hence have negative effects on electricity production. Climate changed-induced decreases in water availability will affect water supply for growing populations; though water use for hydropower is non-consumptive, electricity generation and operation of dams will have to be balanced with the increasing water demands of other sectors. This also refers to water needs of neighbouring countries and hence may increase transboundary water problems (see articles in part II of this volume).

7 Private sector involvement in hydropower generation

In order to overcome the financial problems for investing in hydraulic energy, the country has embarked on alternative financial models involving the private sector such as Build-Operate-Transfer (BOT), Build-Own-Operate (BOO) and Transfer of Operating Rights (TOOR) schemes since the 1980s (see Baskan in this volume).

Prior to 1980 large dam construction in Turkey was supported by public investment without any private sector involvement. However, in 2008 more than 90 percent of the present 14,119 MW installed capacity had been implemented by DSI (DSI 2010). Although DSI is responsible for the construction of hydropower plants, operation has been handed over to the Turkish Electricity Authority (TEK) since its establishment in 1970. From 1993 onwards instead of one single authority (i.e. TEK), two new establishments came into operation: the Turkish Electricity Generation-Transmission Cooperation (TEAS) and the Turkish Electricity Distribution Company (TEDAS). TEK held the monopoly until the 1980s when private sector participation started the generation, transmission and distribution of electricity under BOT, BOO and TOOR schemes. Currently Turkey's energy policy is mainly concentrated on the assurance of liberalization and privatization activities in the energy sector.

As of 2009, a total of 17 HEPP with a total installed capacity of 963 MW and with a production potential of 7,132 GWh will be constructed and taken into operation by the private sector under the BOT model (MENR 2009).

8 Critiques on the privatisation process

During the privatization process there was a conflict of interest between public agencies responsible for natural resources, the private sector and industrial companies. Experience has shown that the currently operating BOT power stations are not satisfactory for the national energy policy, because they disappointed the users by increasing energy prices. End users suffered from price increases, and public cost triggered by increased prices was so high that investigation reports prepared by several state organs recorded this failure (Tutus 2006).

The Chamber of Electrical Engineers filed a law suit objecting to the competence of the Energy Market Regulatory Authority (EMRA) when issuing production licenses to private companies. There is still bureaucratic and technical vagueness as far as competence and financial background of the companies are concerned. Each company presents detailed feasibility reports firstly to DSI and then to the EMRA regarding financial aspects. Unfortunately, many of the presented investment proposals lack comprehensive feasibility studies and are weak investment projects (Keloglu 2006).

Unfortunately, bureaucratic slackness in the assessment of proposed projects and in the final approval occurred. A stable, reliable, efficient and standardized procedure for final approval needs to be established in order to provide a data base for future interconnected management of the HEPP network throughout Turkey. Such an approach will also enable the authorities to accelerate approval, and to control planning, construction, operation, management and maintenance. It must also be ensured that private investment plans consider sustainability aspects in the river basins.

Despite some problems and delays, the liberalization of the market and the relatively new focus on the exploitation of renewable sources indicate a positive development in Turkey. Available human resources and the use of new technologies are among the advantages of private investments (Basmaci 2006). Due to the high costs involved in order to access the market and for the construction of the physical infrastructure, state support to private companies might be justified for various schemes in order to encourage new investments (Turkyilmaz 2006).

9 Concluding remarks

Water is an essential and basic commodity for the survival and sustenance of human life and environmental activities. It is, therefore, necessary to plan, construct, operate, manage and maintain water resources especially in the form of HEPPs, as in the future dams will have to be constructed more carefully and safely. The water problems in society will bring together many specialists including politicians, environmentalists, commercial and trade interests, water scientists and technicians in addition to technocrats. However, these specialists will be unable to solve the problem but they can present the possible deficiencies and shortcomings. The overall solutions and desirably holistic solutions will need to comprehend and deal with more complex issues including economics, financial support, social, environmental and legal issues. Social, economic and environmental integration of a society cannot be obtained without freshwater resources. Especially, in the twenty first century water is expected to be the most necessary commodity for running industries, providing energy and growing food. In order to meet the increasing demand, Turkey must produce continuous, high quality, reliable and economical electricity and make necessary investments by also taking into consideration the environmental effects. Turkey with its energy deficit has to complete hydropower projects because major domestic sources are hydro and lignite. However, the country encounters financial problems, partially because the most suitable locations for dam construction regarding geology and topography are no longer available. Thus only difficult and expensive projects are possible in the near future.

Turkey, with its intercontinental position near the Middle East, has water resources which have not yet been fully developed for the complete hydroelectric energy generation and proper water supply and irrigation distribution systems. It is clear that the only clean but not self-sufficient local energy is hydroelectric power and only about one-third of this potential has been developed. Turkey will seek to construct additional dams in the next several decades, so as to benefit from local sources and to cover the local energy demand to a maximum. This will hopefully also have an impact on the revitalization of the regions' economies, speeding up land development, agricultural expansion and irrigation efficiency, which will also increase employment and revenues.

References

Basmaci E (2006) Su Politikalari ve Hidroelektrik Santraller. TMMOB Su Politikalari Kongresi (21-23 Mart 2006, Ankara) Bildiriler Kitabi 1.Cilt, Insaat Muhendisleri Odasi, Ankara, pp 298-307

DSI (2010) http://www.dsi.gov.tr/hizmet/enerji.htm Accessed 10 January 2010

EIE (Elektrik Isleri Etud Idaresi) (2006) First National Communication Group Report. EIE, Ankara

IPCC (Intergovernmental Panel on Climate Change) (2007) Summary for policymakers. In: Solomon S, Qin D, Manning M, Chen Z, Marquis M, Averyt KB, Tignor M, Miller HL (eds) Climate Change 2007: The physical science basis. Contribution of Working Group I to the Fourth Assessment Report of the Intergovernmental Panel on Climate Change. Cambridge University Press, Cambridge, United Kingdom and New York, NY, USA

Keloglu N (2006) Turkiye Su Kaynaklarinin Gelecekteki Kullanimi ve Hidroelektrik Enerji Uretimi. TMMOB Su Politikalari Kongresi (21-23 Mart 2006, Ankara) Bildiriler Kitabi 1. Cilt, Insaat Muhendisleri Odasi, Ankara, pp 13-25

MENR (Ministry of Energy and Natural Resources) (2006) Energy scenarios report. UNDP GEF FNC Project, Ankara

MENR (Ministry of Energy and Natural Resources) (2009) MENR website http://www.emo.org.tr/ekler/c3cf452d3da8402_ek.pdf Accessed 29 August 2010

Ministry of Environment and Forestry (2008) Environmental indicators 2008. General Directorate of EIA and Planning. http://www.cedgm.gov.tr/CED/Files/eng_2008_son.pdf Accessed 23 August 2010

Rumeli A (2006) Personal communication. Middle East Technical University, Electric Engineering Department

Sen Z (2003) Su Bilimi ve Yontemleri. Turkish Water Foundation Publication

Sen Z (2008) Global warming threat on water resources and environment: a review. Environ Geol (2009) 57:321–329. doi:10.1007/s00254-008-1569-5

Tutus A (2006) Turkiye'de Elektrik Enerjisinin Tarihsel Gelismi ve Yeni Piyasa Duzeni Icerisinde Hidroelektrik Enerjinin Yeri. TMMOB Su Politikalari Kongresi (21-23 Mart 2006, Ankara) Bildiriler Kitabi 1.Cilt, Insaat Mühendisleri Odasi, Ankara, pp 318-320

Turkyilmaz O (2006) Enerji Politikalari - Yerli, Yeni ve Yenilenebilir Enerji Kaynaklari Raporu. Makine Muhendisleri Odasi, Ankara, pp 58-60

Liberalization of Turkey's Hydroelectricity Sector

Argun Baskan

1 Introduction

The 1980s which were dominated by the liberal policies of then Prime Minister Turgut Ozal brought dramatic changes in the political and economic thinking in Turkey. Liberalization and deregulation of the national energy sector with its subsectors including hydroelectricity production marked a clear departure from the mentality of the earlier decades which was characterised and dominated by public investments, with a growing tendency especially in the 1970s, when large dams were constructed without the participation of the private sector (Tutus 2006, 318-320). Neoliberal policies and globalization starting in the 1980s have clearly influenced the liberalization of the electricity market and macroeconomic policies in Turkey. Privatization of the electricity markets was on the agenda of the General Agreement on Tariffs and Trade (GATT) talks in this period. In addition, Turkey's relations with the World Bank has shaped the neoliberal thinking of the Turkish governments for more than two decades,[1] and the Organization for Economic Cooperation and Development (OECD) has contributed further towards this economic mentality. The European Union's (EU) impact was less direct. However, the Renewable Energy Law of 2005 is a genuine example of how the EU's environmental policy influenced national energy planning and the liberalization of the hydroelectricity production sector, as the EU's energy and environmental policy has been promoting the increase of the renewable sources in the national energy production profiles. The EU's competition and trade policy principles are also accelerating the process of private sector involvement in the hydroelectricity business (Ozguler 2006).

[1]The close connection of national electricity policies with the international financing institutions like the World Bank often receive harsh "anti-imperialist" criticisms in public debates and publications; see for example Makine Muhendisleri Odasi 2006, 23-24.

A. Baskan (✉)
Department of International Relations, Ege University, Izmir, Turkey
e-mail: argunbaskan@gmail.com

A. Kibaroglu et al. (eds.), *Turkey's Water Policy*,
DOI 10.1007/978-3-642-19636-2_5, © Springer-Verlag Berlin Heidelberg 2011

2 Institutional and legal transformation

Established in 1970, the Turkish Electricity Authority (TEK) became a statutory monopoly in 1984. From 1984 onwards the private sector was allowed to participate in the generation, transmission and distribution of electricity under three models namely Build-Operate-Transfer (BOT), Build-Own-Operate (BOO) and Transfer of Operating Rights (TOOR) (see Box 1). In 1993, TEK was split into two new bodies, namely the Turkish Electricity Generation-Transmission Corporation (TEAS) and the Turkish Electricity Distribution Corporation (TEDAS). After the introduction of the Electricity Market Law in 2001, TEAS was further unbundled into three companies, i.e. Electricity Production Corporation (EUAS) (generation), Turkish Electricity Transmission Corporation (TEIAS) (transmission) and Turkish Electricity Trade and Contract Company (TETAS) (wholesale). However this structure has yet not taken its final form (International Energy Agency 2005, 131-132).

Box 1 Models for private sector participation

BOT contracts are designed to use private capital for the construction of new public plants or equipment. Under a BOT, the private firm is responsible for financing, building and operating a new facility or infrastructure network for a specified period of time to meet certain performance standards demanded by the government. BOTs are typically for 10 to 20 years, long enough to allow the private company to pay off its costs and make a profit. The government still retains ownership of the constructed plant and other infrastructure facilities, and functions as both the customer and the regulator. The government is obliged to purchase a minimum level of the output from the private company regardless of the level of end-consumer demands (Plummer and Gentry 2002, 201). The official BOT policy dramatically changed after the economic crisis in 2001. Turkey announced the ending of sovereign guarantees to finance future BOTs and the guarantee period for cost recoveries were reduced to 10 instead of 20 years. Turkish courts have also ruled that BOTs could be subject to legal reviews as they should be accepted as concessions. These developments practically lowered the attractiveness of the sector for private companies (GENI 2008).

BOO, a private company builds a new plant or other infrastructure facilities at its own risk, and owns and operates the facility again at its own risk. The government usually offers purchase guarantees through long-term take-or-pay contracts for big supply facilities, or minimum revenue guarantees (World Bank). BOO schemes are favoured by the private sector as they are not subject to time limitations (GENI 2008).

TOOR, as practiced in Turkey, under a TOOR contract the private company is responsible for operating, and, if necessary, rehabilitating an existing publicly owned plant or other facilities through a lease type arrangement for

15 to 30 years and includes take-or-pay arrangements which secure government purchase of the related service (Hoekman 2005, 191). This approach is criticised by the International Monetary Fund because it would impose restrictions on operating rights (GENI 2008).

The Law on the Utilization of Renewable Energy Resources for the Purpose of Generating Electrical Energy (No. 5346) which entered into force in May 2005, is regarded as a revolutionary step in the hydroelectricity business. The Law entails "the guarantee of purchase" principle which guarantees the purchase of a company's service by the government which acts as a strong incentive for private investments.

3 Practices and criticisms

Before the adoption of the Law on the Utilization of Renewable Energy in 2005, legal aspects of the private sector's involvement in hydroelectricity production business were rather disputed. The core of the debate were objections made by the Chamber of Electrical Engineers (CEE) of Turkey who demanded the protection of the common good and public interests through the maintenance of the public monopoly as the best way of serving public interests. Law suits issued by the CEE in the Council of State (Danistay, in Turkish) caused an interruption of the technical and financial process that would enable the entrance of private companies in the hydroelectricity production business. The Council of State's decisions were a legal victory for the CEE at that time because it ruled that the liberalization process should be suspended. The major objection against the privatization process was based on the claim that the extension of the right to use public natural resources by the private sector through agreements between government bodies and private companies were simply in contradiction with the public interest. Additionally, the CEE filed another law suit objecting to the competence of the Energy Market Regulatory Authority (EMRA) over issuing production licences to private companies.

The Council of State's decision in this second case was also in favour of the CEE' arguments. Again the privatization process was halted. The Renewable Energy Law (No. 5346) of 2005 appeared as a solution to overcome these legal obstacles (see Box 2 comprising principal legislation regulating the energy/electricity sector). The Law made a critical change in the official management of the process and appointed the General Directorate of State Hydraulic Works (DSI) as the main governmental body with the legal competence to implement the initial steps of the privatization process and to organize tenders. What came into existence was a partial change of legal competence from an autonomous agency, that is the EMRA, to another government agency, DSI. Currently private companies do not face any legal obstacles if they wish to construct and run hydroelectricity plants, but the CEE has undertaken some legal actions to stop the process. It is not easy to make an overall assessment of the commercialization process involving the construction of private hydroelectricity plants because the whole process is still in its early stages (Yildiz 2006).

Box 2 Principal legislation in the energy/electricity sector

2001 Electricity Market Law (No. 4628)
2003 Regulation on the Procedure and Principle of Signing a Water Use Right Agreement to Make Production in the Electricity Market
2004 Strategy Paper Concerning Electricity Market Reform and Privatization Programme
2005 Law on the Utilization of Renewable Energy Resources for the Purpose of Generating Electrical Energy (No. 5346)
2007 Energy Efficiency Law (No. 5627)

Yet, a retrospective look towards early experiences in the sector may result in pessimistic assumptions. One major and almost classical problem regarding the legacy of national energy policies has been the lack of sufficient public investment to increase and renew the production capacity, especially in the field of electricity production. The BOT model was introduced from the 1980s onwards to open the way for private investments. Participating companies were provided with purchase guarantees by the Treasury. Today there are still BOT power stations operating. However, generally speaking, the experience of BOT type power stations does not match the performance expectations of the government and other interested parties like the CEE. The management of BOT power stations increased energy prices in comparison with earlier periods, which end users have to bear. Public spending based on the Treasury's purchase guarantees were as high as reported in investigation reports prepared by several state organs. This meant that public money was not spent properly as a result of favouring private companies' interests even when those companies did not fulfil their obligations (USIAD 2004, 24-26, Ulusaler 2006). Between 1980s and mid-1990s the respective governments tried to overcome legal, practical and fiscal problems in the liberalization process of the electricity market but supervision of the overall system was weak (Yildiz 2006).The process led to the introduction of the current model introduced by the Electricity Market Law of 2001.

When it comes to the clearance process, which is actually the market entrance for private companies, one of the most important challenges is the persistence of a number of technical and bureaucratic ambiguities. The process requires a wide examination of the technical competence and financial background of the applicant companies. According to the procedures, companies must first apply to DSI in order to sign Water Use Rights Agreements. In line with the Regulation on the Procedure and Principle of Signing Water Use Right Agreement to Make Production in the Electricity Market (2003), the agreement grants the production license to the private company. Private companies apply for those projects they select from EMRA's pre-determined list of potential projects. The companies start to work after they get EMRA's approval.

There are good reasons for a unified mechanism for assessing the applications in order to achieve a more comprehensive policy. On the technical side, due to the yet unsatisfactory production performance of the companies, gaps may develop between today's production targets and the actual electricity power output. Even today there are signs pointing to poorly prepared projects by private companies, Therefore, the private sector may not be able to meet overall expectations (Keloglu 2006, 23), and might realize only about half of the targeted 10,000 megawatts. A further concern is that river development plans do not take due consideration of environmental issues; private investments should be implemented in line with recommendations made by Environmental Impact Assessment studies to be approved by the Ministry of Environment and Forestry. High entrance costs into the market and high initial costs for constructing the physical infrastructure are seen as a justification for granting state support through various schemes to encourage private companies to finance new investments (Turkyilmaz et al. 2006, 58-60). The acceleration of the bureaucratic procedure for assessing and approving applications would be a facilitating factor in the liberalization process. Articulation of more comprehensive data on Turkey's hydroelectrical production potential would enable the government bodies to make more accurate planning and investment decisions (Yildiz 2006).

A more competitive business environment is a necessity for the proper functioning of the hydroelectricity production system. The small number of companies limits the range of customer choice (International Energy Agency 2005, 152-153). Liberalization of this sort in the national electricity sector requires a well designed preparation process that would precede other steps – the realization and auditing of privatized systems. Legal obstacles and law suits as witnessed in the last few years do not seem to pose major problems for the moment but they may discourage future investments (The Economist Intelligence Unit 2005). There are legitimate doubts that Turkey's bureaucracy is yet ready to undertake this task properly. The overall bureaucratic system needs to be restructured to develop transparent procedures in order to achieve higher productivity as well as developing more sophisticated technical competence. In the simplest sense, Turkish bureaucracy's performance and success is actually dependent on the quality and success of its internal structures and processes (Committee of Water Resources Planning and Electricity Power of the Ankara Branch of the Chamber of Construction Engineers 2006, 308-309).

The overall policy making process which welcomes the participation of the private actors, foreigners in particular, was criticised with a protectionist rhetoric. There are calls for developing a "national" energy policy and for restricting the role of foreign players using a strong "nationalist" and "anti-imperialist" rhetoric (one being that Turkey sells its rivers) (Makine Muhendisleri Odasi 2006). This can be understood as a call for protectionism and economic nationalism to avoid the challenge which foreign companies bring into the Turkish market. Entry of foreign capital into almost all national sectors, especially energy and telecommunications, are perceived as risky not only for the national economy but even for the independence and security of the country. Problems related to the EU accession process also feed suspicious sentiments towards foreign capital involvement. Other objections question not only the management of this liberalization process but the whole notion of private sector involvement in the hydroelectricity production business. The main reasoning behind this kind of objection lies in the belief that renewable energy sources should be developed by public bodies assuming that only they are able to act and decide in the public interest. An accompanying argument points to the failure of similar experiences in European countries especially in the United Kingdom (Ulusaler 2006).

Of course, not all aspects of the privatization process rely solely on official policies and bureaucratic performance steps. For example, companies willing to enter the private hydroelectricity production business face some difficulties in obtaining credits under acceptable terms, which is an issue that needs to be primarily solved between the companies and private financial bodies with as little governmental intervention as possible. However, companies call for official intervention to obtain credits under more favourable financial conditions. As revealed by a series of joint meetings between the General Directorate of DSI and applicant companies, problems of private sector participation are reported as follows,

- delays in the announcements regarding the specifications of calls for applications
- lengthy correspondence between different official bodies and slow decision-making
- obstacles caused by the poor physical state of Turkey's general transportation infrastructure
- price levels of electricity
- protests by environmentalists
- financial and technical difficulties.

Assessments made by particular firms further highlight macroeconomic risks, price levels of natural gas, shareholder relations etc. caused by Turkey's risky macroeconomic climate and business environment. Most sensitive issues that arise are the level of prices and production costs. Bureaucratic delays and "red tapism" of the governmental bodies are also mentioned as obstacles which make it more difficult for private companies to make reliable and timely business plans (Eroglu 2006; Goker and Yurdagul 2006).

Again, considering the licensing process from the view point of environmental protection, it is not surprising to see that the idea of private hydroelectricity production faces local public protests especially in the Black Sea region. Local population has been worrying about the risks hydroelectricity stations pose to the natural beauty and life of the region. It is assumed that these fears are somewhat linked to the ongoing impacts of the Chernobyl disaster of 1986 which still shows impacts on the environment and on human health in the region. There are further worries by the society at large that the exploitation of rivers for hydroelectricity production – be it public or private – may cause harm to nature (Ozguler 2006).

According to the national energy profile projections up to the 2020s, Turkey's energy imports used to produce electricity will grow and private hydroelectricity power stations can be useful in limiting the dependence on foreign sources (Yildiz 2006). From such a perspective, participation of the national energy companies in the electricity sector gains more importance. More specifically, private power stations have the potential to produce electricity that could well match the public sector's production. A comparison of the private and public production capabilities reveal that six of the largest public dams have a production capacity of 7,442 MW whereas private sector's hydroelectricity stations could produce 10,594 MW (Eroglu 2006). Besides the production capacity, use of new technologies in the electricity production is another advantage of the private sector involvement in this field (Basmaci 2006, 308-309). However, concerns about environmental protection should be well addressed in the process. Environmental protection can be a real source of weakness for both the private and public actors.

Despite risky problems and delays, the liberalization of the hydroelectricity market and the relatively new focus on the development of renewable energy resources are positive factors for Turkey's integration into the EU. The increase of renewable resources including hydropower in the composition of the national energy production profile would even be a source of prestige in the EU context. Turkey's national targets for renewable energy production are not far from those of the EU, as the Union tries to promote clean renewable energy use in the member and candidate countries in connection with its energy and environment policies. However, it has to be noted that even under the best conditions - full exploitation of all available renewable resources in the country - hydroelectricity production is not the final answer to the country's growing energy demand. Yet, we assume that the development of the hydroelectricity capacity with the introduction of the privately operated power stations would make a positive contribution to the national energy production profile. The future of the hydroelectricity liberalization process will depend on the level of liberalization to be introduced to the sector by the public authorities (Ozguler 2006). Positive developments in Turkey's accession process to the EU would facilitate integration into and harmonization of Turkey's national energy policy with the EU's energy and climate policies (Isik 2004).

Depending on its level of production, Turkey is interested in becoming an electricity exporter or even a regional leader in the South European electricity production and transmission system (International Energy Agency 2005). EU-Turkey approachment in energy policy issues could also make contributions to the general political relations between the EU and Turkey. Transformation of the Turkish hydroelectricity market would improve Turkey's growing image as a key transit country or even an energy hub in this region, in terms not only of hydropower but of oil and natural gas trade as well. Despite the wider political and economic challenges of Turkey-EU relationship, Turkey's importance in regional energy geopolitics is also acknowledged by the EU. Turkey is keen to use its promising national energy profile as a bargaining chip in the EU accession talks. Again considering the more specific picture of hydroelectricity politics, another expectation is that foreign investors participate in hydroelectricity delivery systems. The production side of the hydroelectricity business would probably remain the domain of national investors (Yildiz 2006). The official interest in opening the hydroelectricity market for private participation is also part of the global climate change agenda (Turkey ratified the Kyoto Protocol in 2009; International Water Power and Dam Construction, 16 January 2008), and also suits commercial interests of the national hydropower business that wishes to participate in (voluntary and formal) emission trading schemes.

4 Conclusion

Liberalization and deregulation in Turkey's national energy policy, and more specifically its hydroelectricity production policy, are expected to be stronger trends due to the growing dynamism of the private sector, adoption of international business norms and Turkey's increasing integration within international markets. Earlier international experiences reveal that liberalization and deregulation processes may not necessarily be flawless neither in the preparation nor in the application stages. Procedural and practical gaps and weaknesses can be expected in the quality of the process especially in relation to the environmental protection.

Strong criticism has already been voiced by interested third parties such as the CEE. Public authorities need to develop more comprehensive and reliable schemes to harmonize the interests of the private companies and the final users of their services. Inclusion of third parties like the non-governmental organizations can help to develop a better framework to promote the hydroelectricity production capacity of Turkey by more transparent and environmental-friendly means.

References

Basmaci E (2006) Su Politikalari ve Hidroelektrik Santraller. TMMOB Su Politikalari Kongresi Bildiriler Kitabi 1.Cilt 21-23 Mar 2006, Ankara. Insaat Muhendisleri Odasi, Ankara, pp 306-317

Committee of Water Resources Planning and Electricity Power of the Ankara Branch of the Chamber of Civil Engineers (2006) Su Kaynaklarimizdan Enerji Uretimi Politikamiz ve Gelecek Projeksiyonlari. TMMOB Su Politikalari Kongresi Bildiriler Kitabi 1.Cilt 21-23 Mar 2006, Ankara. Insaat Muhendisleri Odasi, Ankara, pp 306-317

Eroglu V (2006) Hidroelektrik Enerji Talebi ve Cozumleri Sunumu 2006. General Directorate of State Hydraulic Works, http://www.dsi.gov.tr/enerji_2006.pdf Accessed 4 February 2007

Global Energy Network Institute (GENI) (no year) An energy overview of the Republic of Turkey. Global Energy Network Institute, San Diego CA. http://www.geni.org/globalenergy/library/national_energy_grid/turkey/EnergyOverviewofTurkey.shtml. Accessed 15 April 2008
Goker O, Yurdagul B (2006) Ak Enerji Arastirma Raporu (April 2006). Ak Enerji, Turkey. http://www.akenerji.com.tr/media/AKENR-Nis06.pdf. Accessed 6 February 2007
Hoekman BM (2005) Turkey: Economic reform and accession to the European Union. World Bank, Herndon, USA
International Energy Agency (2005) Energy policies of IEA countries: Turkey 2005 Review. IEA Publications Service, Paris
International Water Power and Dam Construction (2008) Carbon drive for Turkey hydro. International Water Power and Dam Construction Journal, 16 January 2008. http://www.waterpowermagazine.com/story.asp?storyCode=2048365 Accessed 16 April 2008
Isik Y (2004) Turkey's energy prospects in the EU-Turkey context. EU-Turkey Working Papers, no 9. Centre for European Policy Studies, Brussels. http://www.ceps.be. Accessed 4 February 2007
Keloglu N (2006) Turkiye Su Kaynaklarinin Gelecekteki Kullanimi ve Hidroelektrik Enerji Uretimi. TMMOB Su Politikalari Kongresi Bildiriler Kitabi 1.Cilt. 21-23 Mar 2006, Ankara. Insaat Muhendisleri Odasi, Ankara, pp 13-24
Makine Muhendisleri Odasi (2006) Turkiye'nin Dogal Gaz Temin ve Tuketim Politikalarinin Degerlendirilmesi Raporu. Makine Muhendisleri Odasi, Ankara
Plummer J, Gentry B (2002) Establishing appropriate organisational and contractual arrangements. In: Plummer J (ed) Focusing partnerships: A sourcebook for municipal capacity building in Public-Private Partnerships. Earthscan Publications Limited, London, UK, pp 179-212
The Economist Intelligence Unit (2005) Turkey Country Profile 2005. The Economist Intelligence Unit Website. www.economist.com Accessed 13 September 2006
Turkyilmaz O et al. (2006) Enerji Politikalari - Yerli, Yeni ve Yenilenebilir Enerji Kaynaklari Raporu. Makine Muhendisleri Odasi, Ankara
Tutus A (2006) Turkiye'de Elektrik Enerjisinin Tarihsel Gelisimi ve Yeni Piyasa Duzeni Icerisinde Hidroelektrik Enerjinin Yeri. TMMOB Su Politikalari Kongresi Bildiriler Kitabi 1.Cilt. 21-23 Mar 2006, Ankara. Insaat Muhendisleri Odasi, Ankara, pp 318-330
Ulusaler K (2006) Personal communication. Chairman of the Chamber of Electricity Engineers
USIAD (2004) Elektrik Enerjisinde Ulusal Politika Ihtiyacimiz: Elektrik Enerjisinde Bugun ve Gelecegimiz Raporu. USIAD Website. http://www.usiad.net Accessed 13 September 2006
World Bank (no year) Private participation in infrastructure projects database glossary. World Bank Website. http://ppi.worldbank.org/resources/ppi_glossary.aspx Accessed 24 April 2007
Yildiz D (2006) Personal communication. General Directorate of State Hydraulic Works, Ankara

Water for Agriculture: A Major and Inefficient Consumer

Sevilay Topcu

1 Introduction

The history of irrigated agriculture in Turkey dates back to as early as 6000 BC. Throughout history, Anatolia, located on the crossroads of many civilizations, has played an important role as a trade bridge between western and eastern countries. From the beginning of the Turkish Republic the agricultural sector was crucial for the economic development of the country in terms of producing food and fibre, supplying raw material for industry, preventing migration from rural to urban areas, and creating employment. Because of the unreliable and erratic precipitation regime, Turkish agriculture depends heavily on irrigation, an exception being the Eastern Black Sea Region. Real advancements in irrigated agriculture in the country started therefore with the development of land and water resources projects 60 years ago. In 2008, irrigated areas covered about 5.3 million hectares, in 1950 it was only 0.15 million hectares.

Irrigated agriculture is by far the greatest consumer of freshwater which raises significant issues for water resource management with regard to water scarcity, competing demands from other sectors such as industry and domestic use, irrigation service delivery, irrigation system management, and water use efficiency. The challenges in the forthcoming years and even decades will be to balance water supply and demand among users to ensure adequate water for agriculture in addition to satisfying other needs. Although benefits from irrigated agriculture are obvious, irrigated agriculture, if not properly managed, can also be detrimental to the environment, and endanger land and water resources.

The performance of irrigation systems is low in many respects. Options for improving water use efficiency in irrigated agriculture comprise a large variety of means including engineering, agronomic, environmental and socio-political

S. Topcu (✉)
Department of Farm Structures and Irrigation, Faculty of Agriculture, Cukurova University, Adana, Turkey
e-mail: stopcu@cu.edu.tr

A. Kibaroglu et al. (eds.), *Turkey's Water Policy*,
DOI 10.1007/978-3-642-19636-2_6, © Springer-Verlag Berlin Heidelberg 2011

aspects. While some approaches aim at increasing the outputs per unit of water at catchment level, others focus on reducing losses of water and preventing water pollution or reallocating water to higher priority uses. However, physical, socio-economical, institutional, technical as well as financial factors which account for the low efficiency of irrigation projects in a catchment or region are difficult to overcome.

The chapter reviews the relative importance of the agricultural sector, the main water user in Turkey. It covers trends in irrigation and water use, and then discusses constraints for and limits to increasing efficient water use in agriculture. Moreover it describes the impacts irrigated agriculture has on the availability as well as the quality of water resources. Finally, it attempts to outline options for improving irrigation efficiency at both field and basin level.

2 The natural resource base

Good agricultural harvest has always been dependent on climate, land and water resources. A growing population, man-made climate change and environmental degradation are major threats to natural resources, and consequently for sustainable agricultural production.

2.1 Changing climate and its impacts on resources

Climatic conditions show significant differences from one region to another in Turkey. In the coastal areas, Mediterranean climate is dominant with long, hot and dry summers, and short, mild and rainy winters. Whereas, in the Anatolian plateau winter months are cold and the summer months are extremely hot, with limited rainfall throughout the year. The long-term (1965-2007) average annual precipitation in Turkey was about 620mm with considerable spatial and temporal variations from year to year (DMI 2009). The driest regions are Central Anatolia and South-eastern Turkey with an annual precipitation which is frequently less than 300mm. Throughout the country, approximately 70 percent of total precipitation falls from October to March, but rains during the summer months are insufficient for crop production. Besides the limited rainfall, evapotranspiration rates are high due to prevailing high temperatures during summer (Fig. 1). Consequently, as shown on the drought map for August (Fig. 2), most of the country is very dry during summer and classified as desert. Therefore, agricultural production, particularly in regions with fertile lands such as Western, Southern and South-eastern Turkey, can only be achieved by supplying water for irrigation.

Additionally, global climate change will likely cause lower and more erratic rainfall combined with increased temperature, resulting in higher rates of evaporation and increasing water requirements for crop production. Mediterranean countries are

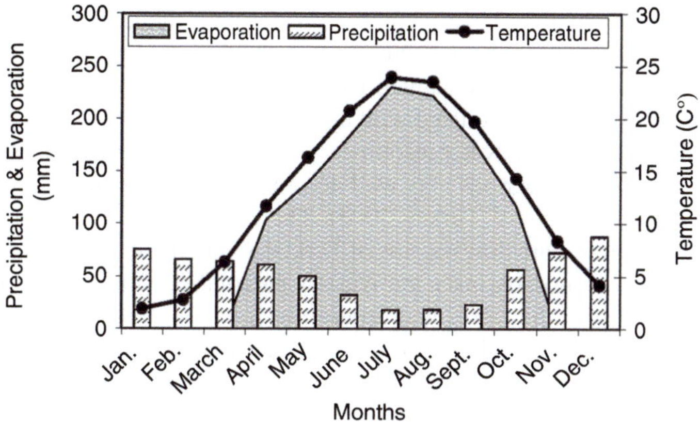

Fig. 1 Long-term observations on monthly temperature, precipitation and evaporation in Turkey (based on data from Turkish State Meteorological Services)

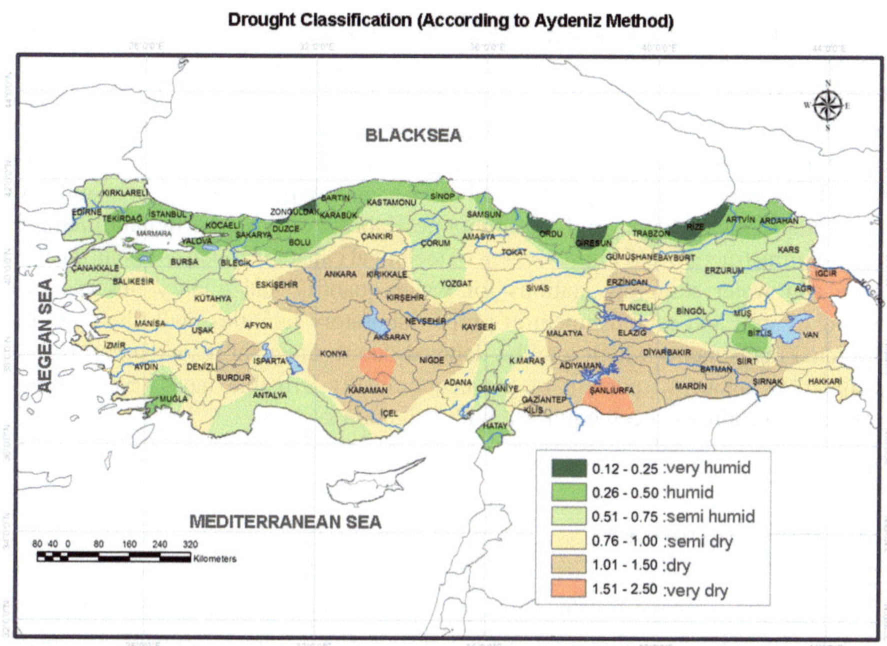

Fig. 2 Drought map of Turkey (1971–2000) (Turkish State Meteorological Services http://www.meteor.gov.tr. Accessed 20 September 2009)

highly vulnerable to climate change, in particular to extremely hot and dry summers while a higher probability of extreme weather events such as floods and/or droughts is to be expected (Giorgi 2006). Recent analyses of Turkey's climate data show that minimum temperatures in winter and minimum and maximum temperatures in spring and summer months have increased. While only moderate decreases in the total amount of precipitation have been reported, the temporal distribution pattern of winter rainfall has significantly changed. The decreasing trend in precipitation, particularly in winter months, is correlated with drought events (Turkes 2003). The precise impact of climate change on water resources is uncertain, however, it is expected that stream flows at low-flow periods may well decrease and cause a further deterioration of water quality (Republic of Turkey 2003). The examination of annual minimum, maximum and mean stream flow records of Turkish rivers showed significant decreasing trends throughout most basins in the Western part of the country (Topaloglu 2006). It is likely that, due to the lesser and erratic precipitation, less water will infiltrate the soil and consequently recharge of groundwater may be diminished. Thus, providing irrigation water will be adversely affected by decreasing water resources.

2.2 *Land resources*

Turkey's total cultivable land area is about 28 million hectares of which 90 percent is irrigable. Considering the water resources potential given by the State Hydraulic Works (DSI), 12.5 million hectares can be irrigated, but due to technical and economic constraints only 8.5 million hectares are planned to be irrigated by the year 2030, 7.9 and 0.6 million hectares from surface and groundwater resources respectively (DSI 2009). However, this goal may not be realized because of the decreasing availability of water, financial constraints and the economic viability of projects. Rehabilitation of existing infrastructure and improvement of technical capacities such as installing flow gauge and water sampling stations for water quantity and quality observations, training of staff and establishment of proper laboratory facilities, will likely require additional investments which may be delayed due to financial resource constraints. As of 2008, about 62 percent of the economically irrigable lands (5.32 million hectares) have already been equipped with irrigation systems.

2.3 *Water resources*

Comprehensive surveys on the country's water resources were started about 50 years ago by DSI, the main public agency responsible for land and water resources development in the country. Since some relevant hydro-meteorological parameters such as precipitation and stream flows show significant decreasing trends over the last several decades, as mentioned above, it is likely that using data given by DSI lead to an overestimation of the present status of water availability. The impact of climate change on natural resources has been investigated by governmental organizations including DSI, but research has not yet been completed.

The gross renewable surface water potential is estimated at 193 billion cubic meters (BCM) (DSI 2006). Besides climatic variations, e.g. precipitation anomalies, it is impossible to harness the entire potential of both surface and groundwater due to topographic, geologic, technological and economic constraints. Thus, only 51 percent (98 BCM) and 34 percent (14 BCM) (DSI 2006) of the surface and groundwater resources respectively can be used technically and economically.

Turkey has 25 river basins with a wide range of catchment sizes and river discharges. The Euphrates and Tigris rivers system for example contribute over a fourth to the total surface water resources. Some regions such as Thrace, the Aegean Sea coastline, Central Anatolia include about 20 percent of Turkey's cultivable areas but only 10 percent of the country's surface water resources. Thus, these regions have agricultural production potentials and higher water demand but scarce resources. According to the classification of the International Commission of Large Dams, 673 large dams have already been constructed and put into in operation, and additionally 143 dams are planned to control and utilize the surface water resources in Turkey (DSI 2009).

Total freshwater consumption in Turkey was 46 BCM in 2008. Industry consumed the smallest portion with a share of 11 percent (5 BCM), while 15 percent (7 BCM) of total consumption was used for domestic purposes. The largest share of freshwater resources is utilized by the agricultural sector: the sector uses 75 percent of the available surface (only 25 percent remains for other sectors) and 66 percent of available groundwater resources. Similar conditions are observed for groundwater resources. These are average national figures, though, and different proportions apply when regions are considered separately.

Per capita freshwater use for domestic and drinking purposes has increased from 98 l/day to 270 l/day during the last two decades (DSI 2009). Rapid urbanization coupled with fast developing industry and an expanding tourism industry will affect the amount of water allocated to agriculture in the future. However, the Turkish government's goal is to develop all water resources by the year of 2030. According to these plans, 64 percent of the total economically exploitable annual water potential of 112 BCM are planned to be utilized in the agricultural sector at the final stage, while the remaining 16 percent and 20 percent are to be available for domestic and industrial use, respectively (DSI 2009).

3 Irrigated agriculture and its role for the Turkish economy

Agriculture in Turkey is still a substantial productive economic sector, a major source of economic growth and employment (Republic of Turkey 2003), and a large contributor to export revenues not only generated from agricultural products directly but also as producer of raw materials for industrial export commodities. Within the agricultural sector, irrigation is often the dominant contributor to value added, employment and exports. Thus, agricultural policies affecting the irrigation sector, particularly those related to trade, commodity prices and investments form an important part of overall agricultural policy.

3.1 Development of irrigated agriculture in Turkey

Archaeological remains of irrigation structures, such as barrages, channels and deep wells, date back to as early as 4000 to 6000 BC in Anatolia. Also the Ottoman Empire constructed many water structures, part of which are still in operation. The first modern irrigation and drainage project, the Cumra Irrigation and Drainage Project in Central Anatolia, was constructed from 1908-1914. During the 1950s, in parallel to the so-called Green Revolution, the share of national investments allocated to irrigation projects increased in Turkey as in other parts of the world. In the 1980s, particular attention was paid to the South-eastern Anatolia Project (GAP, in Turkish acronym) under which 1.83 million hectares are planned to be irrigated from the Euphrates and Tigris rivers. If all goes in accordance with the GAP Master Plan, 20 percent of Turkey's economically irrigable land will be irrigated by the year 2030 and 22 percent of the country's viable hydropower potential will be produced by harnessing the water of the GAP region.

During the 1990s, Turkey was confronted with serious economic crises, which resulted in decreasing funds being allocated to the agricultural and other water-using sectors as compared to previous years (Fig. 3). Thus the share of investments for irrigation within the State's overall investment declined from 30 percent to less than 10 percent (Erdogan et al. 2003). The integrated regional development project GAP is one of the affected projects that could not be implemented due to financial limitations. DSI's recent feasibility reports show that only 1.06 million hectares from the originally planned 1.83 million hectares can be irrigated economically in the GAP area. Although 75 percent of the GAP's energy projects have already been

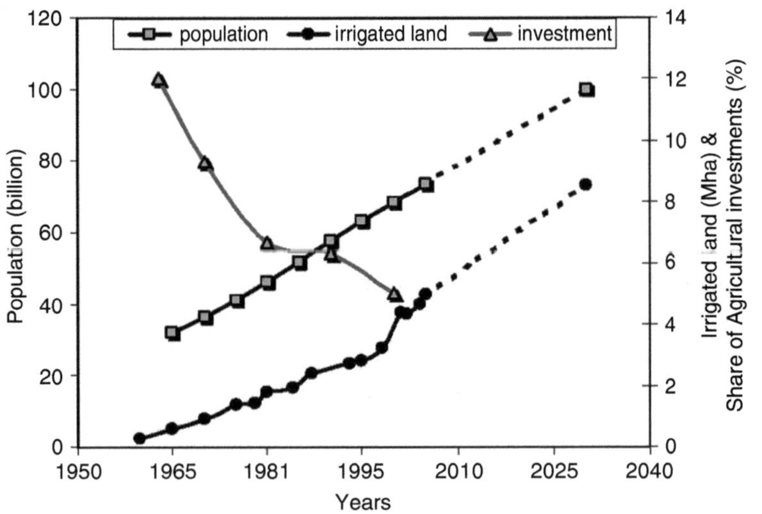

Fig. 3 Development of population, irrigated areas and share of agricultural investments (based on data from Turkish Statistical Institute and Pamuk 2006)

realized, 74 percent of the GAP's irrigation projects are awaiting completion, projected to be by 2030. In June 2008, the Turkish Government issued a new GAP Action Plan covering the period 2008-2012 which targets at speeding up basic infrastructure development projects paying particular attention to irrigation; in accordance with the GAP Action Plan, the allocated budget for irrigation investments in 2009 has been increased 10 times if compared to 2008 (DSI 2009).

3.2 *Current irrigation systems*

Currently, about 75 to 80 percent of Turkey's irrigable area is irrigated using surface water, and the remaining 20 to 25 percent using groundwater. In those areas irrigated with surface water, water is mainly transported by means of gravity-fed systems, whereas pumping systems are used on only 15 percent (DSI 2006). Investment costs for gravity-fed systems are lower if compared with pumping systems which have high operation costs because of high electricity prices that can constitute up to 80 percent of operation expenses.

Water delivery systems in irrigation schemes comprise classic trapezoidal open canals (44 percent) and canalettes, which are small precast concrete half ellipsoidal open canals (42 percent) as well as piped systems (14 percent). Water losses due to leakage and evaporation in open canals are high if compared with closed systems. Therefore DSI increased the share of piped systems to up to 55 percent in new schemes which are already under construction, and foresees public investments to replace canals and canalettes with pipes in existing schemes. Surface irrigation methods such as furrow, basin, border and even wild flooding, are widely applied in the country (in approximately 90 percent of the total irrigated area), while conventional hand-move sprinkler systems, and drip irrigation systems are used in the remaining 10 percent of total irrigated lands. Farmers are using drip irrigation in particular in the Aegean and Mediterranean regions, and partly in the Konya closed basin. Sprinklers are widely used (62 percent of the irrigated areas) in the Marmara Region, particularly in Bursa (Wasamed 2003). However, during the last two years, droughts in addition to attractive loans of banks speeded up the use of drip irrigation systems in many regions. Application efficiency (the percentage of water delivered to the field that is used by the crop) of surface irrigation methods including furrow, basin and flooding is low, approximately 50 to 60 percent, compared to those of modern irrigation methods such as drip and sprinkler (80 to 90 percent). Water loss via deep percolation and runoff is by far the highest in surface irrigation methods.

Extension activities by governmental and non-governmental organizations for raising public awareness about water scarcity and the need for saving water gained momentum in recent days. DSI and the Ministry of Agriculture and Rural Affairs (MARA) encourage farmers to use sprinkler and drip irrigation methods, particularly in areas with high climatic vulnerability (e.g. Aegean and Mediterranean Region). These areas suffer from scarce resources due to overexploitation of groundwater (e.g. Middle Anatolia, particularly in Konya Closed Basin). After the drought experienced in 2007, the government has given priority to apply short term measures to compensate for the losses and to introduce long term measures for tackling droughts by using economic instruments. Farmers' debts repayments to government banks were postponed for one year, interest-free loans and subsidies for implementing water-saving irrigation technology were provided, along with direct payments to farmers for mitigating drought effects. Within the first three months following this initiative, more than 8,000 farmers had changed their irrigation practices from flooding to pressurized systems such as drip and sprinkler. The number of farmers changing their irrigation practices from furrow to drip increased after the subsequent drought in 2008. Undoubtedly, subsidies for water-saving irrigation technologies like drip and sprinkler will not only have a substantial impact on irrigation efficiency but will also contribute to a more sustainable resource use in general. Although drip irrigation is one of the few products that Turkish farmers have seen dropping in price – due to competition in the irrigation business – over the last years, it is still a considerable investment for Turkish farmers. Besides being a relatively high capital investment, the reasons for low incidence of drip irrigation usage among farmers are various; such as higher energy costs compared to furrow and flooding methods, lack of public awareness about water-saving technologies, knowledge and information deficit on the drip system and its use etc. Furthermore, the small and dispersed form of field plots in addition to changing cropping patterns year by year are other factors affecting the wide adoption of drip systems in the country.

The most common crops in terms of percentage of irrigated area in Turkey are cereals, maize and cotton. However the crop pattern shows regional variance depending on land-soil-climate conditions, as well as economic and social factors such as input and output market prices and farmer habits (DSI 2006). Sprinkler irrigation is mainly used for sugar beet, cereals, sunflowers, melons and vegetables in the Aegean as well as in west and northwest Anatolia. Drip irrigation however is preferred for citrus, vineyards, vegetables and strawberries in the western and southern parts of the country (DSI 2005).

3.3 The economic relevance of irrigation

Agricultural statistics (covering the years from 1880 to 2000) reveal that the total agricultural output increased about nine fold during this period and the Gross Domestic Product (GDP) per capita about six fold, while the increase in population was about four fold (Pamuk 2006). Increases in agricultural output contributed to growing agricultural exports from 1880 until the 1980s, which, however, was mainly stimulated by the expansion of the cultivated area. Similarly, the agricultural labour force also showed an increasing trend until the 1980s. With a rapid growth in other sectors, notably manufacturing, trade and services, agriculture's significance in the economy in terms of total employment has declined from 80 percent in the 1990s to 35 percent in 2000 (Pamuk 2006). The significant decrease of the agricultural contribution to GDP (Fig. 3) in the last 50 years is a consequence of the declining share of agriculture within national capital investment (Boz and Volkan 2006). Although the share of agriculture in the Turkish economy has been falling, it still accounts for a relatively larger share of total output and employment as compared to industrialized European Union (EU) and North American countries (Dogruel et al. 2003).

Agricultural production in Turkey has a distinctive regional distribution based on geographic and climatic conditions. Crop production, with 77 percent of total production, commands the largest share, while the remaining part consists of livestock production. Crop production mainly comprises cereals, pulses (annual leguminous crops), industrial plants, perishables (fruits, vegetables, etc.) and fodder crops. Turkey is largely self-sufficient in food production and among the top ten producer countries with regard to the production of cereals (barley, rye and wheat), pulses (lentils and chick-peas, green beans), industrial crops (cotton, sugar beet and tobacco), vegetables (watermelon, cucumber, tomato, onion, eggplant, spinach) as well as fruits and orchards (cherries, apricots, figs, hazelnuts, chestnuts, walnuts, citrus, olives, pistachios, apples, grapes, tea, strawberries) (FAO 2005).

As the largest producer and exporter of agricultural products in the Near East and North African Region, the country's dominating export commodities are hazelnuts, dried apricots and figs, tomato paste, wheat flour and pastry, besides soft wheat and mixed grain, lentils, cotton lint and scrap, tobacco leaf and scrap, as well as sunflower and cotton seed oil (Cagatay and Guzel 2003). Irrigation is essential for most of the above listed crops.

Turkey has a long tradition of supporting the agricultural sector, either directly through price subsidies or indirectly through agricultural credits, virtual non-taxation of rural incomes and guaranteed state purchases of strategic commodities (Dogruel et al. 2003). In addition to capital grants (i.e. for factories for processing, storage and packaging of agricultural products, greenhouses, animal husbandries, pressurized irrigation systems), fertilizer, irrigation, seed and pesticide subsidies, the most widely used instruments in agricultural policy over the years have been output price support followed by input subsidies. However, these traditional agricultural support policies fall into the category of so-called distributional policies (Cakmak 2003) that only target towards improving short-term economic conditions of the rural community. Structural and institutional weakness, low productivity mainly caused by land fragmentation, small farm size and low educational level are among the reasons for weak impacts of subsidies, in particular for low-income small farm holdings.

To adapt Turkey's agricultural policy to the Common Agricultural Policy of the EU, Turkish governments have had to carry through some institutional, technical and socio-economical reforms such as the abolition of administered prices and input and credit subsidies, the introduction of a Direct Income Support scheme, restructuring of agricultural small and medium enterprises etc. For this purpose, the Agricultural Reform Implementation Project was planned by the government for the period 2001-2005, as a result of the financial support from the World Bank for Agricultural Reform Implementation Project, and relevant commitments from e.g.

the World Trade Organization, the EU, and the International Monetary Fund, so that the project was started in 2001. In accordance with this programme, output price support, input and credit subsidies have been eliminated and Direct Income Support- (which is not contingent to input use or output production decisions of farmers but to area) has been implemented instead. One idea behind the introduction of Direct Income Support was to reduce over-produced goods and to realize the transition to alternative products (Cakmak 2003; Olhan 2006). After the completion of Agricultural Reform Implementation Project, a new agricultural strategy paper covering the period 2006-2010 is prepared with a view to continuing the reforms of the agricultural sector. This strategy paper set priorities to ensure development of a sustainable agricultural production, product quality, food security and safety, competitiveness of agricultural holdings, markets and marketing, rural development and producers' organizations. In accordance with the aims of the Agricultural Strategy Paper, area-based support including Direct Income Support payment was discontinued and production-based supports was underlined for 2010.

4 Management of irrigation systems in Turkey

Land and water resources development in the country has been carried out by public sector agencies and by the private sector - both individual farmers and groups of farmers.

The public sector is represented by the DSI, which nowadays comes under the Ministry of Environment and Forestry (MoEF); the General Directorate of Rural Services (GDRS) and the General Directorate of Agrarian Reform, which are both under the MARA DSI is responsible for planning, designing, constructing and managing multipurpose water infrastructure. With respect to groundwater resources, DSI is mandated to issue permits for the exploration and use of groundwater resources, and decides on the number, location, depth and pumping rates of wells. The GRDS, on the other hand, is responsible for the development of small-scale irrigation schemes and small reservoirs, rural roads and water supply to rural areas. Land consolidation and on-farm development works in all irrigation projects such as on-farm drainage were under GDRS's jurisdiction until 2005, when GDRS was abolished and its responsibility transferred to Special Provincial Administrations. Special Provincial Administrations, however, report to the Ministry of Interior (Directorate General of Local Authorities) and their responsibilities are more diversified compared to the work previously done by GDRS. Besides agricultural services such as irrigation, drainage, land levelling, prevention of erosion, protection of soil and construction of artificial ponds; also duties related to industry, trade and health were added to the Special Provincial Administrations' task load. After the abolishment of GDRS, only the General Directorate of Agrarian Reform (GDAR) continued to be responsible for carrying out land consolidation. Together with its regional directorates, the agency plans and implements projects, not just for consolidation but also for land-use planning, land reclamation, irrigation and drainage works in consolidated and/or newly developed agricultural areas. Unlike DSI, GDAR and Special Provincial Administrations plan and carry out the design and construction of irrigation and drainage projects as well as infrastructure for drinking water for urban and rural areas, but do not participate in management.

According to its establishment law in 1953, DSI is entitled to transfer the operation and maintenance (O&M) of irrigation systems to different organizations such as private entities, cooperatives, municipalities, village administrations and irrigation associations. Due to a shortage of funds, DSI deferred maintenance of off-farm irrigation and drainage infrastructure, leading to a serious reduction in water deliveries. Since the early 1960s, DSI encouraged farmers to establish irrigators groups (*sulayici gruplari*) in large-scale irrigation schemes that participated in O&M activities of tertiary irrigation canals, more precisely in controlling water distribution, as well as maintaining and repairing tertiary irrigation canals. The irrigator groups spread slowly but surely up until 1993. The village headmen as their representatives entered into contracts and agreements with DSI, which defined their responsibilities and rewards. As a result, the irrigation ratio, which represents the ratio of actually irrigated area to the command area, increased in some regions of the country (Scheumann and ul-Hassan 2001). However, transfer of O&M responsibilities was limited to tertiary canals and solved neither financial constraints nor managerial deficiencies. Therefore, DSI started with the Accelerated Transfer Programme in 1993 with the financial support of the World Bank.

While in 1993, DSI was still operating 95.2 percent of all irrigation schemes and only 72,000 hectares of the irrigation command areas had been transferred to irrigation associations, the Accelerated Transfer Program took off very quickly, and areas operated by irrigation associations increased to about 1 million ha (61 percent of total irrigated land) in 1995 (DSI 2004). Besides irrigation associations which carry out the O&M services in 90 percent of the transferred schemes, DSI transferred a relatively tiny portion to irrigation cooperatives, village authorities and municipalities, 5, 3 and 2 percent respectively.

The total irrigated area in Turkey is about 5.3 million hectares of which 3.06 million hectares were developed by DSI. Almost 2.1 million hectares have been transferred to irrigation associations and the like, and thus O&M will be carried out by those organizations. Irrigation networks in about 26 percent of the country's total irrigated land were developed by GDRS. Before its abolition in 2005, GDRS transferred schemes under its control, informally to village authorities and irrigation cooperatives. The balance of the total number of irrigated projects (or simply: the other irrigation projects) in the country is constructed and still operated by the private sector, i.e. farmers (DSI 2005 and 2009). However, it is worth noting that in reality there are considerable amounts of unregistered groundwater wells developed and used by farmers, without having taken the permits from the DSI. For example recent surveys of DSI showed that in the Konya [closed] basin, 68,000 wells out of 90,000 were unregistered (Berke 2008).

Efforts to increase user participation have been induced by the poor performance - in terms of efficiency, accountability but utmost cost recovery - of large-scale irrigation systems managed by the DSI. Greater participation by farmers, through irrigation associations, has helped to overcome some of the problems such as collection of irrigation fees, maintaining the secondary and tertiary canals and operation of the irrigation network. A comparison of the DSI expenditure for maintenance and operation before (1991) and after transfer (2005) shows a substantial decrease (from 100 to 16 percent), thus, the economic target of the transfer process was achieved. Yet despite the economic success, irrigation efficiency decreased slightly after the transfer: from 44 percent in 1995 to 42 percent in 2005. While there may be various reasons for a decline in irrigation efficiency, institutional aspect often has a major influence on efficiency. Besides lack of personnel, knowledge and experience, most of the irrigation associations are suffering from bad financial situations. In addition, there is a lack of collaboration between irrigation associations within an irrigation scheme. There is also inconsistency between projected and submitted irrigation water demand before the season and the actual water intake during the season, which is generally caused by changing cropping patterns affected by weather and/or market conditions of the year. For instance, if the farmers fail to sow winter wheat because of unsuitable weather conditions, they tend to grow first-crop corn, sunflower or cotton instead of second crop after wheat in the western and southern parts of the country. Or after harvesting the winter wheat, some plots are left fallow during the irrigation season. Changes in proposed crop pattern sometimes result in less water intake to the fields and diverted water from the main canal flows to the drainage without being used. This however results in reduced irrigation efficiency in the sub-basin and/or basin. DSI had been confronted with similar problems before the transfer process. Furthermore, DSI and irrigation associations only measure the inflows into the irrigation canals but outflows via drainage canals are not considered. This causes mistakes in calculation of water use efficiency and also in evaluating irrigation water management. Measuring inflows and outflows, despite high construction and installation costs, would enable the irrigation associations to assess, improve and/or verify their practices. Evaluation of determined irrigation efficiencies of all associations within a scheme could help to avoid conflicts between head-end-tail-end, upstream-downstream as well as between irrigation associations and DSI, and finally controlling inflow and outflow would sustain the efficient use of resources.

4.1 Financing irrigation services

According to DSI's establishment law No. 6200 water has no unit price. However, water tariffs are levied against the expenses for operation, maintenance and repair (O&M&R), and to a certain degree for capital costs. These costs have to be borne by the farmers using the irrigation and drainage infrastructure. O&M charges for the schemes operated by DSI are predetermined each year by the General Directorate and proposed via the MoEF, which are then formalized in a ministerial decree. The charges are calculated annually so as to recover the majority of the previous year's O&M expenditure as well as a small part of the investment costs, however without adjusting it to inflation.

Although DSI still takes care of O&M&R of such structures which are jointly used by many irrigation associations and as such are labelled "shared facilities", the sum of expenses for the shared facilities will be charged to all irrigation associations using these same facilities. Also those parts of the irrigation network exceeding the boundary of one irrigation association, or for which the Association refused to take responsibility for will be operated by DSI and again related O&M costs will be collected from the Associations after the irrigation season. However, depending on the economic capacity of an association, payments can be collected in installments.

In schemes managed by irrigation associations, each association sets its tariff independently and uses different methods depending on region and scheme: the "*area-crop–based*" charging system, in which the charge per hectare is determined by the type of irrigated crop, is used mostly for gravity irrigation, while volumetric pricing is widely used in pumping schemes. Furthermore, in many irrigation schemes, the irrigation technologies such as water-saving technologies like drip and sprinkler irrigation are also considered in determining the "*area-crop–based*" pricing system. Thus, producers may pay lower costs per hectare when using water-efficient irrigation technologies.

O&M charges vary not only according to crops but also depending on whether it is a gravity or pumping scheme. However, these charges also differ from one association to another and from one region to another. They are generally higher in pumping schemes compared to schemes where water is supplied by gravity. Irrigation charges for pumping schemes also have great variations: for instance, in DSI-managed schemes, O&M charges for wheat, maize and cotton are 89 TL/ha, 180 TL/ha and 280 TL/ha, respectively for gravity irrigation. However, those are much higher (e.g. 340 TL/ha for wheat, 2,187 TL/ha for maize and 1,096 TL/ha for cotton) for pumped irrigation in DSI-managed pumping schemes in 2009 (DSI 2009).

The amount associations charge for irrigating wheat, maize and cotton in gravity irrigation schemes were 55 TL/ha, 155 TL/ha and 155 TL/ha respectively, whereas the irrigation charges for all other kinds of crops were determined at 1,000 TL/ha in pumping schemes under the associations' management for 2009.

Irrigation charges changed drastically in the course of the last decade. Their share makes up about 3-7 percent of agricultural production costs. However, during the last few years also depending on increased input and decreased output prices caused by the economic crisis, farmers in the Cukurova region complained about rising irrigation charges, which together with irrigation labour costs correspond to an average increase for different crops of about 9 percent of agricultural production costs (Cetin et al. 2009). Nevertheless, small farm holdings already have problems with the timely payment of the irrigation charges.

If association members do not pay they will be subjected to penalties e.g. first a warning, then a fine, followed by the disconnection of services and a lawsuit. There is evidence, although with increasing trend, that some associations have filed default cases with the courts. However, so far none of them has refused service to its members (Scheumann and ul-Hassan 2001). Under DSI management, the collection rate of O&M charges was about 41 percent (Ozlu et al. 2002a); this percentage has been increased to over 80 to 90 percent after transfer of management responsibilities because the associations have effectively penalized non-payment (see Ozlu et al. 2002b; Yildirim and Cakmak 2004).

5 Agriculture, an inefficient water consumer

Many irrigated areas in Turkey have lower productivity than originally planned due to low cropping intensity and inefficient water use. Water use efficiency (see Box 1) is a combination of water productivity (a biological factor) and irrigation efficiency (a physical factor) (Alizadeh and Keshavarz 2005). Long-term average for the cropping intensity, so called irrigation ratio, in DSI-managed schemes is about 65 percent (DSI 2009); it varies between 45 and 90 percent in association-managed areas (Yavuz et al. 2006; Cakmak et al. 2009). Low irrigation ratio or cropping density means that a considerable amount of land within an irrigation scheme is not irrigated although they have access to the irrigation networks (see Box 1).

Another important performance indicator for irrigation projects is irrigation efficiency: according to the latest records, the irrigation efficiency was around 42 percent in irrigated areas managed by the irrigation associations (DSI 2005), meaning that 58 percent of diverted water is lost. Considering the period before

Box 1 Definitions

Cropping intensity: The actual rate of irrigated area as a percentage of the irrigable area (= irrigation command area).

Irrigation ratio: The share of actual irrigated crop area within the total command area of the irrigation scheme.

Conveyance efficiency: The ratio of the water that is delivered to a field compared to the total water diverted or pumped into the conveyance system at the upstream end.

Irrigation efficiency: The ratio between estimated irrigation water requirements and actual irrigation water withdrawal.

Application efficiency: The percentage of water delivered to the field that is used by the crop.

Water use efficiency: The ratio of the amount of water actually used for crop growth in relation to the amount of water withdrawn from the source.

Water productivity: Harvested crop yield against water consumptively used in evapotranspiration.

and after transfer to irrigation associations - there is no significant change in irrigation efficiency.

The first performance study in the country has been carried out in the Lower Seyhan Irrigation System in Adana in the East Mediterranean Region. The system has been evaluated according to its technical aspects, such as water conveyance, seepage and tail water losses, irrigation efficiencies, consequently all system design and operating efficiency. Results reveal that only 50 percent of water diverted from the reservoir, Seyhan Dam, was available for plants in the field. However, this was mainly a consequence of administrative, financial and management problems (Benli et al. 1987). Recent studies summarized by Kanber et al. (2004) revealed that both conveyance (Ec) and application (Ea) efficiencies vary depending on regional conditions as well as irrigation methods. The average application efficiencies have been recorded as 84 percent, 80 percent and 55 percent for drip, sprinkler and surface irrigation methods, respectively, while conveyance efficiencies (Ec) vary between 87 percent and 97 percent across the country (Kanber et al. 2004).

A survey of irrigation networks carried out in 90 different developing countries in 1998 showed that water-use efficiency was as low as 38 percent (Shiklomanov 1999). Similar surveys from Europe indicate that in open canal distribution networks, the water losses are estimated at up to 40 percent in unlined ditches and up to 25 percent in lined canals. The irrigation application efficiency ranged from 45 percent to a maximum of 60 percent in areas where water was distributed via open canals. However, higher irrigation efficiencies have been achieved in piped systems (Phocaides 2007). In irrigation schemes where water is diverted using open canals and field plots are irrigated with surface irrigation methods such as flooding, border and furrow, irrigation water use efficiency is about 40 percent in China (Deng, et al. 2004), Pakistan (Pasha 2002) and in Syria (Abeer 2007). Both, conveyance and application efficiencies are around 60 percent in Mexico (Palacios-Vélez 1994). In northern Spain, application efficiency is also low, with an average of 49 percent, whereas the basin wide irrigation efficiency only reaches reasonable values when the system operates under water scarcity (49 percent in 2000 versus 66 percent in 2001) (Lecina et al. 2005).

According to a comparative study (open canals with surface irrigation versus pressurized pipes with sprinkler or drip irrigation systems) the results show that, irrigation efficiencies vary between 42-53 percent for open surface canal with surface irrigation, whereas efficiency rose up to 70 percent for pressurized pipe systems in the Jordan Valley (Battikhi and Abu-Hammad 1994).

Since the irrigation sector is the largest user of freshwater with a share of 75 percent, it is necessary to optimize water use and promote conservation, and improve irrigation efficiencies to cope with scarce resources in Turkey. Such optimization however requires modern and viable management operation systems for planning, water distribution, O&M including engineering, agronomic, social, economic and financial aspects (ICID 1999).

6 Constraints and limits to efficient water use

One of the most limiting factors for the efficient use of water on agricultural land is the low irrigation ratio. That is to say: despite access to the irrigation networks, some land owners are not irrigating the land for different reasons. Suboptimal social and economic conditions, the continuing practice of rain-fed farming and of leaving lands fallow instead of practicing crop rotation are the main factors explaining the persistence of non-irrigated areas within irrigation schemes. According to DSI evaluation reports, the causes and their shares for different years vary, as shown in Table 1.

In addition to low irrigation ratio, constraints such as physical, agronomic, environmental, institutional, managerial as well as financial are resulting in a low irrigation efficiency which is about 40 percent. However it ranges spatially and temporally from 10 to 70 percent (DSI 1998, 2005). These highlighted problems are described more in detail below.

Physical (engineering) problems

Almost 90 percent of irrigation canals are concrete lined, increasing the conveyance efficiency in the system. However, transporting and diverting water via open canals leads to substantial conveyance losses. Conveyance losses of 5 percent in main canals and 5 percent in branch canals occur mainly due to evaporation, overflows and leakage (DSI 2006). Surface irrigation methods like flood, border and furrow are dominating, whereas water-saving piped and pressurized irrigation systems and

Table 1 Limiting factors and shares (percentage) of non-irrigated areas within the irrigation schemes managed by irrigation associations. Compiled from DSI Evaluation Reports 1997, 2001 and 2005 (DSI 1998, 2005)

Limiting factors	Years		
	1996	2000	2004
Inadequacy of water resources	2.0	1.4	2.6
Insufficient irrigation structures	1.5	2.3	2.0
High water table	1.4	1.1	0.8
Salinity and alkalinity problems	0.9	1.1	0.8
Inadequate repair and maintenance	0.4	0.4	0.7
Unsuitable topography	1.3	1.4	1.3
Rain-fed agriculture	7.4	9.7	8.6
Fallow	2.9	3.3	4.3
Social-economic factors	4.8	5.8	6.6
Other factors*	4.4	6.1	5.6
Total share of non-irrigated areas within the irrigation schemes (%)	**27**	**33**	**33**

*Water pollution, marketing problems, agricultural areas which converted from agriculture into industrial and settlement areas and graze lands, pastures etc.

methods such as sprinkler and drip account for less than 10 percent of the overall irrigated area.

Technically necessary drainage facilities have not yet been constructed in many irrigation schemes. Even in new developed areas in the GAP region, 40,000 hectares of land have already been degraded due to high water tables.

Topography is a limiting factor for efficient irrigation in many areas, and land levelling works are insufficient. Almost 70 percent of the producers have less than 5 hectares lands are fragmented; lands are owned by people who are not living in rural areas, thus resulting in low efficiency and productivity. Also irrigation structures such as conveyance and distribution systems are insufficient and old in many schemes.

Agronomic problems

Input and output prices, years of experience and habits are important in influencing the farmers' decisions regarding crop selection. Most of the farmers are not aware of new drought resistance cultivars. Fertilizer applications, in particular, nitrogen fertilizer application rates are high which stimulates more plant vigour and increased water use than required. Dry land (or rain-fed) farming is only preferred by small farm holdings due to high input costs for irrigated agriculture.

Environmental problems

Irrigated agriculture can cause environmental degradation in many respects: over exploitation of water resources, discharging pollutants and sediment into surface and/or groundwater, soil erosion due to poor agricultural practices, salinisation and water logging of irrigated land. However, agriculture suffers from inadequate water supply and the low quality of surface and ground water too. In Turkey, due to excess water application and unconscious irrigation in addition to lack of drainage facilities, about 3 and 1.5 million hectares of irrigated lands are at risk from water logging and salinisation, respectively, or are already affected (Sengun 2006). Due to excessive use of agrochemicals such as fertilizers and pesticides in irrigated agriculture, surface and groundwater resources can be contaminated with heavy metals and nitrates. In West Turkey (Bornova-Izmir) and in Middle Anatolia (Nigde) groundwater resources are polluted while Thrace Region the Ergene River is polluted with agrochemicals in particular with nitrate concentration (Harmancioglu et al. 2001; Gidisoglu et al. 1996). Also, poor maintenance of drainage systems reduces the effective removal of excess water. Also to note that drainage systems are not only used for removing excess irrigation return flows from agricultural lands but they are also used to discharge the treated and/or untreated wastewater from industrial areas and municipalities (Scheumann and ul-Hassan 2001).

Institutional and managerial problems

A number of organizations such as DSI, municipalities, irrigation associations, irrigation cooperatives, private sector are responsible for water management in the country. Due to their special laws and rules, each institution works independently and without any efficient coordination with authorities. Irrigation associations are established under the Municipality Law which is not adequate for the management of relatively complex irrigation and drainage systems. Thus there is a lack of comprehensive water law and a new legislation for irrigation associations. However, the MoEF in particular DSI drafted a new law for irrigation associations in summer 2008 which is currently in the legalization progress. There is also a lacking legal basis for forming federations of associations for joint purchasing and supplying "lumpy" services such as equipment maintenance.

Although DSI organised seminars and workshops for training technicians and general secretaries at the beginning of the transfer process and still provide technical support if requested and/or needed, there is a lack of regular and formal training programs (Kibaroglu et al. 2009). Since irrigation associations only date back a decade, staff are therefore not well experienced and need routine training. Extension services in particular are not as effective as required, and collaboration between staff of the irrigation associations and extension services is weak.

Financial problems

Water for irrigation has always been subsidized in this way promoting irrigated agriculture. If ever, pricing policies have aimed at recovering service costs (O&M&R), and to a lesser extent, capital costs. If levied per area and crop, irrigation costs are only a proxy of the amount of water used, and incentives for saving water are difficult to introduce. Given the technical difficulties of volumetric measuring, it is not easy to implement charges as a stimulant for water savings.

While transferring O&M of irrigation systems to irrigation association has improved O&M cost recovery, state financial transfers are still needed for rehabilitation and maintenance of irrigation and drainage systems.

7 Conclusion

Future growth in irrigated agriculture of the country will be limited by; i) scarcity of land and water, ii) increasing competition for water, iii) rising cost of development, iv) inadequate management, v) deterioration of existing systems, vi) degradation of

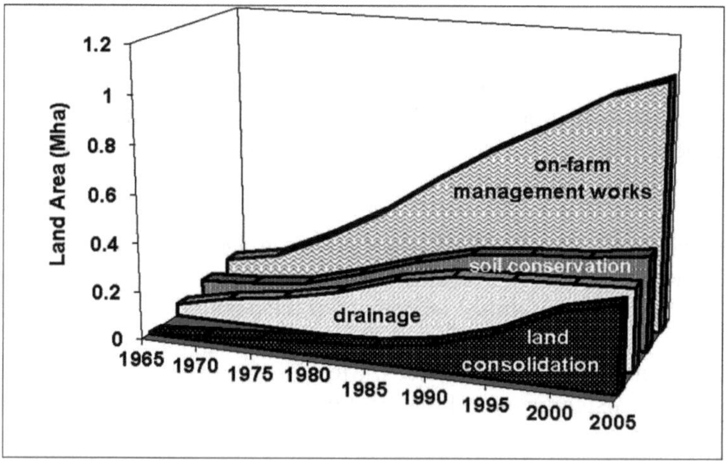

Fig. 4 On-farm land management (based on data from DSI, GDRS and Ministry of Agriculture and Rural Affairs)

the environment, and, finally, vii) alignment with EU legislations related to water and environment.

However, there is potential for improving water use efficiency, and options might cover a combination of physical (engineering), agronomic, environmental, institutional and managerial as well as socio-political measures (Howell 2001).

Physical (engineering) measures include the rehabilitation and modernization of irrigation network, and the completion of on-farm works to increase the irrigation ratio and to decrease conveyance losses. As shown in Figure 4, there are still large areas to be drained and levelled. Using flow meters by diverting and applying irrigation water can prevent water wastage and increase efficiency. Higher water use efficiency can be achieved by improved control in distribution networks, and by combining structures and simple operating rules with automation. Remote operation and control of irrigation schemes using high tech systems can allow adjusting and/or correcting the application failure and misuse of the system.

Agronomic measures comprise the development of varieties of better yielding crops under conditions of water scarcity, the selection of drought tolerant varieties, the replacement of high-water consuming crops, applying intercropping such as growing a deep-rooted crop with a shallow-rooted crop to utilize soil water in an efficient way etc. (Alizadeh and Keshavarz 2005). Additionally, improved water supply management and deficit irrigation can increase water productivity of crops. Methods such as conventional deficit and partial root-zone irrigation (e.g. Topcu et al. 2007) besides surge flow irrigation increase efficiency by reducing drainage losses on the borders of the farm and ensuring more even application, root to shoot signalling and regulating transpiration without limiting photosynthesis, thus limiting losses from deep percolation and evapotranspiration.

Environmental measures mainly consist of environmentally friendly practices such as precision irrigation and utilizing unconventional methods and waters for irrigation including recycled drainage and tail water, and the use of treated wastewater. The efficient use of water and the safe reuse of wastewater are the most economical and often the only sources of additional water, and at the same time, a very effective means of controlling water pollution. During the last decade, wastewater treatment plants have been installed in cities, and resulted in an increased connection of population (almost 66 percent of the population is connected, and ca. 66 percent (1.9 BCM) of wastewater is treated using biological, physical and advanced combined treatment methods. Most of this water is flowing into rivers and lakes, while a significant volume of untreated wastewater is discharging into lakes, ponds and onto lands. Usage of wastewater for irrigation is as yet limited to the arid/semi-arid regions like the central and south-eastern regions in the country. About 200,000 hectares of land is irrigated using wastewater in various provinces (Gokcay 2004). Irrigation return flows can also be reused in the lower part of the irrigation schemes like in the Seyhan and Harran Plains, where water is not sufficient and/or irrigation infrastructure has not yet been completed. However, application of low-quality water requires controlled management and expert knowledge in order to conserve soil resources.

Coordination between universities, stakeholders and end users needs to be encouraged by institutional and managerial measures. Thus, knowledge gained from long-term laboratory and/or field works by universities and relevant authorities can be transformed to massive applications by the end users in irrigation districts. There is no doubt that establishing technical assistance programs, as well as water consulting services and training the trainers from extension services and irrigation associations can promote water use efficiency. Routine trainings and enhancement of communication between associations can help to disseminate successful practices from one group and/or region to the others. Also providing water delivery planning and irrigation scheduling services can improve water productivity.

Economical and socio-political measures include the transfer of successful policies; reducing irrigation subsidies; planning dynamic education and dissemination activities for farmers and training the trainers; fostering rural infrastructure development through the private sector; dissemination of suitable technology; and supporting environmental sound management techniques and technologies. Stabilized market prices and production policies can encourage farmers to use fallow lands consequently increase irrigation ratio. Application of water saving techniques can be promoted, and particularly facilitating farmers the acquisition of pressurized and/or micro irrigation systems through soft credit may help poor farmers. Reallocating of water from lower to higher value uses can also increase overall water use efficiency.

References

Abeer MH (2007) Water use efficiency in Syrian agriculture. Working paper no 26, Syrian Ministry of Agriculture and Agrarian Reform. National Agricultural Policy Centre. http://www.napcsyr.org/dwnldfiles/working papers/en/26_water_eff_am_en.pdf. Accessed 23 August 2010

Alizadeh A, Keshavarz A (2005) Status of agricultural water use in Iran. Water Conservation, Reuse, and Recycling: Proceedings of an Iranian-American Workshop, pp 94-105

Battikhi AM, Abu-Hammad AH (1994) Comparison between the efficiencies of surface and pressurized irrigation systems in Jordan. Irrigation and Drainage Systems, 8(2):109-121

Benli E, Tekinel O, Yazar A, Cevik B, Kanber R (1987) Ex-post evaluation of the Lower Seyhan Project (Technical Aspects). FAO Project of TCP/TUR/6652, Univ. of Cukurova, Faculty of Agriculture, Irrigation and Drainage Eng. Dep., Adana

Berke MO (2008) Water Management in Turkey. Presentation held in Bursa on 20th November 2008. http://www.wwfegitim.org/pdf/2.pdf Accessed 23 August 2010

Boz B, Volkan F (2006) Financial problems and relevant solutions to improve infrastructure of irrigated agriculture in Turkey. TMMOB Water Policies Congress, 21-23 Mart 2006, Ankara, pp 403-4120

Cagatay S, Guzel A (2003) Review of agriculture and trade policies in Turkey. BSEC Trade Facilitation Project, FAO TCP/REP/2901

Cakmak E (2003) Evaluation of the past and future agricultural policies in Turkey: are they capable to achieve sustainability? Options Méditerranéennes, Ser. A/no 52., CIHEAM. http://ressources.ciheam.org/om/pdf/a52/03400062.pdf Accessed 23 August 2010

Cakmak B, Polat HE, Kendirli B, Gokalp Z (2009) Evaluation of irrigation performance of Asartepe Irrigation Association: A case study from Turkey. Akdeniz University Agricultural Faculty Journal, 22(1), 1–8

Cetin, M. Ibrikci, H., Topcu, S, Budak, F. Dingil, M, Oztekin, E., Karnez, E., Oguz, H., (2009) Annual Report, QUALIWATER: Diagnosis and Control of Salinity and Nitrate Pollution in Mediterranean Irrigated Agriculture, INCO-CT QUALIWATER (2005-015031).

Deng XP, Shan L, Zhang H, Turner NC (2004) Improving agricultural water use efficiency in arid and semiarid areas of China. 4th International Crop Science Congress, September 2004 Brisbane, Australia

DMI (2009) State Meteorological Service website. http://www.dmi.gov.tr/veridegerlendirme/yillik-toplam-yagis-verileri.aspx Accessed 20 September 2009

Dogruel F, Dogruel AS, Yelden E (2003) Macroeconomics of Turkey's agricultural reforms: An intertemporal computable general equilibrium analysis. Journal of Policy Modeling 25 (6-7):617-637

DSI (1998) Evaluation report for 1997 of irrigation schemes operated by DSI and transferred. General Administration of the State Hydraulics Works, Operation and Maintenance Unit, Ankara

DSI (2004) Irrigation management transfer (IMT) in Turkey. Operation & Maintenance Department, General Directorate of Hydraulic Works (DSI). Capacity Building Symposium on Integrated Water Management and Irrigation. 11-14 October 2004, Istanbul

DSI (2005) Evaluation report for 2004 of irrigation schemes operated by DSI and transferred. General Administration of the State Hydraulics Works, Operation and Maintenance Unit, Ankara

DSI (2006) 52 Years General Directorate of State Hydraulic Works, 1954-2006. DSI, Ankara, http://www.dsi.gov.tr/english/english_brief.pdf Accessed 23 August 2010

DSI (2009) Activity report of 2008. DSI, Ankara, http://www.dsi.gov.tr/faaliyet_raporlari/raporlar.htm Accessed 20 September 2009

DSI website (2009) General Directorate of State Hydraulic Works, Ankara. http://www.dsi.gov.tr Accessed 20 September 2009

Erdogan N, Aytac A, Kirbiyik N, Aytac H, Unlu F, Ozgen G, Sen NF (2003) Added value in agricultural sector created with irrigation systems developed by DSI. 2. National Irrigation Congress, 16-19 October 2003, Aydin, pp 181-189

FAO (2005) Major food and agricultural commodities and producers. FAO website, http://www.fao.org/es/ess/top/topproduction.html?lang=en&country=223&year=2005 Accessed 15 September 2009

Gidisoglu A, Cakr R, Tok HH, Ekinci H, Yuksel O (1996) Determination of Ergene river pollution and effects on soil. Soil Water Research Yearbook, 1996. Publication no 102, General Directorate of Rural services, Division of APK, Ankara. pp 308–321

Giorgi F (2006) Climate change projections over the Mediterranean region. Geophysical Research Abstracts, vol 8, 10946

Gokcay C (2004) Evaluation of the Turkish reuse standards and the compliance status. International Workshop on Implementation and Operation of Municipal Wastewater Reuse Plants. 11-12 March 2004, Thessaloniki, Greece

Harmancioglu N, Alpaslan N, Boelee E (2001) Irrigation, health and environment: A review of literature from Turkey. IWMI working paper 6. International Water Management Institute (IWMI), Colombo, Sri Lanka

Howell TA (2001) Enhancing water use efficiency in irrigated agriculture. Agron. J. 93:281-289

ICID (1999) ICID Bali Declaration, http://www.icid.org/decl.html 15 March 2009

Kanber R, Unlu M, Cakmak EH, Tuzun E (2004) Irrigation systems performance. Country report: Turkey. WASAMED, Ankara

Kibaroglu A, Baskan A, Alp (2009) Neo-Liberal transitions in hydropower and irrigation water management in Turkey: main actors and opposition groups. In: Huitema D, Meijerink S (eds) Water policy entrepreneurs. A research companion to water transitions around the globe. Edward Elgar, Cheltenham, UK, pp 294-300

Lecina S, Playán E, Isidoro D, Dechmi F, Causapé J, Faci JM (2005) Irrigation evaluation and simulation at the Irrigation District V of Bardenas (Spain). Agricultural Water Management, 73 (3):223-245

Olhan E (2006) The impact of the reforms: impoverished Turkish agriculture. 96th EAAE Seminar "Causes and Impacts of Agricultural Structures" 10-11 January 2006, Taenikon, Switzerland

Ozlu H, Erdogan FC, Doker E (2002a) Irrigation Management Transfer (IMT): Benefits and arising problems. Follow-up Seminar: Towards A Sustainable Agricultural Development - New Approaches. 15-21 April 2001, Antalya

Palacios-Vélez E (1994) Water use efficiency in irrigation districts. In: Garduño H, Arreguín-Cortés F (eds) Water Use Efficiency. Proceedings of the International Seminar on Efficient Water Use, October 21-25, 1991, Mexico City. http://www.unesco.org.uy/phi/libros/efficient_water/tapaefus.html Accessed 23 August 2010

Pamuk S (2006) Agricultural output and productivity growth in Turkey since 1880. International Economic History Congress, 21-25 August 2006, Helsinki

Pasha M R (2002) Inefficient water management its impact on economic growth. The Journal NIPA Karachi, 7:4, 87-98

Phocaides A (2007) Handbook on pressurized irrigation techniques. FAO, Rome

Republic of Turkey (2003) Turkey Country Report. Prepared for the Third World Water Forum. World Water Council http://www.worldwatercouncil.org/fileadmin/wwc/Library/Publications_and_reports/country_reports/report_Turkey.pdf Accessed 23 August 2010

Scheumann W, ul-Hassan M (2001) Irrigation Management Transfer: Irrigation associations in the Lower Seyhan Plain and the Gediz river basin (Turkey). Study tour at Irrigation Systems in the Provinces of Adana and Izmir. Project Report, Germany

Sengun MN (2006) The mistakes that were made by the institutional structure about land consolidation on-farm management works and economical usage of water. TMMOB, Water Policies Congress. 21-23 Mart 2006, Ankara

Shiklomanov LA (1999) World water resources at the beginning of the 21st century. IHP Report. UNESCO, Paris, France

Topaloglu F (2006) Trend detection of stream flow variables in Turkey. Fresenius Environmental Bulletin, 15(6):644-653

Topcu S, Kirda C, Dasgan Y, Kaman H, Cetin M, Yazici A, Bacon MA (2007) Yield response and N-fertiliser recovery of tomato grown under deficit irrigation. Europ. J. Agronomy 26:64–70

Turkes M (2001) Global Climate Change: Impacts on agriculture and water resources. Proceedings Panel on Climate Change impacts on agriculture, Ministry of Agriculture and Rural Affairs, Ankara, pp 91-128

Yavuz MY, Kavdr I, Delice NL (2006) Performance evaluation of water user associations: A case study of the lower Seyhan basin. J. Agric. Fac. HR.U. 10(3/4):35-45

Yldrm E, Cakmak B (2004) Participatory irrigation management in Turkey. Water Resources Development, 20(2):219–228

Wasamed (2003) Country Report on water user associations in Turkey. WASAMED Workshop, 15-19 December 2003, Sanliurfa, Turkey

Turkey's Policy for Combating Water Pollution

Gokhan Orhan and Waltina Scheumann

1 Introduction

Water is becoming scarce in most regions of Turkey not only due to growing demands and climate change but due to pollution stemming from point and non-point sources, which threatens water availability for agriculture, households and industry - which in turn are also the major polluters.

Water pollution control as a subject of Turkey's environmental policy has been stimulated by the Stockholm Conference (1972) leading to legislative and institutional reforms in the 1980s (OECD 1999, 19). Furthermore Turkey's membership in a number of international regimes and her efforts towards European Union (EU) accession in particular has given impetus to combating water pollution and further the protection of its natural resources by implementing for instance the EU Urban Wastewater Directive (2006). Despite these endeavours, the pressure on Turkey's water resources is intense, and Turkey's governments face the challenge of developing and protecting the country's water resources (Republic of Turkey 2003, 97-98).

Turkey has made some progress in combating water pollution, but achievements are not sufficient. Turkey's water pollution control policy still has a long way to go until pollution is curtailed, thus creating healthy living conditions for its population. Failure to establish sufficient treatment plants and to apply best agricultural practices has the potential of creating serious harm to public health and the environment. Inadequate achievements in this sector might be attributed to a variety of factors such as financial constraints, regulatory deficiencies, inappropriate approaches and instruments, limited administrative capacity for monitoring and

G. Orhan (✉)
Department of Public Administration, University of Balikesir, Bandirma, Turkey
e-mail: gorhan@balikesir.edu.tr

W. Scheumann
Deutsches Institut fuer Entwicklungspolitik, Bonn, Germany
e-mail: waltina.scheumann@die-gdi.de

A. Kibaroglu et al. (eds.), *Turkey's Water Policy*,
DOI 10.1007/978-3-642-19636-2_7, © Springer-Verlag Berlin Heidelberg 2011

enforcing legislation (Hamamci et al. 1992; Hamamci and Emre 1993; Erim et al. 1996; Algan 2000; Yazgan 2000; Arat 2000).

This chapter focuses on a particular aspect of Turkey's water pollution control policy, namely wastewater treatment in urban areas and the industrial sector.[1] It will first set the stage for Turkey's environmental policy, report on the state of water pollution and the achievements made; then proceeds to describe the regulatory framework for combating water pollution and dealing with wastewater services. It aims at analyzing the bottlenecks (financial, institutional); the regulatory approach followed, and looks at the inherent challenges for Turkey's administration.

2 Setting the context of Turkey's environmental policy

Turkey, like other developing countries, faces the dilemma of coupling economic and social development with protecting its natural resources. As an Organization for Economic Cooperation and Development (OECD) report has put it, "Turkey confronts the challenge of ensuring that economic growth is associated with environmental and social progress, namely sustainable development. It has experienced increasing environmental pressures from energy, industry, agriculture, transport and tourism. They translate in a range of environmental challenges concerning air quality, water services, water resources, waste management, soil erosion and nature protection, as well as marine issues" (OECD 2008, 14).

Until the 1970s the Turkish political and bureaucratic elite had not attempted to establish a separate specialist branch to address environmental issues. However, environmental problems started to surface and came to the attention of public authorities during the late 1960s and early 1970s. Initially, these problems were either ignored or dealt with by sectoral branches of the public administration and municipalities. Environmental policies were at best characterized as reactive and ad-hoc and were not national in scope (NEAP 1999, 11).

Although there has been legislation prior to 1970s, environmental policy reached a new stage after 1972, when it became institutionalized and broad environmental legislation was enacted. The Stockholm Conference, at which Turkey participated, had stimulated these early developments (Algan and Mengi 2005, 95-96). After the Stockholm Conference environmental policy as a distinct sphere of public policy was introduced during the 1970s by the establishment of an environment administration, and developed during the 1980s and 1990s with the gradual introduction of relevant environmental legislation. In 1972, the Environmental Problems Coordination Council was established and environmental issues entered into the Third Five Year Development Plan (1973-1977) and the 1973 government programme. In 1974, a Permanent Board of Consultants for Environmental Problems was established. However, modern organizations dealing with

[1] The chapter does not cover diffuse land-based pollution (for a discussion of land-based pollution see Avaz et al. 2007).

environmental issues and a legal framework regulating environmental matters came later when the Prime Ministry Undersecretariat of Environment was established in 1978. A framework Environment Law was enacted in 1983, and by-laws have been introduced throughout the 1980s and 1990s.

From 1978 to 1991, the Prime Ministry Undersecretariat of Environment attached to an unspecified Minister of State was responsible for the coordination of all national and international activities and administrations pertaining to the environment. The Undersecretariat was expected to set environmental objectives, prepare regulations and cooperate with other ministries and agencies as deemed necessary for ensuring sound environmental management (OECD 1992, 25). The Ministry of Environment established in 1991 took on the functions of the Undersecretariat of Environment. Finally, the Ministry of Environment and the Ministry of Forestry were merged in 2003 and named the Ministry of Environment and Forestry (MoEF). As it stands, MoEF sets the strategic priorities of environmental policies in coordination with other ministries and public authorities, and shares with them the responsibility of implementation, monitoring and enforcing regulations.

Meanwhile, a legal and institutional framework has been established and strategic documents like Agenda 21 (1996), the National Environmental Action Plan (1998) and in particular the Environmental Chapter of the EU Accession Partnership Document (2008) was prepared to address, reverse and prevent environmental degradation. Turkey's government has started aligning with the EU's environment and water acquis, including the directives addressing water pollution (e.g. the EU Urban Wastewater Treatment Directive, the EU Dangerous Substances Directive), and the Water Framework Directive (WFD) which is key in defining the overall environmental objectives to be achieved in the end (see Sumer and Muluk in this volume).

3 Water pollution issues, with a specific focus on domestic and industrial wastewater

In the early years of environmental policy formulation in Turkey, deforestation and soil erosion had much more publicity than water quality problems. Soil erosion, as a consequence of deforestation had been perceived as early as the 1950s as a major threat for Turkey's agriculture (Cevre Bakanligi 1997, 103-118). Nowadays, soil pollution stemming from other sources such as industrial production, unsafe waste disposal and intensive use of pesticides and fertilizers in agriculture receives increasing attention.

Until recently, water policies were designed to accomplish those sectors' objectives which are perceived as key to Turkey's economic growth, e.g. industry and agriculture, and public authorities did not pay due attention to their impact on environment in general and water resources in particular (Orhan 2003, 233-238). Sector specific objectives such as providing drinking water for households, water for irrigating agricultural land and for hydropower generation have been priority targets while pollution of natural water bodies and its impacts on related ecosystems have not been a major concern. Meanwhile, since the By-Law Water Pollution Control of 1988, the topic is permanently fixed on the political agenda but must compete with the manifold water-related problems Turkey is facing, one of them being flood control which gained prominence during the 2000s due to disastrous events as for example in the Meric river basin (see Kramer and Schellig in this volume).

Rapid growth of population and the large influx of rural migrants into large cities has intensified the pressure on water resources and also contributed to their pollution by greater production of waste and wastewater. The migration from rural to urban areas of about 1.4 million people per year shows adverse effects, which are felt heavily in Ankara, Bursa, Istanbul, Izmir, Adiyaman, Antalya, Diyarbakir, Batman and Icel (OECD 2008, 159). Settlements in metropolitan cities are left without proper infrastructure facilities including sewage collectors and treatment plants (Cevre Bakanligi 1997, 377-387).[2] Industry and tourism alike, which are the fastest growing sectors of the Turkish economy produce a sizable amount of the wastewater pollution load. Approximately half of the 190,000 small industrial enterprises in Turkey are active in highly polluting industries (textile / clothing / leather; metal products / machinery / equipment; food / beverages / tobacco; forest products / furniture), and of those only about 1.4 percent are established in organized industrial zones with reliable and environmentally sound infrastructure.

The state of Environment Report of the MoEF (MoEF 2007, 66) categorizes water pollution sources as follows;

1. Discharge of untreated or only partially treated domestic and industrial wastewater into surface waters
2. Pollution of groundwater stemming from leakages in sewer systems and waste disposal sites
3. Diffusion of pesticides and fertilizers into surface waters and aquifers, and soils
4. Deforestation and agricultural practices that facilitate soil erosion and sedimentation in lakes and dams

Although not mentioned, large-scale interventions such as dam construction also affect water quality by changing a river's thermal regime, water chemistry and its organisms and biodiversity (McCartney 2009; OECD 2008, 92) (see Scheumann in this volume).

Assessments based on data of the General Directorate of State Hydraulic Works (DSI) indicate that water quality has been deteriorating in most of Turkey's inland water bodies, and has reached alarming levels in surface waters situated in the vicinity of large metropolitan cities (Baltaci et al. 2008, 321; OECD 2008, 61). Most of those rivers flowing into the eastern Mediterranean Sea, the Marmara, Aegean and Black Sea are in class III ('polluted') and IV ('highly polluted') (MoEF 2007). Furthermore, analyses of groundwater resources beneath some cities, as early as 1998, indicate that groundwater resources are polluted and not suitable

[2]See also EFT (1995) and OECD (1998; 2008) for a thorough documentation of Turkey's environmental problems.

for domestic use any more (Karaguzel and Irlayici 1998; Muhammetoglu and Yalcin 2003).

In rivers which are not polluted by industrial and domestic wastewater, pollution stems from agricultural chemicals (fertilizers and pesticides) intensively used in irrigated agriculture (Cevre Bakanligi 1997). Although the most important problem with regard to agricultural irrigation in Turkey is the scarcity of water and, in some regions, lack of drainage systems, the infiltration of fertilizers and plant protection chemicals in groundwater and surface water bodies poses serious threats to water quality (MoEF 2006, 12; Karaguzel and Irlayici 1998). Nutrient run-off from farmland affects in particular the lakes of Apolyont (Marmara Region), Golcuk and Koycegiz (Aegean Region), while domestic and industrial discharges are risk factors for the lakes of Kus and Sapanca (Marmara Region) and Van (Eastern Anatolia Region) along with pollutants stemming from farm run-off (OECD 1999, 52).

If compared with the 1999 OECD environmental performance report, progress in rehabilitating these water bodies has been slow (OECD 1999; 2008, 73; NEAP 1999, 31-38).

4 Sewage system connection rates increase between 1994 and 2008

In the late 1990s, financing the extension of water supply systems had priority, but funding for sewerage networks and wastewater treatment plants increased too. Alaton et al. (2006, 118) observe a distinct pattern in providing basic sanitation and wastewater treatment services in Turkey: the percentage of provinces served is high in the Western parts of the country which have zones of high industrial and/or tourism activity and/or urbanization; connection rates are higher in Metropolitan municipalities with a high population density and in tourism regions as well as being higher in urban than in rural areas. Actually population in most metropolitan cities is connected to wastewater treatment facilities, and the country-wide connection rate has increased from 10 percent in the mid 1990s to 46 percent in 2008 (see Table 1) with the number of treatment plants constructed having almost doubled.

Referring to the Turkish Statistics Institute (TUIK) which provides regular and updated information on the municipalities' water supply and sanitation services, changes between 1994 and 2008 - the year where the latest statistics were published (in 28 April 2010) - are as follows:

- The volume of wastewater discharged from municipal sewer networks[3] increased from 1.5 million cubic meters (MCM)/year in 1994 to 3.2 MCM/ year in 2008 (these figures do not include wastewater from industry outside municipal borders and from enterprises not connected to municipal systems).

[3]Sewer networks receive domestic wastewater or a mixture of domestic with industrial wastewater and / or run-off storm water.

Table 1 Municipal wastewater figures (1994 – 2008) (TUIK website 2010)

	1994	1996	1998	2001	2004	2008
Number of municipalities	2,663	2,750	2,757	3,227	3,225	3,225
Total municipal population	47,597,657	47,843,698	47,862,511	53,407,613	53,935,050	58,581,515
Population included in municipal survey	45,658,019	45,920,465	46,529,408	53,377,431	53,903,955	58,581,515
Number of municipalities connected to sewerage systems	1,188	1,383	1,647	2,003	2,226	2,421
Percentage of total population connected to sewerage systems	52	55	59	64	68	73
Percentage of municipal population connected to sewerage systems	69	72	78	81	86	88
Volume of wastewater discharged (million m^3/year)	1.510	1.679	2.091	2.301	2.923	3.261
Volume of wastewater treated (million m^3/year)	150	202	590	1.194	1.901	2.251
Percentage of total population connected to treatment plants	10	10	17	27	36	46

- Of this volume, 0.150 MCM/year were treated in 1994 and 2.2 MCM/year in 2008 (these figures do not include wastewater from industry outside municipal borders and from enterprises not connected to municipal systems).
- In 2008, biological[4] treatment was applied to 43.3 percent; physical treatment to 33.4 percent and advanced treatment to 23.3 percent (in 2006, it was 10 percent only (Kauffmann 2011, 87)).
- Out of 3,225 municipalities only 2,421 are connected to sewage systems but not necessarily to treatment plants.
- percent of the municipal population was connected to sewerage systems in 1994 (52 percent of the total population), and 88 percent in 2008 (73 percent of the total population).

Although industrial wastewater makes up less than one percent of the total wastewater generated and discharged, it contains highly poisonous substances such as mercury, lead, chromium and zinc. Within the period under review by the OECD (1998-2008), the discharge of untreated industrial wastewater decreased by 10 percent. Out of the total wastewater generated by the manufacturing industry, 36 percent is treated by the industry's own plants and 7 percent by municipal facilities (OECD 2008, 68), the rest being released untreated into receiving bodies.

4,030 establishments within the scope of the Manufacturing Industry Waste Inventory Survey of 2004 discharged 1.145 billion cubic meters (BCM)/year of wastewater. 2,112 of them discharged 760 MCM/year without being treated; 1,918 industrial facilities discharged 385 MCM/year of treated wastewater. That means, approximately 66 percent of total wastewater originating from the manufacturing industry was discharged without any treatment (MoEF 2006, 33), the greatest amount coming from metallurgical industry (48 percent), followed by food and beverages (13 percent), textile (12 percent) and chemical (9 percent) industry (OECD 2008, 68).

These figures do not include pollution stemming from small- and medium-sized industries, but their contribution to water pollution should not be underestimated. According to "Turkey – National Study" (Burak 2007, 706), Turkey has registered 216 Organized Industrial Districts, of which only 17 have wastewater treatment plants which discharge directly into a receiving medium (as of 2004); treatment plants were under construction for an additional 18 Organized Industrial Districts, and nine Organized Industrial Districts are connected to municipal sewers, leaving

[4]Physical or primary treatment means treatment of urban wastewater by a physical and/or chemical process involving settlement of suspended solids (SSD), or other processes in which BOD5 are reduced by at 20 percent discharge and the SSD of the incoming wastewater are reduced by at least 50 percent. Biological or secondary treatment means treatment of urban wastewater by a process involving biological treatment with a secondary settlement or other process in which the requirements established in the EU Urban Wastewater Treatment Directive are respected. Advanced or tertiary treatment refers to several types of treatments with the purpose of reaching a treatment level of higher quality than that which one could normally expect from secondary treatment. It can aim for a higher level of removal for conventional parameters such as SSD or also concentrate on certain parameters for which there is little removal in secondary treatment such as phosphorus (European Commission, International Office for Water 2001).

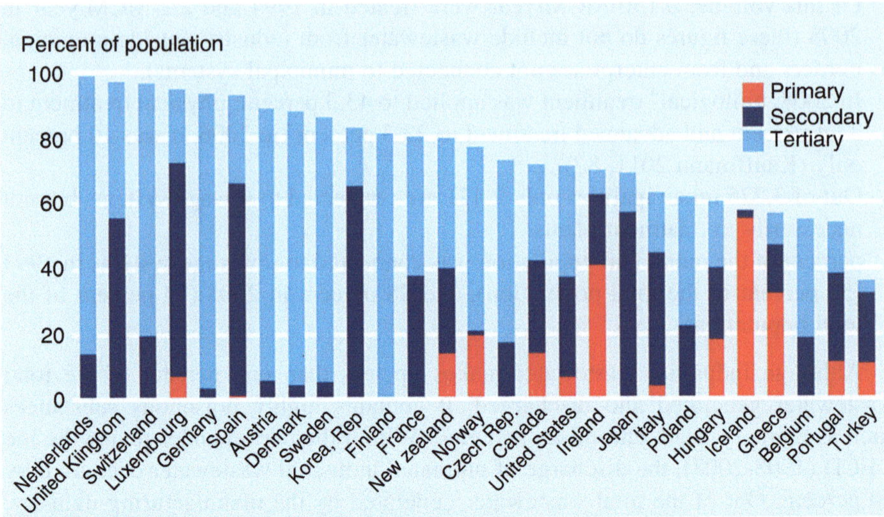

Fig. 1 Levels and types of wastewater treatment in OECD countries and selected European countries, as of 2006 (UNESCO 2009, 142)

172 without any treatment. According to GlobeNet reporting on investment opportunities in Turkey's water and wastewater industry, only 3,000 industrial establishments have treatment facilities of their own, out of about 190,000 small to medium-sized industrial enterprises (GlobeNet 2009).

According to 2008 figures, the public sector spent 8.98 billion TL on environmental investments of which 55 percent were spent on water supply services and 24 percent on wastewater treatment services. Municipalities accounted for 72.43 percent, Special Provincial Administrations for 2.05 percent and other public sector institutions for 25.52 percent of environmental investments. The public share of environmental investments is 0.35 percent of the Gross National Product (GNP) (TUIK 2010).

If the proportion of the population connected to sewage treatment plants serves as an indicator for the attempts made to combat water pollution, Turkey has progressed. However, if compared with OECD countries, Turkey ranks last (see figure 1).[5]

The OECD estimates that about "3,000 new plants remain to be built in towns with a population over 2,000" to follow the EU Urban Wastewater Treatment Directive for which EUR 18 billion is needed for investment in and rehabilitation of treatment facilities (OECD 2008, 66). EU funds are expected to support 40 percent, while local administrations will have to co-finance these projects through credits from the Bank of Provinces.

[5]See Kuks (2004, 26-28) for connection rates of households in the Netherlands, Switzerland, France, Italy, Spain and Belgium.

5 Regulatory framework for protecting water quality and combating pollution

Turkey's legislative commitment to environmental policy targets, in particular, the protection of water resources from pollution, is reflected in a detailed regulatory system. Regulations date back to the early years of the Republican period, the 1920s and 1930s, namely the Village Law (No. 442) of 1924, the Municipality Law (No. 1580) of 1930 and the General Hygiene Law (No. 1593) of 1930 which mandated municipalities and village administrations to provide clean drinking water and to protect it from pollution. However, modern water legislation and a new legal structure for environmental protection and water management emerged from 1980s onwards.

Hierarchically, the Turkish Constitution of 1982 is the major source of reference, followed by laws that require parliamentary majority, and by-laws and statutory decrees. Considering the chapter's specific topic, each category will be presented selectively following a hierarchical and chronological order (see Table 2).

Table 2 Framework for water pollution control (own compilation)

Year*	Law No.	
		Water pollution control
1988	19919	By-Law for Water Pollution Control
2004	25687	
2008	26786	
2004	25377	By-Law on the Protection of Waters from Agricultural Nitrate Pollution
2005	26005; 26040	By-Law on the Control of Water and Environmental Pollution Stemming from Hazardous Substances
2006	26047	By-Law on Urban Wastewater Treatment
		Water quality standards
1991	20748	Technical Procedures of the By-Law for Water Pollution Control on Irrigation Water Quality**
2005	25999	By-Law for the Quality of Surface Waters Used for Drinking Water Purposes
2006	26048	By-Law on Bathing (Recreational) Water Quality
		Mandating service providers
1924	442	Village Law
1930	1593	General Hygiene Law
1930	1580	Municipality Law
2005	5393	
2004	5126	Metropolitan Municipality Law
2005		
2005	5302	Special Provincial Administration Law

(*) The year of enactment is mentioned first, followed by the last ordinance.
(**) In Turkish: Teknik Usuller Tebligi.

5.1 *Constitutional and parliamentary ruling*

The **Turkish Constitution** introduced environmental rights as a human right in Article 56: "Everyone has the right to live in a healthy, balanced environment. It is the duty of the state and citizens to improve the natural environment, and to prevent environmental pollution." The Turkish Constitution assigns stewardship of the country's water resources to the public domain, where only public institutions may grant water use-rights to both public and private parties as long as they accommodate 'public benefit' or 'common wealth' (Republic of Turkey 2003, 94).

The **Turkish Civil Law** of 2001 (No. 4721) regulates ownership of and user rights to water resources in detail and also addresses pollution: If water resources get polluted or spoiled Article 757 regulates matters of compensation and Article 758 regulates issues concerning their restoration.

The **Groundwater Law** of 1960 (No. 167) has put property rights into the public domain. User-rights are subject to licensing upon request (within the safe yield of an aquifer), and can neither be transferred nor sold. State Hydraulic Works issues user licenses, monitors pollution but its mandate does not cover groundwater pollution control.

The **National Parks Law** of 1983 (No. 2873) prohibits all activities, including water pollution, that may contribute to the degradation of the environment and ecological balance in areas designated as National Parks, and asks for strict measures to be applied for activities outside of protected areas that may affect them.

Furthermore, the **Municipality Law** of 2005 (No. 5393), the **Metropolitan Municipality Law** of 2004 (No. 5126) and the **Special Provincial Administration Law** of 2005 (No. 5302) mandates local governments for providing water and sanitation services within their jurisdictions, while Special Provincial Administrations do the same in villages, rural and non-municipal areas of Turkey's 81 provinces.

The **Environment Law** of 1983 (No. 5491) which was revised on 26 April 2006, is a framework document that determines general principles concerning the protection of the environment and the prevention of pollution. It endorses the 'Polluter Pays Principle' and handles environmental issues broadly. The aim of the law, which considers the environment as a whole, is not only to prevent and eliminate environmental pollution, but also to allow for the management of natural and historical values and land in such a way as to utilize and preserve its richness for future generations. According to its basic principles, citizens as well as the State bear responsibility for the protection of the environment. It emphasizes that every effort should be made to minimize and solve environmental problems in economic activities, in particular when determining production methods. Acknowledging the relationship between protecting the environment and economic development, the

Law stipulates that environmental protection efforts must be carried out in harmony with the targets set for economic and social development (EFT 1987, 33-4; OECD 1992, 24).[6]

5.2 *By-Laws specifying water pollution control*

Since the Environment Law (1983) is a framework document, it was assumed that the relevant regulations would be introduced after three years. Instead, it took more than one decade, but recent efforts towards EU alignment have accelerated legislative change.

The **By-Law for Water Pollution Control** of 1988 (No. 19919) revised in 2004 (No. 25687) and 2008 (No. 26786) aims at both conserving the quality of water resources and water-dependent ecosystems, and protecting and improving water quality to meet human demands. It establishes emission limit values (or discharge standards) which define the maximum allowable discharge of pollutants (including priority substances as defined in the EU Dangerous Substances Directive and the EU Nitrate Directive) into receiving natural and artificial water bodies. By means of classifying inland surface waters into four classes,[7] water quality standards of receiving water bodies are defined. In order to prevent further pollution and/or to improve water quality, the MoEF may restrict wastewater discharges until watershed protection plans are prepared, including the designation of watershed protection zones around drinking water reservoirs and wells. Discharge limit values may be modified if current and future usage of the respective water bodies is negatively affected. It further regulates the permit system for direct (into receiving natural water) and indirect dischargers (into municipal sewage systems), and authorizes the Provincial Environment and Forestry Directorate to issue the permits (the Local Environment Commissions act as their advisors). Although MoEF is responsible of inspection, monitoring and controlling, State Hydraulic Works, provincial governors, municipalities and the metropolitan Water and Sewage Administrations are also authorized to the implement this By-Law.

The **By-Law on the Protection of Waters from Agricultural Nitrate Pollution** of 2004 (No. 25377) deals with nitrate and nitrate-based components that originate from agricultural practices (application of fertilizer and animal waste) and

[6]From a critical perspective, the Environment Law makes a distinction between economic development and environmental protection, and prioritises the former over the latter. Although the Environment Law was amended in 2006 and introduced the notion of sustainable development, a distinction between sustainable development and environment remains. This allows economic concerns to be prioritized over environment.

[7]Classification is based on a) physical and inorganic chemical, b) organic, c) inorganic pollution and d) bacteriologic parameters. Class I waters refer to high quality waters, Class II waters refer to minimal pollution, Class III refers to polluted water and Class IV refers to highly polluted water (the number of parameters on which assessments rest was extended to 45 in 2004), see Amendments to the By-Law, Table 1.

pollute groundwater, surface waters and soil. It defines standards for drinking water and criteria for determining regions at risk. The Ministry of Agriculture and Rural Affairs (MARA) is responsible for promoting good agricultural practices, i.e. fertilizer use in agriculture and for developing programmes towards this end.

The **By-Law on Controlling Water and Environmental Pollution Stemming from Dangerous Substances** of 2005 (No. 26040) covers technical and administrative standards for identifying dangerous substances that pollute surface waters, estuaries and regional waters; for preparing pollution reduction programs, for monitoring and preventing pollution; for making an inventory of dangerous substances discharged into water resources; for determining wastewater discharge standards and water quality standards related to fourteen dangerous substances in receiving environments (Appendix 1), and determines environmental quality standards for marine waters and estuaries (not for inland water bodies) for forty less dangerous substances (Appendix 2) that can be released into the water bodies.

The **By-Law for the Quality of Surface Waters that are used for Drinking Water Purposes** of 2005 (No. 25999) categorizes surface water,[8] and authorizes the MoEF to prepare water protection plans, determine quality criteria and means to improve water quality. Major implementing authorities are DSI, the Bank of Provinces, municipalities and the Special Provincial Administrations.

The **By-Law on Urban Wastewater Treatment** of 2006 (No. 26047) regulates the collection, treatment and discharge of urban wastewater to protect the environment from negative effects of untreated industrial and domestic wastewater. It defines technical and administrative principles pertaining to the collection of urban and industrial wastewater that is discharged into sewage systems, their treatment and discharge; it details administrative mandates for monitoring and reporting of wastewater discharge (see Table 3); it specifies a transition period for municipalities to complete their wastewater collectors and treatment plants in parallel with population growth. Discharge into receiving water bodies and connection into sewage systems (pre-treatment might be necessary) are both regulated whereby the Local Environment Commissions advice authorities.

The **By-Law on Bathing Water Quality** of 2006 (No. 26048) aims to prevent recreational and bathing waters from pollution, mainly microbiological, for the sake of human health and the environment. It covers quality criteria for bathing and recreational waters and administrative procedures for monitoring and reporting. The By-Law prohibits the discharge of untreated wastewater and solid waste disposal into bathing waters. Major responsibility for implementing this by-law rests with the MoEF.

Table 3 shows the public institutions involved in providing sanitation services.

---

[8]A1: surface waters that become drinkable after basic physical treatment and disinfections; A2: surface waters that become drinkable after physical treatment, chemical treatment and disinfections; A3: surface waters that become drinkable after intensive physical and chemical treatment, advanced treatment and disinfections.

Table 3 Public institutions involved in providing sanitation services (own compilation; Cinar 2009, 351-2)

Functions / tasks	Public institutions
Agenda setting / policies	
Strategic decisions	Ministry of Environment and Forestry
Setting water quality standards for uses, and emission limit values	Ministry of Environment and Forestry State Hydraulic Works (DSI) Ministry of Health Ministry of Agriculture and Rural Affairs
Wastewater treatment	
National investment plans incl. water and sewage systems	State Planning Organization
Planning and implementation	Metropolitan municipalities State Hydraulic Works Bank of Provinces (3,000 – 100,000 inhabitants) Industry / Organized Industrial Districts Ministry of Culture and Tourism Special Provincial Administrations
Financial services	National treasury subsidizes credits through Bank of Provinces External financing institutions (loans, credits) European Union funds (co-financed)
Operating units	Municipalities Metropolitan municipalities Industrial operators / Industrial Zone Management Organizations International water companies (e.g. ONDEO) under BOT and management contracts
Issuance of permits and licenses, and monitoring	Metropolitan municipalities (connection licenses from Water and Sewage Administrations) Local (provincial) environment commissions Provincial branches of Ministry of Environment and Forestry
Setting tariffs	Non-metropolitan municipalities: proposed by municipal departments, to be endorsed by Municipal Council, and approved by Provincial Governor; Metropolitan municipalities: Water and Sewage Administrations, to be accepted by Municipal Council
Pollution prevention charge	Ministry of Environment and Forestry

6 The command and control approach

The regulatory framework in place has two important features: it defines water quality standards for various uses (drinking, bathing, irrigation) with distinct thresholds for pollutant substances. It further addresses water pollution control by means of emission standards for wastewaters (or, emission limit values). These emission standards define maximum allowable discharges of pollutants (such as Biochemical Oxygen Demand - BOD, Chemical Oxygen Demand - COD) into receiving water bodies. Emission limits are embodied in the pollution permits that each significant source of pollution must possess. Polluters must submit an application for a pollution permit containing a technical analysis of the quality characteristics of the effluents in relation to the standards. In the event of exceeding emission standards, polluters are obliged to treat wastewater, or are subject to penalties. According to OECD: "Industrial and mixed discharges are also required to have a prior quality control permit specifying pollution limit values and quantities. Compliance with permit conditions is low: some estimates indicate that 60 percent of industrial operators (especially small and medium-sized ones) operate without permits" (SOGESID 2055 in OECD 2008, 63). However, if negative consequences from permitted wastewater discharges are observed and impact on present and future uses of the respective water bodies, emission limit values can be tightened and become more stringent.

It has been argued that the use of a command and control approach to combat water pollution does not translate into effective policy implementation due to monitoring and enforcement constraints at the local/municipal level. Applying a "mixed policy" approach utilizing compliance incentives through economic instruments (e.g. tax exemptions, credit support) along with permits (Hummel 2003, 45) would be more effective than command and control instruments alone because, as pinpointed by Barde (1994, 8), "poor enforcement is the weak link of the regulatory chain".

In an attempt to tackle non-adherence to emission standards by industry, the MoEF has imposed a pollution prevention charge on industrial plants that do not operate their wastewater treatment plants for a certain period or are unable to reduce pollution parameters below permitted levels. This charge applies to all irrespective of whether they discharge into public sewers or outside as an incentive to build and operate treatment plants. However, according to the OECD's last performance review "lax enforcement and low collection rates often lead industrial operators to discharge wastewater without treatment" (OECD 2008, 69). Unlike industry, municipalities which are the biggest polluters of freshwater in many townships, are seldom penalized as treatment plans are shelved or delayed due to lack of funding or are lower in priority compared to water supply investment (Hummel 2003, 45).

A number of voluntary agreements ("environmental declarations") have been signed between the MoEF and industrial enterprises to install wastewater treatment plants. The industries, which declared their commitments in 1995 and 1997, include, for instance, the yeast, sugar and paper industries, and the leather industry respectively. Voluntary agreements allow for a transition period for polluting industries to complete their treatment systems and infrastructure in line with emission standards. It is further reported that the Turkish Business Association has coordinated a number of initiatives aiming at cleaner production processes and reducing wastewater generation in order to meet the high environmental standards on a voluntary basis (OECD 2008, 142).

All (domestic and industrial) users who are connected to municipal sewers and treatment plants have to pay charges for sewage services (as well as for water supply). Sewage charges set by individual municipalities (see 7.2), however, are not meant to primarily function as an economic incentive for pollution control but are designed to cover at least the operational expenditure for wastewater collection and treatment - which they however, rarely do.[9] Until the revision of the Law on Environment in 2006 the level of the wastewater charge could not be higher than 50 percent of the charge for supplying drinking water. With the new regulation, rates now can reflect marginal social costs.

Levying effluent charges (effluent charges are borne by those who directly discharge into natural water bodies) aligns with the Polluter Pays Principle and could be an economic pollution control instrument in that they apply to all discharges and are based on the actual pollution load. This instrument aims at providing incentives for dischargers to reduce pollution loads by either treating their wastewater and / or by redesigning production processes to reduce or eliminate pollutants (Barde 1994; OECD 2009). However, financial assistance might be required to help industrial polluters to comply with emission standards (note, municipalities get subsidized credits).

7 Governing and managing water pollution control

Since 1980, following the implementation of the structural adjustment programme, Turkey has undergone substantial economic and institutional reforms. It has abandoned central planning and state dominance in economic life, and has transformed its economy into a dynamic liberal market economy. This has affected the water sector too, although decentralized elements had existed before: providing water and sanitation services have always been the mandate of municipalities (see Table 3).

As a structural characteristic of Turkey's administrative setting, provision of sanitation infrastructure and services and protection of water resources is vested in many government institutions with no one being assigned the lead function. The MoEF which is supposed to coordinate across bureaucracy and administrative levels has not been strong in this respect.

7.1 Central administration: strategic decisions on policies, standards, and financing

The MoEF has a key role in drafting environmental legislation but it shares the authority of strategy and norm setting with many other ministries and institutions. These are amongst others DSI (for irrigation water), the Ministry of Health (MoH, for drinking, bathing and recreational waters), and the MARA (for aquaculture and fishing). All these public institutions have rightful concerns about polluted waters

[9] In most European countries effluent charges recover service costs but do not work as an incentive (European Parliament GD for Research 2001; Rudolph et al. 1998).

but their interests need to be reconciled in an appropriate system of national standards, eventually being part of a national framework water law.

Financing and allocation decisions prioritizing public expenditure is the mandate of the State Planning Organization (SPO). All major public investment projects as well as projects proposed by municipalities which are either to be financed from domestic or foreign resources must be approved by SPO. According to Hummel (2003, 43), however "SPO takes a rather reactive approach to sector development" and is "concerned with reviewing individual projects case-by-case on the basis of technical feasibility studies (...) and does not systematically engage in masterminding future sector development".

Individual wastewater treatment projects approved and included in the annual public investment programmes are then assigned to either DSI or the Bank of Provinces for implementation. Financing sewage systems and wastewater treatment plants is largely provided through the Municipalities Fund of the Bank of Provinces attached to the Ministry of Public Works and Housing, which supports municipalities with less than 100,000 inhabitants to plan, build and finance water supply and sewage systems and wastewater treatment plants. After completion, water supply and sewage systems are handed over to municipalities, and routine management and water quality control becomes their responsibility (OECD 1999, 55).

7.2 *Provincial and municipal level: service providers and inspectors*

According to the Municipality Law municipalities are responsible for providing sanitation. Depending on the size of the non-metropolitan municipalities, municipalities operate under different modes (Cinar 2009, 351-2):

- Municipal public works departments, financed out of the municipal budget, operate in cities with a population of less than 10,000.
- Directorates or water offices of municipalities which are not separate legal entities, operate in cities with population between 10,000 and 50,000.
- Separate operating units established by municipal councils as legal entities assume management responsibility based on their own budget in cities with a population greater than 50,000.

The Water and Sewage Administrations of the 16 metropolitan municipalities[10] (out of 81 provincial capital municipalities) which serve about 45 millions of people (about 50 percent of Turkey's population), are autonomous public entities with an independent budget and are free to set their own water and wastewater tariffs, which need to be endorsed by the municipal council but without formal

[10] Istanbul, Ankara, Izmir, Adana, Bursa, Gaziantep, Konya, Kayseri, Antalya, Diyarbakir, Mersin, Eskisehir, Samsun, Erzurum, Kocaeli and Adapazari.

approval of central government authorities (Cinar 2009). They enjoy a high degree of autonomy in planning, financing, implementing and operating the infrastructure along with competencies in pollution control. Metropolitan municipalities enjoy access to financial services for investing in wastewater treatment projects, and are capable of mobilizing financial sources beyond the mechanism of the Bank of Provinces including foreign loans under the Treasury Guarantee Scheme (see Kibaroglu and Baskan in this volume).

In contrast, non-metropolitan municipal administrations are "firmly integrated into a centralized system of providing the water and sewerage infrastructure" (Hummel 2003, 43) depending on a public sector financing mechanism with planning and implementation assigned to either DSI or the Bank of Provinces. These municipalities rely on the central administration's decisions which are less concerned in addressing local issues, leaving little room of manoeuvring and no institutionalized means of influencing decisions at the central level in order to push forward their investment priorities.

In order to comply with emission limit values as defined by law, investment costs for pollution control equipment are high and seem to severely stress the national budget, exceed the affordability in many of the non-metropolitan municipalities and the economic capacity of water users to recover the investment cost of pollution control equipment (Hummel 2003, 44).

As already mentioned, the regulatory system for wastewater discharge relies on permits which are either granted by metropolitan municipalities, namely the Water and Sewage Administrations, or by the provincial governors' offices. Critical to a regulatory system based on strict standards and permits is reviewing and assessing applications, setting licensing requirements to be met by the polluters, as well as inspection and enforcing regulations at the municipal level where polluters operate. At the provincial level, offices of MoEF are not adequately equipped in manpower and supporting facilities to assess applications and to carry out inspections. Very often the responsibility for industrial wastewater analysis is left to the very industries that apply for the permits and which have not yet been done by inspectorates operating independently. In 2002, a Directorate for Inspections in the Ministry of Environment was established, in this way separating inspections from permit issuing. However, because they are formally subordinate to the provincial governor's office and not to the central inspectorate of the Ministry of Environment, enforcement officers may face pressure to balance economic and environmental protection goals (OECD 2008, 137).

At this level, Local Environment Commissions were established after 1993 which operate under the governors' chairmanship.[11] Concerning pollution control,

[11]The Local Environment Commissions are basically composed of the governor of the province, the mayor (or the metropolitan mayor), provincial directors of ministries, a delegate of the Ministry of the Interior, and representatives of the Chambers of Industry, Trade and Agriculture. If necessary, the chairman may invite representatives from other governmental institutions, universities and NGOs to their meetings. The Commissions take decisions for protecting the environment and preventing pollution in line with environmental policies and principles adopted

their major task has been advising the respective authority on how to approve connection permits to municipal sewer systems and how to grant permission for the direct discharge into natural water bodies which industry and settlements may ask for. In 2005, for instance, 102 out of 181 decisions made by the Local Environment Commission of the Balikesir Province were related to discharge permits applied for by housing estates and industrial plants as stipulated in Articles 37 and 38 of the By-Law on Water Pollution Control.[12]

Pointing to the key role local administration would have to play, Hummel observes "(...) a large degree of compromise in enforcing the regulations at the local level. It appears that enforcement at the local level takes very much the form of a bargaining process meant to prevent, mitigate or intermediate environmental conflicts, balancing the interests of government and polluters" (Hummel 2003, 45-46). This is due to various factors, among them the belief of local governments that enforcing tight environmental regulations tend to drive away industries crucial to the local economy for employment and growth. If industry has difficulties in meeting the permitted emission limits, their representatives bargain with the local authorities for leniency in granting permissions. It is said that their representatives in the Local Environment Commissions are often exerting influence to suppress effective enforcement of on-site pollution prevention measures on the grounds of jeopardy to employment and competitiveness (Hummel 2003).

8 Concluding remarks

Turkey's drive towards EU membership resulted in aligning with European environmental directives from the year 2000 onwards, including those addressing water pollution. However, these efforts were not welcomed by everybody. There have always been reservations about the predominance of water quality policies, and it has been argued that country-specific conditions should be given due consideration when aligning with the EU environment and water acquis. Turkey's priority should lie in developing water resources to satisfy its rising water demand for household consumption, food and energy production, and that it should not prioritize water quality issues. Increasing public expenditure for both water supply and sanitation and competing for scarce financial resources reflect this ambiguity and the country's stage of developing and protecting its water resources.

With regard to the regulatory framework, it appears that Turkey lacks a comprehensive water law that would equally cover and balance the development and the protection of water resources; environmental quality standards for specified

by the MoEF and other ministries. They prepare programs, take precautionary measures, ensure implementation, are responsible for reporting provincial environmental problems, propose solutions to the Ministry and are platforms for coordination among public authorities.

[12]Http://www.balikesir-cevreorman.gov.tr/mck_karar.asp. Accessed 26 January 2007.

water bodies; and the harmonization of the national water quality standards for various uses.

As regards the transposition of the regulatory framework into workable policies, environmental quality standards for specific water bodies have yet to be defined and watershed protection and rehabilitation plans are in the process of being developed. This deficit is strategically and programmatically addressed in the EU WFD which aims at achieving a good status of the respective water bodies. Environmental quality standards need to consider the cumulative effects of emissions stemming from all sources on the receiving water bodies taking into account their assimilative capacity.

Taking a closer look at Turkey's sanitation sector, national budget constraints are evident, and affordability of high-cost investments in many of the non-metropolitan municipalities can be questioned as well as the ability of the water users to recover investment costs. Access to international financial services might be a solution but comes at cost, because it increases the municipalities' debts (see Kibaroglu and Baskan in this volume). Risk and hot spot analyses can provide the basis for the better targeting of public investments towards local sanitation needs and achieving environmental objectives for the respective water bodies.

Poor enforcement is the weak link of the regulatory chain, and this is evident when looking at the local / municipal level where pollution occurs. Provincial administrations and municipalities and the Local Environment Commissions play as yet a minor role in the monitoring and the enforcement of regulations (OECD 2008, 136), although the increasing amount of legislation and sharper standards would suggest the opposite. The Provincial Directorates for Inspections that were established separately from units issuing permits, remain a weak element in the regulatory chain because of their subordination under the provincial governor's office.

It has often been argued, that Turkey's regulatory framework would be sufficient and if implemented, an effective instrument to combat water pollution. However, the use of command and control instruments did not translate into effective policy implementation. Instead, a "mixed policy" approach which utilizes economic incentives along with permits would be more effective than command and control instruments alone. Instead of inspecting compliance, that is checking emission values only, authorities should embark on facilitating cleaner manufacturing processes and on the supplement or replacement of end-of-pipe-measures by 'process' solutions and preventive action.

References

Alaton IA, Gurel M, Eremektar G, Ovez S, Tanik A (2006) Urban waste water treatment plants and reuse potential in Turkey: selected examples from the western cities. International Journal Environment and Pollution, vol 28, no 1/2:115-127

Algan N (2000) Turkiye'de Devlet Politikalari Baglaminda Cevre ve Cevre Korumanin Tarihine Kisa Bir Bakis. In: Turkiye'de Cevrenin ve Cevre Korumanin Tarihi Sempozyumu. Turkiye Ekonomik ve Toplumsal Tarih Vakfi, Istanbul, pp 221-234

Algan N, Mengi A (2005) Turkey's sustainable development policies in the EU accession process. European Environmental Law Review 2005, pp 95-108
Arat Z (2000) 1970'lerden Sonra Cevrede Kurumsal Yapinin Gelisimi. In: Turkiye'de Cevrenin ve Cevre Korumanin Tarihi Sempozyumu. Turkiye Ekonomik ve Toplumsal Tarih Vakfi, Istanbul, pp 166-175
Avaz, G. et al. (2007) Development of National Action Plan to Address Pollution from Land Based Activities in Turkey. In: Gonenç EI, Vadineanu A, Wolflin JP, Russo RC (eds) Sustainable Use and Development of Watersheds. NATO Science for Peace and Security Series. Springer, Dordrecht, pp 383-401
Baltaci F, Onur AK, Tahmiscioglu S (2008) Water quality monitoring studies of Turkey with present and probable future constraints and opportunities. Desalination 226:321-327
Barde JP (1994) Economic instruments in environmental policy: Lessons from the OECD experience and their relevance to developing economies. OECD/GD (93), Working Paper no 92, Paris
Burak S (2007) Turkey national study. In: Water demand management, progress and policies: Proceedings of the 3rd Regional Workshop on Water and Sustainable Development in the Mediterranean. Zaragoza, Spain, 19 – 21 March 2007. MAP Technical Reports Series no 168. UNEP/MAP, Athens, http://www.planbleu.org/publications/mts168_eau_Saragosse.pdf Accessed 3 September 2010
Cevre Bakanligi (1997) Turkiye Cevre Atlasi. Ankara
Cinar T (2009) Privatisation of urban water and sewerage services in Turkey: some trends. Development in Practice, vol 18, no 3:350-363
EFT, Environment Foundation of Turkey (1987) Environmental policy of Turkey. EFT, Ankara
EFT, Environment Foundation of Turkey (1995) Environmental profile of Turkey. EFT, Ankara
Erim R, Toroslu N, Cortoglu S, Turgut N (1996) Ulusal Cevre Eylem Plani: Hukuksal Duzenlemeler ve Kurumsal Yapi Calisma Grubu Raporu. Ankara
European Commission, International Office for Water (2001) Guide: Extensive Wastewater Treatment Processes. European Commission, http://www.environment-integration.eu/download/34-Water/ExtensiveWasteWaterTreatmentProcesses.pdf Accessed 28 August 2010
European Parliament GD for Research (2001) Effluent charging systems in the EU Member States. Working Paper, Environment Series, ENVI 104 EN
GlobeNet (2009) Water and wastewater industry in Turkey. GlobeNet, http://www.globe-net.com/articles/2009/may/2/water-and-wastewater-industry-in-turkey.aspx?sub=12 Accessed 27 August 2010
Hamamci C, Emre C (1993) Turkiye Cevre Kurumlari Arastirmasi: Kurumsal Strateji (Ikinci Adim Raporu). AU SBF, Ankara
Hamamci C. et al. (1992) Cevresel Kurumlar ve Duzenlemeler Envanteri: Birinci Asama Raporu AU SBF, Ankara
Hummel M (2003) Effective water and waste water sector regulation and monitoring. The municipal water and waste water sector: Key issues and possible solutions. In: Proceedings of International Workshop, Ankara, April 8-9, 2003
Karaguzel R, Irlayici A (1998) Groundwater pollution in the Isparta Plain, Turkey. Environmental Geology 34 (4):303-308
Kauffmann, C (2011) Financing Water Quality Management. Water Resources Development, vol 27, no 1:83-99
Kuks, Stefan (2004) The Sustainability Performance of National Resource Regimes, in Ingrid Kissling-Naef and Stefan Kuks (eds), The Evolution of National Water Regiems in Europe. Transitions in Water Rights and Water Policies, Kluwer Academic Publishers Dordrecht, Boston, London, pp. 25-48
McCartney M (2009) Living with dams: managing the environmental impacts. Water Policy, 11, Supplement 1 (200):121-139
MoEF, Ministry of Environment and Forestry (2006) EU Integrated Environmental Approximation Strategy (2007 - 2023). MoEF, Ankara

MoEF, Ministry of Environment and Forestry (2007) Turkiye Cevre Durumu Raporu. MoEF, Ankara

Muhammetoglu A, Yalcin OB (2003) An integrated water pollution control approach to protect groundwater of Antalya Plain from diffused sources. Diffuse Pollution Conference Dublin, 7B Groundwater, Antalya

NEAP (1999) National Environmental Action Plan of Turkey. Ankara

OECD (1992) Environmental policies in Turkey. OECD, Paris

OECD (1999) OECD Environmental performance review of Turkey. OECD, Paris

OECD (2008) OECD Environmental performance review of Turkey. OECD, Paris

OECD (2009) Making reform happen in environmental policy. ENV/EPOC/WPNEP(2009) 4/FINAL,http://www.oecd.org/officialdocuments/displaydocumentpdf?cote=ENV/EPOC/WPNEP(2009)4/FINAL&doclanguage=en Accessed 27 August 2010

Orhan G (2003) Leviathan versus democracy: Implementation of sustainable development in Turkey. Unpublished PhD Thesis, Department of Government, University of Essex

Republic of Turkey (2003) Turkey Country Report. Prepared for the Third World Water Forum. World Water Council http://www.worldwatercouncil.org/fileadmin/wwc/Library/Publications_and_reports/country_reports/report_Turkey.pdf Accessed 23 August 2010

Rudolph KU, Kraemer RA, Hansen W, Staffel U (1998) European comparison of sewage charges, Summary. Research project no 30/96 for the Federal Ministry of Economic Affairs and the Federal Ministry for the Environment, Nature Conservation and Nuclear Safety

TUIK (2010) TUIK Haber Bulteni "Kamu Sektorunde Cevresel Istihdam ve Harcamalar, 2008" Sayi 41, 11 March 2008

TUIK website (2010) Haber Bulteni. TUIK, www.tuik.gov.tr Accessed 26 August 2010

UNESCO (2009) Impact of water use on water systems and the environment. In: UNESCO (ed) Water in a changing world. World Water Development Report, http://www.unesco.org/water/wwap/wwdr/wwdr3/pdf/WWDR3_Water_in_a_Changing_World.pdf pp 127-149. Accessed 3 September 2010

Yazgan N (2000) Turkiye'de Son 25 Yilda Cevrenin Degisimi. In: Turkiye'de Cevrenin ve Cevre Korumanin Tarihi Sempozyumu. Turkiye Ekonomik ve Toplumsal Tarih Vakfi, Istanbul, pp 15-20

[illegible]

Environmental Impact Assessment in Turkish Dam Planning

Waltina Scheumann, Vera Baumann, Anna Lena Mueller, Dennis Mutschler, Sylvia Steiner, and Thomas Walenta

1 Introduction

By supplying water and generating hydroelectricity, dams play a prominent role in Turkey's economic and social development. Hydroelectric energy generation, for instance, enjoys high priority in the domestic energy mix, and it factors as one of the core elements in Turkey's climate mitigation strategy because it compares favourably with fossil energies in terms of carbon emissions. As the General Directorate of State Hydraulic Works (DSI) reasons: "(...) hydroelectric power is environment-friendly, clean, renewable, able to meet peak demands, highly efficient (over 90 percent), involves no fuel cost, is a balancer of energy prices, has a long life-span (200 years), its cost recovery is short-run (5-10 years) its operational costs are low (approximately 0.2 cent/kWh), and it is an indigenous source of energy which is (...) natural."[1]

However, the praise for (hydropower) dams is not shared by everybody despite their potential for development (Caspary 2007). While all political parties, Turkish society, and even non-governmental organisations are more or less supportive of the government's endeavours to close the demand-supply gap, electricity as well as water, by means of dams. Criticism has arisen because of their harmful effects on the environment and, in particular, on the populace affected. Turkey's dam politics have harboured the headlines of international and national[2] media, lately. The Ilisu Dam for example, from which international consortia withdrew financial services because local opposition supported by international non-governmental

[1] Http://www.dsi.gov.tr/english/service/enerjie.htm. Accessed 16 May 2008.

[2] E.g. Hürriyet, Radikal, Taraf, and Zaman.

W. Scheumann (✉) • V. Baumann • A.L. Mueller • D. Mutschler • S. Steiner • T. Walenta
Deutsches Institut für Entwicklungspolitik, Bonn, Germany
e-mail: waltina.scheumann@die-gdi.de

A. Kibaroglu et al. (eds.), *Turkey's Water Policy*, 139
DOI 10.1007/978-3-642-19636-2_8, © Springer-Verlag Berlin Heidelberg 2011

organizations (NGO) has pointed to harmful resettlement and environmental practices amongst others.

In March 2009, at the Fifth World Water Forum in Istanbul, the former General Director of DSI, Dogan Altinbilek, took a stand in the ongoing debate on large dams: "The answer on the worldwide discussion on dams is not '*Yes*' or '*No*' but wise planning and implementation." Given that large dams interrupt river flows, affect wetlands and wildlife habitats and cause thousands of people to leave their land and houses and settle elsewhere, this statement might also be subscribed to by dam opponents if environmental and social concerns are given thorough consideration in the planning, constructing and operating stages of dam projects.

This article investigates to what extent environmental aspects have been integrated in a dam project's planning process.[3] It looks into changes over the last decades related to the environmental clearance process and in particular to the use of Environmental Impact Assessment (EIA) in decision-making of individual dam projects located in different regions of Turkey. It inquires whether EIAs were issued or not; whether the public could participate throughout the decision-making process of the EIA; whether EIA was conducive towards protecting nature conservation areas; finally whether the EIAs approved determined conditional mitigation means to be implemented by project developers and on how compliance is monitored.

Information has been gathered by reviewing documents and studies and by conducting interviews with experts since the EIA studies, EIA approval documents and monitoring reports are not publicly accessible, an exception being the EIAs for the Yusufeli and Ilisu dams which are posted on websites.[4]

2 Reconciling water resources development and environmental protection

Turkey is in a stage of rapid and dynamic change and confronted with the challenge of balancing environmental conservation and unprecedented development. As an Organization for Economic Cooperation and Development (OECD) report of 2008 put it (OECD 2008, 14):

> "Turkey confronts the challenge of ensuring that economic growth is associated with environmental and social progress (...) Looking to the future, to face its environmental challenges effectively, it will be necessary ... to (i) strengthen environmental policies and

[3]This article is based on a study carried out between February and May 2009 (Scheumann et al., forthcoming in 2011). Expropriation and resettlement were part of the study, but is not covered in this article.

[4]For Ilisu Dam: http://www.dsi.gov.tr/ilisu/ilisu_yyp_eke.pdf; http://www.dsi.gov.tr/ilisu/coe_reposts.htm; for Yusufeli Dam: http://www.ilisu-wasserkraftwerk.com; http://www.dsi.gov.tr/english/yusufeli_report.htm. Accessed 8 April 2010.

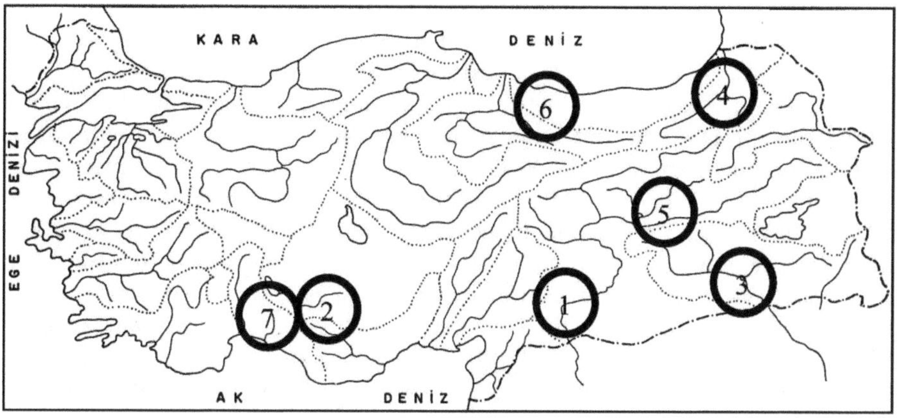

(1) Birecik Dam on Euphrates River
(2) Ermenek Dam on Goksu River
(3) Ilisu Dam on Tigris River
(4) Yusufeli Dam on Coruh River
(5) Konaktepe Dam in Munzur Valley
(6) Series of dams in Firtina Valley
(7) Beskonak Dam on Koprücay River

Fig. 1 Map of Turkey's river basins with location of dams mentioned in the text (own compilation based on DSI. http://www.dsi.gov.tr. Accessed 19 January 2009)

their implementation . . .; (ii) further to integrate environmental concerns into economic and sectoral decisions (. . .)."

Due to the country's high population and urban growth rates, many of its regions are already faced with seasonal or even chronic water shortages. Further, Turkey has an annual increase in energy demand ranging between 6 and 8 percent, with per capita energy consumption nearly 20 percent below the OECD country average. As described by a government official: "In order to meet the increasing demand, Turkey must produce continuous, high quality, reliable energy, and regions are already faced with seasonal or even chronic water economical electricity and make necessary investments by taking into consideration also environmental effects" (Mine 2001, 1).

With the liberalisation and deregulation of the national energy sector, including hydroelectricity generation, which started in the early 1980s and gained momentum with the Electricity Market Law of 2001 and the Renewable Energy Law of 2005, the Turkish government hoped to accelerate the development of the underexploited potential by inviting private investors and private financial service institutions. The classic investment model where planning, financing, construction and operation is carried out by the public sector only, was replaced by private investment models such *Turnkey*, *Build-Operate-Transfer*, and *Licensing* (cf. Baskan in this volume) for hydropower dams and the hydropower component of multipurpose dams. However, the public sector is still supposed to invest in dams for water supply and irrigation.

In Turkey, dams are considered to be important in order to close the demand-supply gap related to electricity as well as to water supply. As of March 2009, 260 large and 413 small dams are in operation, an additional 146 (63 large and

83 small) dams are either under construction or in programme (DSI 2009, 22). The figures are even more impressive if hydroelectric power plants[5] alone are considered: As of 2009, 172 are in operation,[6] 148 under construction, and 1,418 are in planning stage (DSI 2009, 31). According to estimates of the General Directorate of DSI, the operating dams contribute 39.1 percent to the gross national product (WCD 2001). While the theoretical gross hydroelectric potential is 433,000 GWh/year, only 216,000 GWh/year is technically feasible and 127,381 GWh/year is feasible in economic terms (DSI 2009). As of 2009, hydroelectric power generation was 47,871 GWh/year, which indicates that only about one third of the exploitable potential has been developed - a low figure compared with other OECD countries such as the USA with 86 percent.[7] Seen from the perspective of the Turkish government and bureaucracy, the highly underexploited potential and its expected benefits justify its large dam-building programme.

3 Turkey's alignment with the EU EIA Directive

Enactment of the Turkish Environment Law in 1983 and subsequent regulations (e.g. the National Park Law of 1983, the Water Pollution Control Directive of 1988, and the By-Law on Environmental Impact Assessment of 1993) indicate a growing awareness to protect environment and nature. Due to its horizontal cross-sectoral nature, the institutionalization of EIA is regarded as a milestone to serve environmental objectives by integrating environmental criteria into the planning and decision-making process of infrastructure projects, including dams. It is an essential planning instrument which requires the systematic assessment of environmental, social and cultural impacts deriving from infrastructure projects, and the identification of mitigation measures to be implemented by project developers.

However, contrary to common belief, EIA is not an almighty instrument for protecting environment and nature. It is bound by national legislation which determines the environmental goods to be protected. In case of weak legal protection, EIA can not compensate for legal shortcomings. For instance, protection of wetlands, wildlife, flora and fauna and so on are subject to respective regulations to which an EIA has to refer to (see Box 1).

It is, however, the major instrument for "...identifying, predicting, evaluating and mitigating the biophysical, social and other relevant effects of

[5]Including dams with reservoirs and run-off the river type projects.

[6]According to DSI, 24 dams have a generating capacity equal/larger than 100 MW www.dsi.gov.tr/. Accessed 23 March 2010.

[7]According to OECD estimates, only Canada and Turkey have considerable not yet exploited potential (IEA 2008, 19).

Box 1 Environmental clearance according to EIA By-Law (2008)

(1) **EIA Commissions** assess EIA final reports, including public comments, and the mitigation measures proposed.

This assessment is based on national law (laws and regulations referring to National Parks, Wild Life Conservation and Wild Animals Settlement Areas, Cultural and National Heritage Conservation, Special Environment Protection Areas, Wetlands Conservation) and international conventions (e.g. definitions and guidelines for major breeding areas of sea turtles; endangered marine species; wetlands listed under Ramsar Convention).

(2) The EIA Commissions inform and give recommendations to the **Ministry of Environment and Forestry** who in turn finally decides on whether a projects gets environmental clearance or not.

development proposals prior to major decisions being taken and commitment made." (IAIA 2010) It informs decision-makers in order to facilitate balanced decisions in the light not only of an economic but an environmental perspective and its related costs. As it stands now, it is the only institutionalized instrument in Turkey that allows for direct participation of the public in dam decision-making.

Since its adoption in 1993, the Turkish EIA By-Law[8] has been continuously revised by the respective ministry in order to align it with the European Union (EU) EIA Directive of 1985 (85/337/EEC) and its amendments. The four revisions made to date (1997, 2002, 2003, 2008) are owed to the progressive harmonisation of Turkish environmental legislation with the environmental acquis (characteristics of Turkish EIA legislation are briefly portrayed in Box 2).[9]

Box 2 Principal stages of EIA (By-Law 2008)

The Ministry of Environment and Forestry, through its General Directorate of EIA and Planning is the competent authority for EIA and coordinates related matters with other government agencies (e.g. Ministries of Health, Agriculture, Culture and Tourism as well as provincial governors). GD EIA establishes a commission after having received and approved the application file from the project developer. The commission consists of representatives of

(continued)

[8]The By-Law on Environmental Impact Assessment (EIA) was prepared on the basis of Article 10 of Environment Law No. 2872 dated on 09 August 1983.

[9]The environmental acquis communautaire describes the sum of all environment-related EU regulations that must be taken over by EU member states.

relevant institutions and organizations, officials of the Ministry, the project owner and may invite experts, professional associations, trade unions, NGOs etc. as members to its meetings. Its role is to develop the terms of reference and to assess whether they have been met.

Screening (Pre-EIA): While EIA is mandatory for projects listed in Annex I, Annex II projects are subject to screening whether "EIA is necessary" or "EIA is not necessary." Annex I projects are dealt with by the General Directorate of the Ministry. Screening of Annex II projects can be done by the Ministry's Provincial Directorates but are more often than not handled by the General Directorate due to provincial capacity constraints.

Scoping: Should an EIA be necessary the scoping stage identifies which matters are presumably significant in terms of their environmental, social and cultural impacts. Accordingly, terms of reference define the scope of the assessment to which both actors, the project developer and the competent authority, have to agree. Public participation takes place in this phase.

Reviewing: When the developer has prepared the report, it is checked by a commission whether it complies with the terms of reference, and if so, the report is assessed by the commission including the mitigation measures attached and forwarded to the Ministry. Prior to this it is opened to receive public comments.

Decision-making: Considering the commission's report and the comments and recommendations made by the public, the project gets environmental clearance ("EIA is positive") or is rejected ("EIA is negative") by the Ministry. Environmental clearance is tied to conditions specified by the competent authority.

Monitoring: The implementation of obligations attached to a positive EIA decision is to be monitored by the competent authority, and the law requires reports to be submitted by the project developer/operator.

For details, see the respective sections of this chapter and Innanen 2004, Coskun 2005, OECD 2008 and the By-Law EIA 2008.

The process gained momentum in 1999, when Turkey became an accession candidate, and even more in 2001, when the EU Council concluded the Accession Partnership with Turkey. Turkey's National Environment Action Plan of 1998 and its National Programme for the Adoption of the Acquis of 2001 are both committed to adopting EU environmental standards and regulations at a pace feasible for integration with the EU in the long term, and Turkey has declared this a national priority (Innanen 2004, 143; Izci 2005).

One of the short-term priorities of the Environment Chapter of the Progress Report Turkey (2007) to which accession negotiations refer to was to transpose the EU EIA Directive into Turkish law, and consequently the European Commission's annual Progress Reports have reported continuously on and indicated the progress made in this field. The Progress Report of 2007 in particular stated that

> "Turkey needs to ensure that relevant infrastructure investments in the water sector (such as dams) are carried out in line with the requirements of the water acquis[10] (...)" (European Commission 2007, 24).

According to the Progress Report of 2009, the EU EIA Directive "has been transposed to a large degree. However, procedures for public and transboundary consultations have not been fully aligned. (...) Turkey is not party to the Espoo[11] and Aarhus[12] Conventions." (European Commission 2009, 81) The annual Progress Reports further note that mainstreaming of environmental protection into other policy areas and steps to ensure that new investments comply with the environment acquis are in an early stage. The European Commission mentioned that "no progress can be reported on nature protection" (European Commission 2009, 82). The Commission has repeatedly reminded the Turkish Government that there are concerns because of *weaknesses in the implementation and enforcement of EIA* (author's emphasis), and has further called for transposition of the EU Directive on Strategic Environment Assessment (SEA), an instrument to assess the additive and synergistic effects deriving from policies, plans and programmes. OECD Turkey Environmental Performance Review of 2008 mentions that "the regulatory framework for environmental impact assessment of projects has been strengthened and steps launched for the integration of strategic environmental assessment of policies" (OECD 2008, 22).

4 EIA requirements for dam projects: legal loopholes

In Turkey, dams have been subject to EIA requirements ever since the very first EIA By-Law of 1993. While the 1993 provisions for dams were still rather soft, meanwhile stricter criteria are applied. If one compares the Turkish EIA By-Law of December 2003 with the most recent revision made in July 2008, Annex I now makes EIA mandatory for dam projects with a reservoir volume of 10 million cubic meters (MCM) and more, and for run-off the river type projects of 25 MW and more (see Table 1). This is in line with the EU Directive, Annex I (15.) which covers: "Dams and other installations designed for the holding back or permanent storage of water, where a new or additional amount of water held back or stored exceeds 10 million cubic meters," whereas Annex-II (g) mentions: "Dams and other installations designed to hold water or store it on a long-term basis (projects not included in Annex I)."[13]

---

[10] The water acquis comprises foremost the EU Water Framework Directive, the Urban Wastewater Directive, and the Directive of Protecting Waters against Nitrate Pollution from Agricultural Sources.

[11] UNECE Convention on Environmental Impact Assessment in a Transboundary Context, 1991.

[12] UNECE Convention on Access to Information, Public Participation in Decision-making and Access to Justice in Environmental Matters, 1998.

[13] Http://ec.europa.eu/environment/eia/full-legal-text/9711.htm. Accessed 24 March 2010.

Table 1 EIA criteria for dams (2003 and 2008) (Enerji Dunyasi 2008, 46)

Type	Annex I (2008)	Annex II (2008)	Annex I (2003)	Annex II (2003)
Dam with reservoir	Reservoir volume equal / higher than 10 MCM	Reservoir volume equal / higher than 5 MCM	Reservoir volume equal / higher than 100 MCM or 15 km^2 reservoir surface area	Reservoir volume equal / higher than 10 MCM
Run-off river	25 MW and higher	0.5 MW and higher	50 MW and higher	10 MW and higher

Information available on the Ministry of Environment and Forestry's (MoEF 2009) website indicate that 75 dam projects[14] passed the EIA procedure between 3 December 1993, when the EIA By-Law came into force, and 18 March 2010.[15] Comparing this number with the number of dams constructed in about the same period of time between 1991 and 2008 (in total 92 dams[16]), it can be seen that EIAs have not been issued for seventeen dams being constructed. However, information has not been available on whether their size (reservoir volume, surface area and generating capacity) was below the threshold for EIA requirement or whether EIAs were not issued at all.

Provisional Article 3 of EIA By-Law allows dam projects to be exempted from the EIA requirement in a wide range of eventualities: for instance, if a project is approved prior to the enactment of the EIA By-Law of 7 February 1993; if diverse permits and licenses are granted, if expropriation decisions are made, if a project is included in annual investment programmes prior to February 1993.[17]

Exempting projects from the EIA requirement is justified in cases where planning and approval took place prior to the By-Law's enactment date and construction had started shortly after (e.g. construction of Birecik Dam started in 1994; see Table 2). However, it is not unusual that more often than not, dam projects take decades from reconnaissance studies to master plans for river basins, project feasibility studies and project design, approval, arranging for financing to the actual start of construction. Ten to twenty years may pass because financing is difficult to acquire, and has not been secured until today for large dams (e.g. Yusufeli and Ilisu). Currently, the respective project developer can easily refer to Provisional Article 3 in cases where projects fall under any of the mentioned eventualities, but

[14]projects got environmental clearance ("EIA is positive"), one was not approved ("EIA is negative").

[15]Http://www2.cedgm.gov.tr/dosya/cedsonuckarar/cedsonuc.htm. Accessed 22 March 2010.

[16]Http://www.dsi.gov.tr/. Accessed 21 November 2008.

[17]TC Republic of Turkey, Official Gazette No: 26939, 7 July 2008, Provisional Article 3 (1): "As per the projects whose application projects have been approved, for which required approval, permit, license, or expropriation decision has been taken, which have been included in the investment program, or whose local zoning plans have been approved before the By-law on Environmental Impact Assessment published in the Official Gazette No 21489 dated 7 February 1993 and those with documented proof of production initiation and / or operation stage inception before this date the provisions of this Bylaw shall not apply [. . .]." The EU EIA Directive too exempts projects from EIA requirements if they serve "national defense purposes" (Art 1 (4)), and if the Directive's objectives "including that of supplying information, are achieved through the legislative process" (Art 1 (5)).

where construction has yet not started, Article 3 leaves broad scope for exemptions which are no longer justifiable.

Exemptions subject to ministerial approval are contentious issues, and there are a number of cases where individuals and civil society organizations sued the respective ministries for applying the Provisional Article 3. Court verdicts[18] have stopped, for instance, the construction of the Konkatepe Dam in the Munzur Valley, which was declared a National Park in 1971. In July 2005, the court held that Provisional Article 3 did not apply to this project because its application design (one of the mentioned eventualities) had not been prepared before 7 February 1993, the cut-off date specified for exemption. In the absence of an EIA report, the court held that the project lacked legality, all the more as it is located inside the boundaries of a National Park. This decision was overruled by the General Assembly of Administrative Chambers of the Council of State ('Danistay' in Turkish),[19] who argued that the design of the dam and the hydropower plant had already been decided in 1984 – before the EIA Directive came into force. However, the General Assembly ruled that upon request of international financiers, an EIA in line with World Bank standards should be prepared.[20] The license for the Konaktepe Dam and hydropower plant was awarded to Konaktepe Electricity Generation A.S. on 28 January 2010 (Tigrek 2010). Local people are trying to carry the case to the European Human Rights Court. In a similar vein, the Trabzon Administrative Court in 1999 cancelled the ministerial permit as well as the EIA approval report for a hydropower project composed of five dams, six diversion weirs and ten hydropower plants to be built in the Firtina River Valley in the Eastern Black Sea Kackar Mountains, a protected National Park (Coskun 2005, 62). In 2001 the Council of State finally prohibited construction in the Firtina Valley.

Turkish authorities had not required an EIA for the Yusufeli Dam (see Table 2) on the Coruh River, a large dam in every respect, arguing that the EIA By-Law was not applicable because the reconnaissance study (1969), the master plan (1982), the feasibility study (1986) and final project design (1990) were dated prior to EIA By-Law. In the 1990s, DSI granted a construction license to a private consortium led by Spie Batignolles TP, involving private companies from France, the UK, Belgium and Spain. In 2008, the Yusufeli Dam became part of the investment plan. In February 2009 the Turkish government called for tenders for the final design; the bid was won by DOLSAR, a Turkish engineering and consultancy company. However, construction works have not begun because financing has not yet been secured. Due to the involvement of export credit

[18]Both the Environment Law and the EIA By-Law rule that the judicial path can be used if somebody is not satisfied with the decisions taken during the EIA process after having exhausted administrative appeal steps (OECD 2008, 167).

[19]Danistay is the highest administrative court whose ruling is final in the sense that there is no other institution to appeal.

[20]An EIA has not been undertaken according to the Istanbul Water Tribunal (2009).

agencies (ECA), an EIA was carried out later by DOLSAR on their request. The Yusufeli Dam is expected to have a considerable negative impact on the river's regime and the regional ecosystem: it will inundate 1,460 hectares (out of a total of 23,200) of the Coruh Valley Wildlife Protection Area (Encon 2006, ES23), and is expected to seriously affect a large number of endemic species. As a result of the EIA, DSI is preparing a study on how to protect these species, among them an endemic wild goat species and fish ladders are planned to secure the survival of two endangered fish species.

5 When dam projects affect protected areas and resources

As a rule, getting approval for dam projects *located in areas* protected by national law and protected by international conventions is difficult to obtain. The EIA Commissions established by the MoEF, part of which are representatives of concerned conservation authorities, have to consult national laws listed in EIA Annex-V, such as Natural Parks, Nature Conservation Areas, Wild Life Conservation and Wild Animals Settlement Areas, Cultural and Natural Heritage, Archaeological Protected Areas and Special Environment Protection Areas (see Box 1). In some cases dam projects are *located outside but affect* National Parks, strict environmental measures are required by law (Yucel 2009).

Out of the 75 EIAs issued between December 1993 and March 2010, the MoEF had not given environmental clearance for the Beskonak Dam on the Koprucay River,[21] because the dam was planned to be located in the Koprulu Canyon National Park. The EIA non-approval decision led to the cancellation of the project.

However, as has been already mentioned, dam projects may be exempted from EIA requirements although they are located in or affect National Parks. In these instances, individuals and environmental NGOs, civil society groups and chambers (e.g. from engineering profession) addressed the judicial system to defend environmental goods, and asked for issuing EIAs in order to prove a project's 'public benefit' or 'common welfare,' as was done for planned projects located in the Firtina Valley which were stopped by court verdicts.

Interview partners raised general concerns over the legal protection of wetlands which would not be sufficient, even for wetlands designated as of international importance and protected under the Ramsar Convention.[22] Gunes and Elvan (2010)

[21]Http://www2.cedgm.gov.tr/dosya/cedsonuckarar/cedsonuc.htm. Accessed 22 March 2010.

[22]After approving a number of international agreements including the ratification of the Ramsar Convention in 1994, Turkey increased the number of wetlands classified as "Wetlands of International Importance" according to Ramsar criteria progressively to 135, of which 12 are designated Ramsar Sites (Karadeniz et al. 2009, 1108). In 2002, the MoEF issued the Regulation for the Protection of Wetlands (No. 25818, revised on 17 May 2005), set up a National Wetlands Commission and Provincial Local Wetlands Commissions, and prepared the National Wetlands Strategy 2003 - 2008. [23] Management plans are being developed for fourteen Specially

perceive Turkey's wetland policy as being "ambiguous and vague" one reason being the yet to be changed mandate of DSI who might drain wetlands for e.g. agricultural purposes.[23] Karadeniz et al. (2009, 1114) similarly point to the not yet environmentally mainstreamed DSI's establishment law. However, while conflicting interests are almost always present in the course of EIA processes, the major concern is whether the terms of reference extend the scope of an EIA study, as a rule and principle, further downstream to include river deltas and coasts.

Dams with reservoirs have numerous impacts on a river's flow regime, on organisms, biodiversity and fish species because they interrupt the continuum of rivers (McCartney 2009). Ecologists and engineers have developed technical interventions to maintain a river's ecological transmissibility (an objective which is shared by the EU Water Framework Directive). One such means are fish passes and fish ladders which are yet not mandatory in Turkey (DSI 6th Regional Directorate 2009) and are only considered if technically feasible (MoEF 2009). This issue has surfaced for a series of dams under construction and planned on the Goksu River and its tributaries,[24] with the multipurpose Ermenek Dam as the major infrastructure (see Table 2). According to interviewees, the EIA approval report has demanded a minimal discharge of 10 percent of the total river flow for maintaining the ecology of the 8km river stretch between the dam site and the outlet of the pressure tunnel.[25] World Wide Fund for Nature Turkey (WWF-Turkey) on the other hand, points to the lack of specific studies on flora and fauna, and considers the determined minimal discharge to be insufficient to maintain the current ecological condition. (WWF-Turkey 2009) In spring 2010, discussion was ongoing as to whether to increase the minimum in-stream flow requirement to 20 percent. For the time being, 10 percent have to be released, with specifications to be made project-wise (BM Group 2009).[26]

Since development plans on the Goksu and Coruh rivers foresee a large number of dams (reservoir and run-off the river type), accumulated environmental impacts from all types of infrastructure need to be assessed, not only in the vicinity of dam/plant sites but downstream impacts on river deltas and river flow-dependent wetlands. About elevens dams alone are planned on the main stem of the Coruh River, not counting the projects on its tributaries (see Klaphake and Scheumann in this volume). Here, due to hydro-electricity generation maintaining an (annual average) in-stream flow might not be the issue. However, it is not clear whether

Protected Areas and Biological Diversity in the Mediterranean under the Barcelona Convention of which the Goksu Delta under the Authority for the Protection of Special Areas is one.

[23]Gunes and Elvan 2010, http://www.fao.org/DOCREP/ARTICLE/WFC/XII/0165-B2.HTM. Accessed 08 April 2010.

[24]Most of them are run-off the river type.

[25]Http://www.wirth-erkelenz.de/fileadmin/resources/pdf/Tunneling/Fullface_TBM/Ermenek_TUN12.pdf. Accessed 31 March 2010.

[26]Other interviewees were less certain about this obligation, and it could not be verified whether there is a minimum in-stream requirement.

Table 2 Technical information on the dam projects investigated (own compilation)

Dam project characteristics	EIA	Comments
Birecik Dam 1994-2001 Hydroelectricity: 672 MW Volume: 1.22 BCM Surface area: 56 km^2	Exempted from EIA requirement	Protection of bald ibis by NGO in cooperation with DSI after operation started. Minimum downstream flow of 500m^3/s is agreed upon in the 1987 Turkish-Syrian Protocol.
Ermenek Dam 2002-(start in 2013) Hydroelectricity: 302 MW Volume: 4.582 BCM Surface area: 58.74 km^2	EIA approved on 03 Dec. 1999	EIA approved, with conditional minimum in-stream flow for river stretch between dam site and outlet of pressure tunnel DSI management plan to protect coast line in cooperation with local Authority for Protection of Special Areas
Ilisu Dam Construction not yet started Hydroelectricity: 1,200 MW Volume: 10.4 BCM Surface area: 313 km^2	Exempted from EIA requirement but conducted on ECAs' request	EIA report revised in 2002, again updated in 2006 Environmental Monitoring and Management Plan
Yusufeli Dam Construction not yet started Hydroelectricity: 540 MW Volume: 2.13 BCM Surface area: 33 km^2	Exempted from EIA requirement but conducted on ECAs' request	Two options but not officially acknowledged and appraised (two dams* (government DSI) vs. three dams (local association)) Protective means for endemic wild goat and fish species

*The two-dam option includes Artvin and Yusufeli dams.

the terms of reference required for assessing national downstream impacts, imply that the operator of hydropower plants has to consider that downstream flows should not be to the detriment of a river's ecology and wetlands.

6 Status of EIA in dam project decision-making

Article 6(3) of EIA By-Law of 2008 defines the status of EIA in project decision-making and reads as follows: "No incentive, approval, permission, construction and usage license can be given, no investment can be initiated, nor any tender awarded for projects subject to this By-Law unless 'Environmental Impact Assessment is Positive' decision or 'No Environmental Impact Assessment is Required' decision is made." In this respect, however, fundamental criticisms are raised by both WWF-Turkey and the Industrial Development Bank of Turkey (TSKB) the latter providing financial services. They have pointed to the rather late stage of the project cycle at which the public is able to raise its voice as part of the EIA process (WWF-Turkey 2003; TSKB 2009). WWF-Turkey criticises in particular that the EIA process in general starts only after planning and project design is almost completed, thus leaving little room for modifications (WWF-Turkey 2009).

Until recently, in private investment models, EIAs were issued only after projects are approved and construction licenses are granted (see Box 3).

Box 3 EIA in private investment models

In a *Turnkey* project, different private entities – usually organised in a consortium – are responsible for the financing and construction of the dam, while DSI remains in charge of project design and operation. After construction, the project is turned over in a ready to use condition to the public sector for operation. The EIA is issued by the private consortium, which usually subcontracts an authorised consulting company.

Build-Operate-Transfer (BOT) is a form of project financing, where private investors establish a company which receives a concession from the public sector (DSI, EIE[27]) to design, finance, construct as well as to operate a hydropower plant. The government still retains ownership and acts as a regulator. The EIA procedure is issued as in the Turnkey model.

In the *Licensing Model* project planning and implementation differ: The decision of DSI to allow a private company to construct a hydropower plant is based on the feasibility study (which includes a mandatory chapter on potential environmental impacts[28]). A *Water Use-Right Agreement*[29] signed between DSI and the applying project developer, forms the basis on which the electricity generation license is then issued by the Energy Market Regulatory Authority. Only then does the company start to prepare the EIA report to be submitted to MoEF.

This procedure was changed in early 2010: now, a project developer only gets the license from the Energy Market Regulatory Authority when the related EIA was approved by the MoEF (BM Group 2010) (see Box 4). This procedural change has enhanced the overall status of EIA in the decision-making process, and at the same time has reduced risks for the respective project developer that the project may not actually take-off, or be considerably delayed. However, it has been reported that project design can significantly be changed while verifying the feasibility study and seeking optimization of cost-output relation.[30]

[27]Elektrik Isleri Etud Idaresi (Electrical Investigation Administration).

[28]DSI/EIE Projelerinde Istenecek (On)Fizibilite Raporunda Yer Alacak Ana Basliklar, Bolum -7. Cevresel Etkiler (Genel).

[29]According to the Turkish Constitution of 1982, water resources are vested in the State domain, and the government only can transfer / assign user rights. These transfers have inadequately been coined as "selling the rivers." However, a subsequent problem of water use-right transfers is whether the state maintains control over private activities – one being monitoring the implementation of environmental requirements (see Baskan in this volume).

[30]See Ferrari (2010, 81-86) for the Alara hydropower project.

Box 4 EIA in project decision-making as of 2010

Water Use-Rights Agreement

Signed between DSI and project applicant

Based on (pre-) feasibility study

Key licenses / clearances

Environmental Impact Assessment (EIA)

Forestry and agriculture clearance

Expropriation decision made

Final license for projects from Electricity Market Regulatory Authority for 49 years after key licenses (e.g. EIA) are granted. If EIA decision is not positive, Water Use-Rights Agreements nullify.[31]

7 The public's right to participate

The revisions to the Turkish EIA By-Law made since 1993 have incrementally improved the means for the public to participate in the EIA procedure, and although Turkey has not yet signed the Aarhus Convention, Turkey's National Programme for the Adoption of the Acquis envisions the improvement of access to environmental information in order to seek alignment with this convention (SPO 2001, 403ff.). As amended, Article 30 of Environment Law No. 2872 meets the criteria of Article 9 of the Aarhus Convention on the right to access environmental information; the Right to Information Act which came into force in April 2004, "seems to give more room for environmental NGOs (. . .)" (Izci 2005, 96).

As it stands now, the Turkish By-Law of 2008 provides for public participation in the scoping phase and for a public meeting (Art.10) and written submissions by the public (Art. 12) concerning the final EIA report. The term 'public', as used in the Turkish By-Law, allows in principle everybody to participate and is thus broader than the requirement of the EU EIA Directive which restricts the right of participation to the 'public concerned'. In Turkey, the EIA reports are supposed to be made available to the public, including non-technical summaries, by displaying them at the provincial governors' offices or the MoEF Provincial Directorates. During participation meetings the public may ask questions, and the EIA commission and the project developer are obliged to answer them. All comments made by the public must be recorded and taken into account during the subsequent stages of the EIA process (Turgut 2003, 167). Written submissions by the public, that is, comments, questions and recommendations, can be sent to the MoEF or its

[31] Yonetmelik Hidroelektrik Santrallarin Su Kullanim Anlasmasi, No. 25150, 26 June 2003, Madde 12, http://www.dsi.gov.tr/ska/yonetmelik_tamami.htm. Accessed 30 March 2010.

Provincial Directorates, and the EIA commissions and the MoEF need to take into account the reservations and recommendations made and to inform the public of its overall decision.

According to the EIA Department of MoEF the public frequently writes comments and submits questions. However, the EIA Dept. of MoEF qualifies "Public participation in the final EIA (as) a checking procedure, no important issue comes up anymore" (MoEF EIA Department 2009).

The EIA process for the Ermenek Dam shows how the public made use of the EIA instrument. The mayor of the town of Ermenek was invited by the MoEF to the EIA commission's meeting, where he put four items on the agenda: (i) treatment of sewage water that will flow into the reservoir; (ii) effects on farmland, 30 percent of which will be inundated; (iii) a historical 700-year old bridge that ought to be relocated; and (iv) permission for farmers to use water from the reservoir for irrigation. The local public participation meeting, which was announced in a local newspaper as well as by speakers in the town, was attended by the surprisingly high number of 350 people, none of whom expressed reservations towards the dam, but supported it, despite the mayor's requests. As a result, a wastewater treatment plant was built (financed by DSI), and expropriation rates for farmland were eight times higher than required By-Law (paid by the private company) (Mayor of Ermenek 2009).

As a general observation made by an EIA expert from the Cukurova University of Adana, Professor Muzaffer Yucel, public participation is rather limited and only few people would attend meetings if projects affect flora and fauna only. Therefore, EIA experts working in academia and consultancy services have been asking the respective authorities to actively support participation, and emphasized the need to create more transparency in all phases of EIA (Yucel 2009). A search on the MoEF's website shows that public disclosure of information is rather poor, and environmental clearances are not in the public domain. This refers to

- final EIA reports,
- the EIA commissions' recommendations,
- the Ministry's approval document and whether and how the public's concerns have been considered or not (as requested in the EIA By-Law's Article 14 (2)),
- the conditional requirements (mitigation measures) which ought to be implemented by project developers,
- reports on monitoring compliance to be undertaken by the EIA Department of MoEF, and reporting by project developer.

8 Civil society organisations mobilize for EIAs

Environmental awareness in the Turkish society is rather low, and participation in the EIA process gains high interest only if it concerns social impacts such as expropriation and resettlement issues. However, there is a growing environmental movement in Turkey, starting with the liberalization of the economy and followed by the democratization of political and social life. Starting from 1980s and 1990s onwards, issue-based environmental policies began to emerge, but the growing awareness for environmental (and social) impacts from growth-based economic activities have rarely questioned the development paradigm, but rather the agent of development (market or state) (Aydin 2005, 53-70; Adem 2005, 71-86).

A variety of civil society organizations such as country-wide operating environmental NGOs and locally-based, ad-hoc established associations can be found. Doga Dernegi (i.e. Nature Association) and WWF-Turkey are two prominent environmental NGOs participating in the national dam discourse, and are as such, using the EIA process as a platform.

Being exempted from EIA requirements, Ilisu Dam and Yusufeli Dam are two recent examples where NGOs together with local groups organized protests and allied with mayors from affected communities. Attorneys, residents and the *Save Hasankeyf Initiative* – comprising 70 organizations – filed lawsuits against the relocation of the historical city of Hasankeyf. The lawsuits failed but were instrumental in gaining publicity. Local opponents allied with the international NGO *ECA Watch* and with NGOs from those countries, whose governments had agreed to export credit guarantees, in order not to enter into financial arrangements unless conditions were met. The mayors of the towns of Yusufeli and Kilickaya, together with the Local Culture Association, filed a lawsuit against the respective Ministry to stop the realization of a DSI-favoured project option by conducting an EIA with the aim of investigating its social and environmental impacts (Mayor of Yusufeli 2009). The Local Culture Association proposed instead that three dams should be built, which would reduce the number of people to be resettled, and would prevent the town of Yusufeli from being inundated. The plaintiffs were successful in the first instance, when it was unanimously ruled that the construction of the Yusufeli Dam was unlawful without an EIA and that the exemption rule of the 1993 EIA Directive did not apply. Without EIA, the realization of the two-dam option based only on the feasibility study of 1986 would not fulfil the common welfare criteria. However, when it came to an appeal, the plaintiffs lost.

These groups' aim had not been to prohibit dam construction in general but had rather been at achieving sustainable dam development by applying international principles and standards, such as the World Bank safeguard policies, international conventions etc.

9 International organizations call for higher-quality EIAs and strict enforcement of required protective measures

In Turkey, important players in the dam business are no longer multilateral development banks but private companies and financial institutions both national and international that refer to international standards for planning, implementing and operating dams.[32] The liberalization of the energy market has further caused the state (namely DSI and EIE) to refrain from investments in exploiting Turkey's hydropower potential and assigned this to the national private sector which allies with international partners (see Baskan in this volume).

Where international actors are involved, they have expressed their dissatisfaction with the quality of EIA studies.[33] Export Credit Agencies have called for the application of higher environmental standards, improved baseline studies and strict implementation of mitigation means.

In this respect, the Ilisu Dam is informative because EIA studies were not correctly pre-assessing the environmental impacts due to e.g. poor baseline data based on which suitable mitigation measures are to be recommended.

Studies for the Ilisu Dam on the Tigris River started as early as 1954. A preliminary report was released in 1971; the feasibility study and final project design were completed in 1982, when the project was incorporated into the Master Plan of the South-eastern Anatolia Project. Only in 1997 did the dam become part of the investment plan, and the Ilisu Engineering Group was assigned the task of performing an EIA study, which was completed in 2001. The Turkish government twice subcontracted dam and hydropower plant construction to international consortia whose risks were backed by guarantees from export credit agencies. Both consortia – under close scrutiny of NGOs – withdrew because a number of conditions had not been met by the Turkish government. These conditions covered, among others, the construction of upstream water treatment plants capable of ensuring that the water quality in the reservoir is maintained; maintenance of adequate downstream flows at all times and development of a detailed plan to preserve as much of the archaeological heritage of Hasankeyf as possible.

The first EIA for Ilisu was revised in 2002 at the demand of the British Export Credits Guarantee Department, and the second consortium (Austria, Germany and Switzerland) asked for an update of the revised study, which was finally made available in English and Turkish in 2006. Reports of an independent Committee of Experts released in December 2007 and May and November 2008 concluded that the environmental terms of reference had not fully been met by the Turkish implementing agencies. Controversies between the ECAs and the Turkish Government could not be settled, and the ECAs of Austria, Germany and Switzerland finally cancelled their export credit guarantees on 7 July 2009[34] (see Kibaroglu and Scheumann in this volume).

---

[32]See for instance, the OECD's Common Approaches, the International Hydropower Association's hydropower sustainability assessment protocol, and the Equator Principles for the financial industry to manage social and environmental issues in project financing.

[33]If EIAs are done on the request of e.g. ECAs, then they serve as advisory for mitigation means and are not part of a project's environmental clearance process. Whether mitigation means are considered binding by the respective project developer and the authority is rather questionable, and depends on the arrangements agreed upon between ECAs, Turkish authorities and the project developer.

[34]Http://www.oekb.at/de/unternehmen/Presse/pressetexte/Seiten/070709-ausstieg-ilisu.aspx. Accessed 08 April 2010.

10 Monitoring compliance

The Turkish EIA of 2008 includes a general requirement that mitigation measures be identified, but it does yet not require environmental management plans (Coskun 2005, 57). However, a matter of major concern is whether compliance of the project developer is monitored by the MoEF's Department on a regular basis in order to secure that conditions attached to positive EIAs are followed. According to By-Law, Article 18(3), project owners are "obliged to communicate to the Ministry or the Governorate the monitoring reports for initiation, construction, operation and post-operation periods of the investment." However, extension time can be granted, but if conditions are not fulfilled, the project can be stopped (Article 19 (1b)).

The EIA procedure including monitoring is highly centralized in that the General Directorate of MoEF is assigned major responsibility (Article 18(1)). Although the MoEF has recently (in 2009) been strengthening its capacity through an increase of personnel by 200, capacity and financial resources are still perceived as being inadequate (Interview MoEF EIA Dept March 2009; OECD 2008, 100). In particular in the provincial directorates of the Ministry which are close to dam sites. For example, the Provincial Directorate of Balikesir has only three staff members in its EIA department, responsible for about 80 projects (not only dams) where monitoring is required (MoEF Balikesir 2009); the Provincial Directorate of Sanliurfa is equipped even worse (MoEF Sanliurfa 2009). The merging of the two Ministries of Environment and Forestry in 2003 has yet to have a positive impact in this respect, since forestry staff does not have the skills for dealing with overall environmental monitoring.

Due to public closure of documents referring to conditions attached to environmental clearance and, if issued, monitoring reports, the public has no means of controlling either.

11 Findings and recommendations

Turkey's governments have made great efforts in aligning national EIA regulation with EU standards. The respective Ministry has set stricter criteria for dam projects being subject to EIA requirements, and the status of EIA in the project licensing process in particular for investments in the hydropower sector was enhanced (environmental clearance is conditional for getting the license). It has further facilitated the rights of access to information based on the Right to Information Act with positive effects for NGOs to participate in the EIA process. Overall NGOs, civil society groups, political party members are meanwhile engaging in the national discourse about dams, and the media has taken up this issue too. Furthermore, an independent judiciary has been playing its role in a number of disputed cases.

However, Turkey has yet to fully integrate environmental concerns into its dam-related policies. In this respect, there is scope for improvement.

Legal loopholes

In order to prevent an unduly wide, and arbitrary, interpretation of Provisional Article 3, its conditions, which allow for exemption of dam projects from EIA requirement, should be revised. National requirements for issuing an EIA should not lag behind what is international practice and has been requested by international financiers and national banks and companies. Revising Provisional Article 3 is all the more important in that the EIA is the only institutionalized instrument that formally provides for the participation of the public, which otherwise has no means of influencing decision-making. Those affected by involuntary resettlement must have a say through the EIA process about a project's social impacts and should not be deprived of their rights.

If the issuing of EIA studies rely on requests from foreign financial service providers, and is not based on national legislation, EIAs do not pass official clearance, conditional mitigation means may not be binding, and the respective financier may even apply low standards.

Stricter protection of environmental goods

It would appear that environmental assessment studies should specifically target and describe potential impacts that undermine the objectives defined in legislation protecting specific areas (e.g. Natural Parks, Natural Monuments, Nature Conservation Areas, Wild Life Conservation and Wild Animals Settlement Areas, Cultural and Natural Heritage, Archaeological Protected Areas and Special Environment Protection Areas).

Due to the unidirectional effect, upstream interventions have on a river's flow regime, scoping ought to be extended in order to assess downstream impacts even beyond national borders.

Because of the accumulated effects resulting from the large number of projects on one river system, project EIAs show serious limitations. The hydropower development programme along the Coruh River is illustrative in this respect where about elevens dams will be constructed on the main river stem (not counting the projects on its tributaries), new transmission lines will be built, state highways will be relocated to replace the roads being submerged. This calls for alignment with and implementation of the EU Directive on Strategic Environmental Assessment.

Public participation and disclosure of information

While legislation provides for public participation, public disclosure of information is weak. This does in particular relate to the final decisions made by the Ministry, to the conditions attached to the clearance document and to both monitoring by public authorities and reporting by project developers. It seems that the provincial administrative apparatuses should be strengthened in particular - in terms of capacity and financial resources – to improve the EIA process as a whole.

Weaknesses in implementing environmental mitigation means

The By-Law is clear in asking the project developer, who only obtained environmental clearance, once he had committed to the mitigation measures, to report on their implementation, and the law makes it possible to sanction non-compliance in the long-run. However, the actual challenge of monitoring and enforcement remains, although the MoEF was strengthened in terms of personnel. The ministry's capacity is still too low to adequately perform its tasks. This is especially true for the Provincial Directorates. Strengthening their capacity and possibly increasing their range of activities might have positive effects on monitoring implementation of mitigation means by project developers.

References

Adem C (2005) Non-state actors and environmentalism. In: Adaman F, Arsel M (eds) Environmentalism in Turkey Between Democracy and Development? Ashgate Publ. Company, Aldershot, pp 71-86

Aydin Z (2005) The state civil society, and environmentalism In: Adaman F, Arsel M (eds) Environmentalism in Turkey Between Democracy and Development? Ashgate Publ. Company, Company, Aldershot, pp 53-86

BM Group (24 February, 6 March 2009, 18 March 2010) Personal communication, Ankara

Caspary G (2009) Assessing, mitigating and monitoring environmental risks of large infrastructure projects in foreign finance decisions: the case of OECD country public financing for large dams in developing countries. Impact Assessment and project Appraisal, 27(1), March 2009:19-32

Coskun A (2005) An evaluation of the environmental impact assessment system in Turkey. International Journal Environment and Sustainable Development. 4(1):47–66

Doga Dernegi (23 February 2009) Personal communication, Ankara

DSI (2009) Water and DSI. 5th World Water Forum Istanbul, Bridging Divides for Water, Ankara

DSI 6th Regional Directorate (2009) Personal communication, Adana

Encon (2006) Yusufeli Dam and HEPP: Environmental Impact Assessment. Ankara

Enerji Dunyasi (2008) Eylul. Enerji Dunyasi. www.enerji-dunyasi.com. Accessed 13 March 2010

European Commission (Commission of the European Communities) (2007) Turkey 2007 Progress Report. Brussels, 6 November, SEC(2007) 1436

European Commission (Commission of the European Communities) (2009) Turkey 2009 Progress Report. Brussels, 14 October, SEC(2009) 1334/3

Ferrari H (2010) Alara-Wasserkraftwerkprojekt Tuerkei – Ingenieurleistungen im Spannungsfeld des Machbaren. In: Schuettrumpf H (ed) Mitteilungen 158, 15. Deutsches Talsperrensymposium, Talsperren im Wandel, 14-16 April 2010, pp 81-86

Gunes Y, Elvan OD (2010) The impact of statutory controversy on wetland degradation in Turkey. FAO. http://www.fao.org/DOCREP/ARTICLE/WFC/XII/0165-B2.HTM. Accessed 31 March 2010

Innanen S (2004) Environmental Impact Assessment in Turkey: capacity building for the European Union accession. Impact Assessment and Project Appraisal, 22(2): 141-151

IEA (International Energy Agency) (2008) Deploying renewables. Principles for effective policies, in support of the G8 Plan of Action. OECD/IEA, Paris

IAIA (2010) International Association for Impact Assessment. http://www.iaia.org/. Accessed 7 June 2010

Istanbul Water Tribunal (2009) Case: Yusufeli Dam and Hydroelectric Power Plant Project in the Coruh Valley. Handout, Istanbul Water Tribunal

Izci R (2005) The impact of the European Union on environmental policy. In: Adaman F, Arsel M (eds) Environmentalism in Turkey Between Democracy and Development? Ashgate Publ. Company, Aldershot, pp 87-100

Karadeniz N, Tiril A, Baylan E (2009) Wetland management in Turkey: problems, achievements and perspectives. African Journal of Agricultural Research, vol 4(11):1106-1119

Mayor of Ermenek (6 March 2009) Personal communication, Ermenek

Mayor of Yusufeli (19 March 2009) Personal communication, Yusufeli

MoEF Balikesir (25 March 2009) Personal communication, Balikesir

MoEF Sanliurfa (12 March 2009) Personal communication, Sanliurfa

McCartney M (2009) Living with dams: managing the environmental impacts. Water Policy 11, Supplement 1 (200):121-139

Mine O (2001) Challenge of Turkey in dam construction and hydroelectric energy potential development in the 21st century. Paper prepared for the World Energy Council, 18th Congress, Buenos Aires, October.

MoEF EIA Department (March 2009) Personal communication, Ankara

MoEF (2009) Ministry of Environment and Forestry. http://cevreorman.gov.tr. Accessed 15 May 2009

OECD (Organisation for Economic Cooperation and Development) (2008): Environmental performance reviews: Turkey. OECD Publishing, Paris

Scheumann W, Baumann V, Mueller A, Mutschler D, Steiner S, Walenta T (2010) International standards for sustainable dam development: dynamics and conditions enhancing their implementation in Turkey. DIE Study. German Development Institute, Bonn

Screening Report Turkey (2007) Chapter 27 – Environment, 22 June 2007.

SPO (State Planning Organisation) (2001): National Programme for the Adoption of the Acquis. SPO, Ankara

T.C. Republic of Turkey, Resmi Gazette, No. 26939, 7 July 2008, By-Law on Environmental Impact Assessment (EIA)

T.C. Republic of Turkey, Yönetmelik Hidroelektrik Santrallarin Su Kullanim Anlasmasi, No. 25150, 26 June 2003, Madde 12. http://www.dsi.gov.tr/ska/yonetmelik_tamami.htm. Accessed 30 March 2010.

Tigrek S (August 2010) Personal Communication, Middle East Technical University, Ankara

TOBB (Turkiye Odalar ve Borsalar Birligi), http://www.tobb.org.tr/abm/taramaraporlari/Chapter%2027%20-%20Environment.pdf. Accessed 17 February 2010

Turgut N (2003) EIA with reference to the EU Directive. Environmental Policy and Law, 33 (3-4), 163-169

TSKB (23 March 2009) Personal communication, Istanbul

WCD (2001) Response to the Final Report of World Commission on Dams, ICOLD members – Turkey: General Directorate of State Hydraulic Works (DSI). World Commission on Dams. http://www.dams.org/report/reaction/icold_turkey.htm. Accessed 15 May 2009

WWF Turkey (World Wide Fund for Nature Turkey) (2003) Ermenek Dam & Hydropower Project: Evaluation of ecological issues. Research and evaluation report, Istanbul

WWF-Turkey (22 February 2009) Personal communication, Ankara

Yucel M (March 2009) Personal communication, Cukurova University, Adana

NGOs Promote Integrated River Basin Management in Turkey: A Case-Study of the Konya Closed Basin

Buket Bahar Divrak and Filiz Demirayak

1 Introduction

The global freshwater crisis has been on the agenda of the international community for the past 30 years: rapidly increasing water demands due to population growth and economic development have to be met with limited resources that are further compromised by over use and pollution. The water crisis has forced people to develop new approaches for water resources management that could accommodate the ensuing conflicts between economic and environmental concerns.

Numerous international meetings and conferences, aimed at developing recommendations and frameworks for improving the water situation, have taken place. In 1992, crucial steps were taken in the area of freshwater management with the International Conference on Water and Environment (Dublin) and the United Nations (UN) Conference on Environment and Development (Rio de Janeiro). Both events highlighted the need for an integrated approach to the management of water resources that links social and economic development with the protection of natural ecosystems. Also, the participation of all users at all levels: the public, planners and policy makers alike, was greatly emphasised (Global Water Partnership 2000). Moreover, the principle that water should be considered a social and economic commodity was established.

Since then, the boundaries of hydrological basins have been found to be the most appropriate spatial unit for managing water resources, rather than political and administrative boundaries. The water cycle within a river basin is the systematic and continuous movement of water between land, sea and atmosphere in which several ecosystem functions and services are provided. Hence, river basins are not

B.B. Divrak (✉)
WWF-Turkey
e-mail: bdurmaz@wwf.org.tr

F. Demirayak
Natural Resource Management Expert and WWF-Turkey Trustee
e-mail: fdemirayak@gmail.com

A. Kibaroglu et al. (eds.), *Turkey's Water Policy*,
DOI 10.1007/978-3-642-19636-2_9, © Springer-Verlag Berlin Heidelberg 2011

only geographical areas but also consistent systems that fulfill many functions for society as well as for the environment (Mostert et al. 1999). Accordingly, Integrated River Basin Management (IRBM) has been developed as a new approach to resources management. The core of IRBM is the integration of key issues and variables that directly impact, or are dependent on the healthy functioning of the basin. IRBM, therefore, takes a much broader approach than traditional water management and includes significant parts of land-use planning, agricultural policy and erosion control, environmental management and so on. However, IRBM does not attempt to be inclusive of all issues and variables but recognizes variations among basins and acknowledges that there is no standard success formula for implementing IRBM. Each basin will have to develop its own mechanism and processes according to its own dynamics, structures, interrelations and so on.

Implementation of IRBM is a complex process which involves different stakeholders with differing priorities or interests for example: government, municipalities, water regulatory bodies, local communities, academic institutions, industries, farmers, non-governmental organizations (NGOs) etc. IRBM can be conceived as a process through which people can develop a vision, agree on shared values and activities, make informed decisions and act together to manage the natural resources of a river basin (WWF 2002). The challenge is to design and establish a collaborative framework to bring together major groups that should be involved, to agree on transparent procedures that facilitate mutual trust – a key ingredient for cooperation.

Since the 1990s, NGOs have been building a body of experience and knowledge in the practical implementation of sustainable water management initiatives. They have played a key role in demanding more participatory forms for river basin management and developing some good practices on the ground. The functions that NGOs typically take in IRBM are as follows:

- Building bridges between local and central governmental level
- Reporting field/real life/local needs to decision-makers
- Spreading information towards field/real life/local level
- Involving/connecting parties that are willing to cooperate
- Building coalitions with local partners
- Acting as facilitators during the whole IRBM process.

NGOs have skills and characteristics that allow them through especially qualified personnel to implement participatory water management practices on the ground. Many NGOs, in order to pursue their objectives, have established good relationships with the public and developed skills to mobilize public support. Secondly, NGOs – as compared to public administrations, academia or the private sector - are more flexible and are able to respond quickly to emerging needs. Furthermore, they show high committment to the whole planning and implementation process and a good sense to coordination, and are able to act as a catalyst by pulling parties together.

This chapter will firstly give an overview of the status of IRBM and environmental NGOs in Turkey. Against this background, the World Wide Fund for Nature

(WWF) project "Towards Wise Use of the Konya Closed Basin" will be presented in a case study. As the project presents the very first attempt to implement IRBM in Turkey, it can provide important lessons for future initiatives aiming to promote participatory and integrated water resources management.

2 Integrated River Basin Management - status in Turkey

IRBM is slowly gaining acceptance in Turkey's national water policy. Recently, the General Directorate of State Hydraulic Works (DSI), i.e. the main responsible institution for water resources development and river basin planning, declared IRBM as one of the new elements of their vision for their future work (DSI 2008). The government's intention to implement IRBM in Turkey's 25 river basins is the most important and strategic shift in the way the administration thinks about, uses and manages water resources to occur in the last 50 years.

IRBM is, however, a new concept in Turkey and existing political/institutional structures do not facilitate the necessary change in mentality; the implementation of IRBM will therefore be a major challenge. There are three main obstacles behind this: 1) unfinished water resources development, 2) administrative problems and 3) lack of participatory approaches.

Firstly, as a developing country, Turkey's water policy is based mainly on the view that water is a means by which to increase welfare. Government agencies are the key actors who develop water through infrastructure projects, thus following a "hydraulic mission" which aims to fully utilize all water resources for developmental purposes.

Starting from the 1950s, with the establishment of the General Directorate of DSI in 1953, water infrastructure schemes (mainly dam construction and irrigation canals) have been considered as a basis for social and economical development. In recent decades, infrastructure development has been further accelerated to meet the demands of a rapidly increasing population, for the expansion of irrigated lands and for the generation of energy – but has also been used as a political tool, in that political parties have made promises to local voters during election times, especially if they would promote water supply for irrigation. The main drivers in Turkey's water policy are:

- Achieve independency from external energy resources
- Provide food security and increase agricultural production
- Provide water for urban, industrial and rural water needs
- Reduce regional discrepancies in economic and social development (e.g. Southeastern Anatolia Project, GAP)

The existing focus in Turkish water policy on further development of water resources stands in contrast to IRBM, as the latter aims to establish a balance between development and utilization on the one hand, and conservation and protection of water resources on the other.

A second obstacle to IRBM implementation in Turkey is grounded in the institutional set-up: Turkish water resources are governed by 14 governmental institutions in terms of planning, management, development, monitoring and research for domestic, rural, industrial and other sector needs. Each institution has experts working on the ground. However, following different legislation and visions, institutions regularly work individually without coordination and consultation even though they operate in the same region. Various institutions collect data on the same resource, but often apply different standards and methods, so that there is no adequate information base for cooperation even if the institutions attempt to collaborate. In addition, each institution aims to maintain its authority and power in water resources management and to prevent others from intervening in their field. Thus, they often resist sharing information, and behave as data monopolies. Therefore, the exchange of information between different government authorities and also with NGOs is very limited if not absent.

Thirdly, the institutional structure does not provide well-defined mechanisms for the participation of non-governmental stakeholders, or the public in general, in the water sector. Since the water management process is coordinated by governmental authorities, the level of public participation is mostly defined by these authorities. Traditions and the legal set-up of government institutions do, however, not promote cooperation. This gives a pessimistic vision for participative water management as well as for coordinated work among authorities which would be necessary to increase efficiency and hinder the misuse of institutional resources of the country. Up to now, there are but a few examples in Turkey, where cooperative and participatory processes were followed. Hence, there is a lack of knowlegde and experience in participatory methods, ways of negotiation and bottom to top decision making processes among non-governmental stakeholders. Despite these obstacles, there are some efforts where – based on participatory processes – IRBM has been successful. The Turkish government has indicated that it wishes to work towards the integration of ministerial agendas and to promote public involvement in decision-making processes for water management.

3 Environmental non-governmental organizations in Turkey

Conservation activities throughout the last decade have led to the development of many national and regional environmental NGOs in Turkey. Focusing on various environmental problems, the main national NGOs aim to propose efficient solutions and to encourage public participation. Their strategies include creating public awareness and providing a basis as a pressure group in decision-making processes. In Turkey, there are 10 to 15 national NGOs (e.g. Turkish Foundation for Combating Soil Erosion, for Reforestation and Protection of Natural Habitats TEMA, Regional Environment Center REC Turkey, Greenpeace, The Nature Association, Birdlife, Environmental and Woodlands Protection Society, Turkey TURCEK, Water Foundation) dealing with environmental issues such as biodiversity conservation, natural resource management, erosion control, research and capacity building. Most of them are based in Ankara and Istanbul since the government authorities, private sector, and media are mostly located in metropolitan cities. In addition, NGOs often have field offices/branch offices in project locations throughout Anatolia.

As international relations and cooperation have gained momentum with the EU accession process for Turkey, most of the Turkish NGOs have developed partnerships with international organizations such as with WWF, Greenpeace, Birdlife etc., to share experiences and to increase fund-raising opportunities. In fact, these partnerships have provided a good basis for the evolution of Turkish environmental NGOs: from species-oriented conservation efforts towards broader natural resource management approaches. Most of the national NGOs started with the protection of some important plant or animal species, and have now turned to promoting rural development and sustainable use of natural resources, as well as campaigning and capacity building on the national level. Meanwhile, some of the NGOs have played important roles in terms of influencing environmental policies through campaigns. One of the most important examples of success were efforts against the planned selling of forest areas for urbanization, industrialization and tourism: The draft law (known as the "2B Law") prepared by the government was not approved by the Parliament because of the public pressure created by environmental NGOs (Arsel 2005).[1]

Although some mechanisms for public participation exist, such as local environment committees, councils for the environment and forestry, and public meetings in the context of environmental impact assessments (see Scheumann et al. in this volume), public participation is a relatively new, yet not institutionalized process in Turkey. It is important for citizens to be heard when projects are at the proposal/planning stage. Yet, most of the people in Turkey are not aware of the importantance of their participation in the desicion-making process. In fact, this may be a natural result of the very common belief in the Turkish society that all desicions and investments are dependent on political power. Another reason for limited participation is the absence of environmental reporting and the lack of institutionalized access to information for stakeholders. Since public participation requires significant outreach, governmental agencies at all levels should provide concrete, clear procedures for accepting, processing and responding adequately to information requests.

Some of the national and local environmental NGOs have developed expertise and experience in wetland management and are currently involved in preparing wetland management plans for internationally important wetlands in cooperation with the Turkish Ministry of Environment and Forestry (MoEF) following the procedures as laid down under the Ramsar Convention. However, as mentioned above, the most appropriate scale for water resources management is the river basin. Therefore, existing efforts for the conservation of wetlands have to be

---

[1]For an analysis of social movements in Turkey, see Arsel (2005) on social movements opposing gold mining in Bergama.

extended to the river basin scale. The Konya Closed Basin Project that was initiated by WWF-Turkey is the first IRBM project in Turkey.

Environmental NGOs representing civil society in general will need to address a range of issues in order to establish themselves as stimulating and constructive partners for environmental progress. There is also a need to establish networks between the relevant NGOs in Turkey to facilitate and initiate integrated approaches to environmental issues.

4 WWF-Turkey and the Konya closed basin IRBM pilot project

WWF's mission is to stop the degradation of the planet's natural environment and to build a future in which humans live in harmony with nature by conserving the world's biological diversity, ensuring that the use of renewable natural resources is sustainable, and promoting the reduction of pollution and wasteful consumption (WWF 2003).

Considering the many current problems and threats, WWF-Turkey has identified the following strategic action areas:

- Improvement of Turkey's national water policies to secure the wise use and conservation of freshwater ecosystems and resources
- Promoting IRBM as a standard approach to freshwater ecosystem management both at eco-regional and sub-basin level
- Promoting sustainable use of water in different sectors (mainly agriculture).

As an NGO, WWF-Turkey's role is to develop best practices in priority river basins and to promote IRBM as a standard approach in water resources management in Turkey. Current efforts are focusing on building the capacity of related stakeholders, facilitating dialogue among them, implementing pilot projects on the ground and also communicating the necessity of IRBM to the public. The following sections presents a case study of a pilot project that WWF-Turkey has been carrying out since 2003, namely "Towards Wise Use of the Konya Closed Basin".

4.1 Characteristics of the Konya Closed Basin

The Konya Closed Basin is one of the largest closed basins in the world.[2] It is located in the Central Anatolian Plateau and covers an area of 53,000 km^2, roughly 7 percent of the total Turkish surface area. The basin's water flows terminate in stable water bodies, marshes or semi-marshes.

[2]River basins that terminate at lakes and/or inland deltas, with no exit to the sea are referred to as closed basins. Examples of closed basins in the world are the Caspian Basin, the Okavango Basin and the Jordan Basin.

The Konya Closed Basin is of outstanding importance to nature conservation in Turkey and globally, particularly for its wetlands, the extensive areas of remaining salt steppe (the largest and most pristine in Turkey), and for the diversity of its fauna and flora. The Konya Closed Basin is one of the "Global 200" eco-regions identified by WWF International. There are 11 Important Bird Areas (IBA) within the basin, providing breeding ground for 8 out of the 13 globally threatened bird species breeding in Europe. Preliminary research also indicates that the basin holds at least 8 Important Plant Areas, covering many hundreds of thousands of hectares.[3] The upper catchments of the basin including the Taurus, the Anamas and the Sultan mountains are home to various threatened mammal species such as the Anatolian wild sheep, the brown bear, the jackal, the lynx and wolf. Several areas in the Konya Basin were declared protected areas by the Turkish Government in the 1990s: 10 areas attained SIT[4] status. In addition an area of 7414 km^2 was declared a Special Protected Area under the Authority for the Protection of Special Areas (APSA) (see Table 1 for the protected wetlands in the basin).

Nearly three million people live in the Konya basin, of which 45 percent in rural areas (declining) and 55 percent in urban areas (increasing). The main industries around Konya and Aksaray include dairy farms, copper and aluminum plants, leather processing units and food processing industries. In terms of agriculture (irrigated agriculture as well as dry land arable farming), sugar beet, wheat and corn are the most common crops. Animal husbandry (mainly extensive grazing of sheep and goats) is widely practiced throughout the basin. Reed harvesting and fishing are important sources of income mainly in Beyşehir Lake. Meanwhile, the salt industry around Tuz Lake produces 60 percent of Turkey's total salt production.

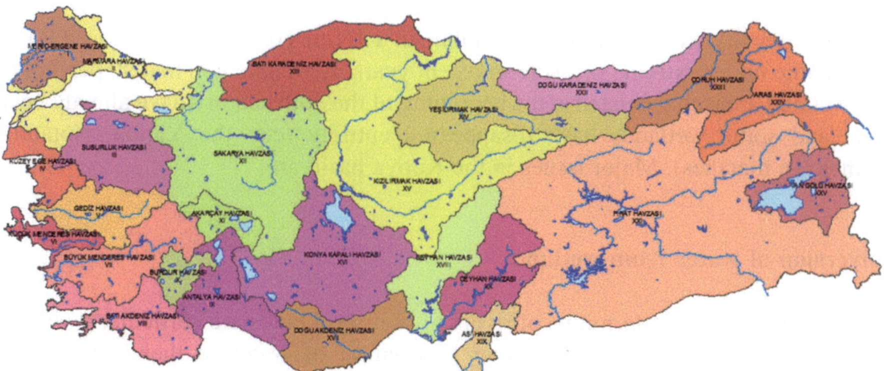

Fig. 1 River basins of Turkey (map prepared by WWF and DSI Regional Directorate for Konya)

[3]IBAs and Important Plant Areas are areas being globally important habitats for birds and plants respectively. However, IBA and Important Plant Areas, as well as Global 200 are no legal site designations but a tool for identifying sites of conservation priority.

[4]SIT is a protection status given by the Turkish Ministry of Culture and Tourism for areas of natural, cultural and historical importance.

Table 1 Protected Wetlands and Lakes in the Konya basin

Wetlands in Konya basin	Original size*	Characteristics	Protection status
Samsam Lake	830 ha	Brackish	SIT (1992)
Kozanli Lake	650 ha	Freshwater, marsh	SIT (1996)
Kulu Lake	860 ha	Brackish	SIT (1992)
Eregli Marshes	37,000 ha	Freshwater, marsh	SIT (1992), Nature Reserve (6,787 ha)
Esmekaya Marshes	11,250 ha	Freshwater and salt lakes, marsh	SIT (1992), Permanent Wildlife Reserve (4,500 ha), lost its SIT status in 2005
Beysehir Lake	73,000 ha	Freshwater lake	SIT (1988/91) Beysehir National Park (88,750 ha), Kizildag National Park (59,400 ha), Drinking Water Reservoir Protection Zone
Sugla Lake	16,500 ha	Freshwater lake	./.
Hotamis Marshes	16,500 ha	Freshwater, small salt lake	SIT (1992)
Bolluk Lake	1,100 ha	Salt lake, saltpans	SIT (1992)
Tersakan Lake	6,400 ha	Salt lake, saltpans	SIT (1992)
Tuz Lake	260,000 ha	Salt lake, steppe	SIT (1992), Special Protected Area (7,414 km^2)

* The figures indicate the original size of the wetlands. It does not take into account the changes due to irrigation projects and reclamation for agriculture.

4.2 Main problems and threats

The biodiversity values and traditional human activities of the Konya Closed Basin have come under extreme pressure in recent years. It is estimated that 90 percent of the wetland areas that existed at the beginning of the 20th century have already been lost, and some wetlands have disappeared entirely (e.g. the Yarma, Hotamis, Esmekaya marshes). Major issues in the Konya basin are:

Diversion of water from wetland

Inland wetlands at the terminus of rivers in semi-arid lands are particularly vulnerable to the impacts of diversions or abstraction of water in their catchments. Farmers hardly (if at all) pay for the water and there is therefore no incentive for the economical use of water; irrigation methods are often inefficient.

Rapidly decreasing groundwater level

Irrigation of drylands and draining of wetlands has had enormous impacts on hydrological functioning of the Konya basin. WWF-Turkey's groundwater research shows that the groundwater level has decreased by more than 14m over the past

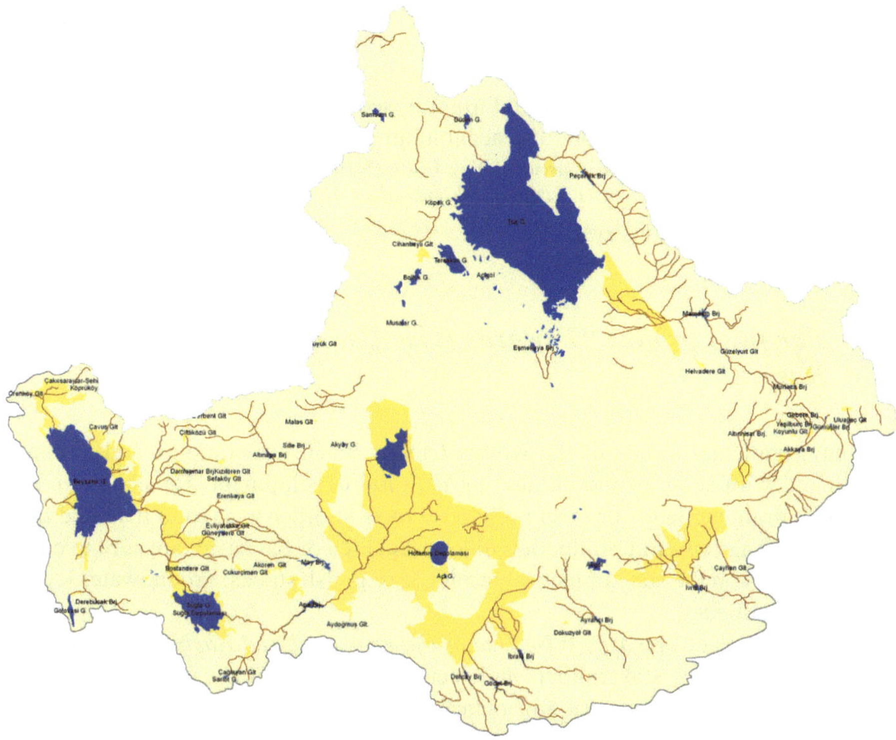

Fig. 2 Konya Closed Basin (map prepared by WWF and Turkish Ministry of the Environment and Forestry)

decades, but mainly during the last 10 years. According to DSI, 68,000 of a total of 94,000 wells were drilled and operate without permission.

Draining of wetlands for arable cultivation and grazing areas

Nearly half of the basin had already been converted into arable land by the 1970s, and today this proportion is even higher. These areas include many former wetland and steppe habitats. Yet, shallowness, salinity structure, thin structure of the soil and erosion limit the fertility of the basin's arable land.

Over-grazing of grasslands

During recent decades ever-increasing stock densities (mainly goats and sheep) have led to severe over-grazing of the rangeland and steppe areas. This reduces habitat quality both for the characteristic wildlife of these areas and also for grazing.

Pollution of water courses and lakes

Untreated effluents from urban areas, and also the disposal of agricultural drainage water from irrigated farmland has led to increased salinity in wetlands, and caused problems of eutrophication and the accumulation of toxic compounds. A horrific example of this is the situation around Tuz Lake (Salt Lake), which receives sewage from the city of Konya (ca. 1 million inhabitants) and untreated wastewater from several hundreds of industries.

4.3 The pilot project "Towards Wise Use of the Konya Closed Basin"

The environmental issues in the Konya Closed Basin are complex and strongly interlinked with socio-economic and agricultural development. Decisions to irrigate land, drain wetlands or even build reservoirs in order to expand irrigated land and speed up agricultural production in a certain area of the basin, are sure to cause reverse effects on the state of natural resources (e.g. fish stocks, groundwater levels, wetlands, reedbeds).

With the project *Towards Wise Use of the Konya Closed Basin*, WWF-Turkey aims to stimulate key stakeholders and decision-makers to make a strategic shift in the way they think about, use and manage water resources in the Konya Closed Basin. This process-oriented project is focused on the empowerment of stakeholder groups and communities in order for them to transform their responsibilities and concerns regarding the environment into concrete actions, long term objectives such as using scarce water in a sustainable and efficient manner, and allocating it to support multiple activities, conserving wetlands and steppes for their rich biodiversity, natural products and functions.

WWF-Turkey sees its role in the Konya Closed Basin primarily as a facilitator of the process towards IRBM. Once this process is set in motion and carried forward by stakeholders, WWF-Turkey hopes to primarily play the role of advocating nature, aiming at the protection and restoration of fragile key habitats within the basin. The project under consideration aims to ignite the IRBM process through:

- Capacity building for effective and sustainable IRBM
- Establishing dialogue between various (inter-sectoral) stakeholders
- Developing and implementing the pilot project
- Communicating the necessity for IRBM to the general public and specific target groups.

4.3.1 Working with stakeholders in the basin

WWF-Turkey has been working with the support and active participation of more than 600 stakeholders in the basin since 2003. The most important stakeholders are as follows:

- **Central authorities**
  MoEF, APSA, General Directorate of Nature Protection and National Parks, DSI, Ministry of Agriculture and Rural Affairs (MARA), Ministry of Culture and Tourism
- **Regional authorities**
  Regional Directorate of DSI, Regional Directorate MARA
- **Local authorities and other local actors**
  Governors of six provinces, the Provincial Directorate of Environment and Forest, the Provincial Directorate of Agriculture and Rural Affairs, municipalities, local NGOs, Chambers of Industry and Trade, Irrigation Associations and Groundwater Cooperatives and Unions, Sugarbeet Cooperatives, Agricultural Credit Cooperatives, local media, Selcuk University, Konya Sugar Factory, salt producers

As the project area covers nearly 53,000 km^2 with eight different provincial boundaries and various stakeholders at the central, regional and local level, a major difficulty has been communication between stakeholders and sustaining durable relations throughout the project. In order to sustain the dialogue's continuity during the project, stakeholders were regularly visited and updated about the progress.

Most of the time, members of local communities do not feel that they are able to influence government decisions and they often lack confidence and knowledge to address pressing issues which directly or indirectly affect their livelihood and/or living conditions. This in turn does not contribute towards the needed "sense of shared responsibility" and unsustainable practices causing pollution and further degradation of natural habitats further threaten the sustainability of the Konya Closed Basin.

The main focus of the project was therefore put on the empowerment of local stakeholder groups, such as local environmental NGOs, irrigation cooperatives, municipalities, private sector and media representatives and farmers. The aim was to enable them to translate their responsibilities and concerns regarding the environment into concrete actions, such as participation in the dialogue and the initiation of new NGOs and IRBM-related projects. An intensive capacity building programme was designed covering a wide range of topics from water resources management to the importance of participation, from policy instruments to implementation techniques on the ground. The underlying goal of these capacity building efforts is to engage local stakeholder groups in the decision making processes and to work together with the government agencies that have proven to be hesitant in joining cross-section dialogues and initiatives. DSI, MoEF and MARA are the key institutions within this process.

4.3.2 Success in influencing policies and planning

The project raised the basin community's awareness of the limited water resources, misuse of water in agriculture and threats for the future. Several capacity building activities on water resources management, IRBM and EU Policies were organized. Water users and responsible bodies came together in several stakeholder meetings. The project further carried out scientific research, which provided important input for the planning process: such as the analyses of agricultural activites, an assessment of the change in groundwater levels and an analysis of stakeholders and socio- economic aspects. Moreover, opportunities and alternatives for a better management and usage were practiced through pilot implementations.

With this strategy, the project managed to influence resources management and to increase stakeholder involvement on several occasions:

- The Tuz Lake (Salt Lake) Management Plan was prepared with APSA through several workhops; implementation of the plan is coordinated by APSA and WWF-Turkey. In addition, the Eregli Marshes Management Plan was prepared in coordination with MoEF and a local NGO entitled The Society for the Protection and Improvement of Eregli Marshes and Akgol
- Four local projects were financed and supported with small grants enabling the development of projects initiated by stakeholder groups and contributing directly to the integration of conservation and sustainable use of the Konya Closed Basin
- The committed work of WWF-Turkey has helped to transform the region in terms of how 'water' is perceived, used and managed and there have been observable changes in the activities of public institutions in the water and agriculture sector. Confronted with the results of the WWF research on changing groundwater levels, the Regional Directorate of State Hydraulic Works initiated a study to identify illegally drilled and operated wells and started to re-assess irrigation projects that had been planned since 1970s. These are steps towards preparing the "Basin Master Plan"
- The project, in cooperation with national and local government authorities supported the implementation of efficient irrigation techniques and training schemes on modern irrigation systems: since 2005, a total of 12 pilot projects covering an area of about 14 hectares have been implemented. In response to the severe drought in 2007 and to WWF-Turkey's pilot implementations in Konya Closed Basin, the Cabinet of Ministers reduced the interest rate for credits used to finance modern irrigation systems to zero. Additionally, the MARA will provide grants for drip or sprinkling irrigation projects. Due to the pilot implementations and training schemes of WWF-Turkey, as well as these financial subsidies, drip irrigation application increased by 400 percent basin-wide and today, drip irrigation system has been adopted in an area of nearly 33.000 hectares in the Konya Basin. *In 2007, the Council of Ministers reduced the rate of interest to zero on agricultural credits supplied through the Bank of Agriculture in order to promote modern irrigation systems* and since then, 52 percent of

the total Bank credit used for modern irrigation all over Turkey is used solely in the Konya province (Konya Regional Directorate of Bank of Agriculture 2007)
- A Drought Action Plan for Konya has been prepared with the collaboration of all state institutions as well as the university, irrigation management organizations and other stakeholders
- An Action Plan for Konya Basin Irrigation Projects (KOP) has been recently finalized as an outcome of the joint effort of the DSI, the Agricultural Directorate and the Directorate of Agricultural Reform. The action plan foresees the transformation of the water distribution infrastructure to a closed system, land consolidation, redistribution of all water rights in the region and use of water meters in all wells
- As a part of the EU regional development strategy, the Mevlana Development Agency (covering the provinces of Konya and Karaman) was established. Even though it is not a local NGO, WWF-Turkey has been invited to become a member of this agency thanks to its committed work in the region

4.3.3 The next step: establishing a stakeholder commission

In the current phase of the project, WWF-Turkey's efforts are mainly directed towards establishing a Konya Basin Platform/Commission as a central body that co-ordinates, integrates, promotes and/or even enforces decisions regarding the use and management of natural resources in the basin. The exact role and responsibilities of this central body will be defined as basin specific. Some of the first tasks within the Commission will be to develop a common vision and declaration for the Konya Basin that is approved by all stakeholders, to build ownership with participating stakeholders, and to develop and implement joint pilot projects. Agricultural organizations (irrigation cooperatives, sugar beet cooperatives, sugar factories), municipalities, local NGOs and local media are the most important groups to be involved in the proposed Konya Basin Commission. The commission will have to address the complex relationships between agriculture, the environment and water resources. Moreover, the socio-economic importance of water resources management, functions and values of wetlands, the linkages between water and poverty relation, and water pricing are the key issues to be mentioned and solved through this Commission. Since the Konya Basin is a very good 'laboratory' in terms of solving the complex problems related to water resources management, the solutions developed for Konya Basin will serve as a model for the rest of the river basins in Turkey.

In order to have an effective and easy functioning commission, the organizational structure should be as simple as possible. Tasks and responsibilities of each body need to be well-defined, in order to reduce the risk of conflicts or problems in the functioning of the commission. Two different models for the commission have been suggested by WWF-Turkey and are being discussed by relevant parties: One of them is developing the Konya Basin Commission through Local Wetland Commissions (LWC). According to the Legislation for the Conservation of

Wetlands, each province with an internationally important wetland within its administrative boundaries, should establish a LWC consisting of 13 members: related governmental bodies, two local NGOs and two academicians. Currently, the cities of Konya and Isparta established a LWC. After the establishment of Aksaray and Karaman LWC, the Konya Basin Commission may be developed through combining these LWCs. The second model would depend on sub-basin organizations, which may be developed as a part of the Salt Lake, Eregli Marshes and Beysehir Lake Management Plan processes. After the establishment of the Beysehir, Eregli and Tuz Lake sub-basin organizations, the Konya Basin Commission could be developed through these structures. Finally, the process should end with establishing a management body/committee which coordinates the implementation of its activities. For Salt Lake Management Plan, this role is under discussion to be delegated to the Union of Municipalities in Salt Lake sub-basin; however this does not function well in practice. Therefore, there are no sub-basin organizations available right now but they need to be developed. The common characteristic of these two alternatives is that both of them provide platforms for different stakeholders and sectoral representatives through technical working groups.

5 Discussion and lessons learned

The Konya Basin reflects the complex structure of water resources management in Turkey. There are several institutions responsible for the planning, development, monitoring and research of water resources. DSI, MoEF, MARA, Municipalities, Regional and Provincial Directorates of these ministries, irrigation cooperatives, sugar beet cooperatives, salt producers, sugar factories, local media, local NGOs, academicians are the most effective groups within the Konya Project. Some of the stakeholders were hesitant to discuss the problems/threats at the beginning of the project. They do not believe in the importance of participation and necessity to involve all parties affected by existing practices. This is mainly because of the heavily centralized structure of the Turkish administrative system. Meanwhile, government institutions are concerned about losing their authority or power in the water resources management system. However, throughout the project, stakeholders got to know each other and managed to establish a dialogue.

Considering the problems of the basin, there are two main groups of actors within the process: water users (mainly farmers, irrigation cooperatives, sugar beet cooperatives) and government authorities (DSI, MoEF, MARA, municipalities). The efforts for the conservation of wetlands, groundwater resources and the other natural resources of the basin can only be achieved with the cooperation of these two groups of actors. While water users are changing to more water efficient practices, the government authorities should develop new policies for the better management of water resources. In addition to this, new subsidy schemes for water saving agricultural techniques, effective control mechanisms for illegal groundwater extraction and capacity building activities for different water users have to be developed. NGOs can guide and facilitate these processes.

Several lessons can be learned from WWF-Turkey's work in the Konya basin in order to guide future efforts in other river basins:

There is a strong need for a holistic view that goes beyond the symptomatic issues to understand and tackle the systemic problems. For this, full and continuously updated knowledge about the region is crucial and access to data is a must.

Building bridges and facilitating coordination among stakeholders is extremely important. In this respect it is better to keep on equal terms with all stakeholders (instead of taking sides), to be neutral and to facilitate the dialogue and process. To maintain transparency, it is vital to always keep stakeholders informed about other stakeholders' actions, plans and projects. Language should be understandable for all parties, therefore the terminology and the language used should be as simple as possible because there is a high range of linguistic capacity among stakeholders.

The greatest challenge in initiating IRBM in a basin is providing a platform where all stakeholders can have their say, towards a coherent and achievable set of targets. Open-ended actions should not be left unattended, next steps should always be stated, documented and declared to stakeholders. It is necessary to build trust, be consistent and keep promises that were made. Keeping contact open between all stakeholders is crucial, so as not to lose a stakeholder, and informing them about the stages of the project regularly is helpful for this purpose. Showing the stakeholders that their voice has been heard and transmitted to others increases the feeling of ownership (Butler 2003).

Encouraging stakeholders to appreciate success in smaller steps, as the way towards achieving bigger goals prevents disappointment, as working on the basin level means working on a very large scale, where progress will take time. Patience is required instead of rushing and becoming disappointed in the end. Even small demonstration projects contributing to wider goals encourage stakeholders and increase their motivation. Motivation is a prerequisite in order to be able to carry on with such a long term process.

It should be kept in mind that for the sustainability of the process, the central role in the project is played by the people living in the basin, not the project team-organization; this message should be transmitted as often as possible.

References

Arsel M (2005) The Bergama Imbroglio. In: Adaman F, Arsel M (eds) Environmentalism in Turkey. Between Democracy and Development? Ashgate Studies in Environmental Policy and Practice, pp 263-275

Butler WH (2003) Planning for Water: Statewide Approaches to Watershed Planning and Management. Unpublished Masters Thesis, Blacksburg, VA

General Directorate of DSI (2008) DSI Annual Report. DSI, Ankara

General Directorate of Electirical Power Resources Survey and Development Administration http://www.eie.gov.tr/turkce/YEK/HES/hidroloji/havzalar.html. Accessed 8 November 2009

Global Water Partnership (2000) Integrated Water Resources Management. Technical Advisory Committee Background Papers no 4

Konya Regional Directorate of Bank of Agriculture (2007) Personal communication, Konya

Mostert E, van Beek E, Bouman NWM, Hey E, Savenije HHG, Thissen WAH (1999) River Basin Management and Planning. Keynote paper for International Workshop on River Basin Management, The Hague 27-29 October 1999. UNESCO, IHP-V, Technical Documents in Hydrology

WWF (2002) WWF College for Conservation Leadership IRBM Module version 2.0 (developed by Ilke Tilders with the support of Cap-Net).

WWF (2003) Managing Rivers Wisely: Lessons from WWF's work for integrated river basin management. www.panda.org/downloads/freshwater/managingriversintroeng.pdf Accessed 8 November 2009

Part II
International Cooperation on Turkey's Transboundary Rivers

The Water Dimension in Turkish Foreign Policy

Ilter Turan

1 Introduction

Much more so than in earlier times, water today constitutes a central concern of international politics. There is a general perception that humanity, as it develops both numerically and economically, is consuming the global water resources exhaustively. Soon the world is expected to reach a stage where there will not be enough water to meet all human needs.[1] Countries will then fight for water to meet the needs of their population at the expense of others. It is not surprising that such anticipations of inter-state conflict are strongest in regions that are not blessed with large amounts of water. Frequent references to 'water wars' both in the world and in the Middle East are but one indicator of this phenomenon.[2]

The conceptualization of water as a resource whose distribution among countries can only be settled by applying conflictual methods is not necessarily correct. Countries may choose cooperative as well as conflictual means, or a combination of both, in striving to meet their water needs. But if such an attitude comes to be widely shared by a group of countries that have common rivers, aquifers or lakes, then a preference for conflict oriented methods is likely to occur. It is, therefore, important to dwell upon the potential of cooperation in the solution of water problems between countries.

In this context, it becomes particularly important that Turkish foreign policy perceives and approaches water as an element of cooperation. Although more a part

[1]Kolars and Mitchell cite six uses of rivers which are the most important source of water in the world. These include in diminishing order of importance irrigation, domestic use, hydropower, industrial use, navigation and fisheries. The first four of these are consumption oriented; the other two are in the nature of using (1991, 78).

[2]A Google search for the term 'water wars' produced 333,000 references (15 March 2009), the less comprehensive Yahoo search 295,000.

I. Turan (✉)
Department of International Relations, Bilgi University, Istanbul, Turkey
e-mail: ituran@bilgi.edu.tr

A. Kibaroglu et al. (eds.), *Turkey's Water Policy*,
DOI 10.1007/978-3-642-19636-2_10, © Springer-Verlag Berlin Heidelberg 2011

of Southern Europe in terms of location and foreign policy orientation, two major rivers in the Middle East, namely the Tigris and the Euphrates, originate in Turkey. Several other rivers, some originating in Turkey, others elsewhere including the Asi (Orontes), the Meric (Maritza) and its tributaries Tunca (Tundja) and Arda, Coruh (Chorokhi), Aras (Araks) and tributaries Arpacay (Ahuryan) and Kura, all display some or all of the characteristics of transboundary rivers.

This article will trace why and when transboundary water issues have gained importance as a matter of foreign policy and appeared on agendas, and the influence non-water issues have had on cooperation. It will conclude with suggestions on how the deadlock can be overcome in highly conflicting environments.

2 Water appears on the Turkish foreign policy agenda

Water has not constituted a major concern of Turkish foreign policy until the last three to four decades (Turan 2004, 191). This derives in part from the fact that the emergence of water questions to central importance in international politics is a relatively recent phenomenon. Indeed, in the past, there were problematical and sometimes conflict-ridden relations between the riparians on certain rivers but they were looked upon as being of local or regional in nature, rather than as a manifestation of a global water problem. But the lack of emphasis on water questions in Turkish foreign policy also owes much to the fact that for a long time Turkey was not interested in using its transboundary rivers for agricultural, i.e. irrigation purposes. Initial Turkish interest in rivers was motivated by generating of hydropower and supplying cities with drinking water. Therefore, river regulation rather than consumption has prevailed as the guiding Turkish concern.

The Treaty of Lausanne which in many ways constitutes the 'birth certificate' of the Turkish Republic, did not devote much attention to Turkey's transboundary rivers except if they constituted the border with her neighbours. Article 109 referred to the protection of established rights and depicted that international arbitrage would be used if conflicts could not be settled by an agreement. There was also a reference that Iraq would be consulted before Turkey could construct hydraulic infrastructure on the Tigris and Euphrates rivers. The Good Neighbourliness Agreement with France in 1926, reflecting a similar approach, made references to the water rights of Syria on the Euphrates while the 1930 Treaty of Aleppo created a mechanism for conflict resolution.[3] Then, in the 1946 Treaty of Friendship and Good Neighbourliness between Iraq and Turkey, the parties agreed that the Euphrates and the Tigris rivers should be regulated: "Iraq agreed to contribute to the expenses of the installations aimed to regulate water, if it also aimed for (sic) the

[3]Beschorner (1992, 39) makes this point in a general way. The 1930 Treaty, however, does not refer specifically to water but creates an international committee for conflict resolution which would naturally include water-related ones.

benefit of Iraq." (See Annex 11) (Inan 1994, 235).[4] The understanding stated in Lausanne that Iraq would be consulted as regards the commencement of development projects on the twin rivers was also reiterated in the treaty (Beschnorner 1992, 39). There were other agreements regarding other rivers as well, such as the Joint Utilization of Border Rivers Agreement signed with the Soviet Union in 1927. Turkey, however, did not give priority to the utilization of the potentials of the transboundary rivers during the early years of the Republic.

Because many of its rivers originate and reach the sea within national boundaries, and because it had hardly exploited the potential of these rivers until relatively recently, Turkey had come to think of its rivers as national waters. It encountered its first major challenge to this perception when it initiated plans to build the Keban Dam on the Euphrates in 1963.[5] Prior to the commencement of construction when Turkey asked for financial support from United States Agency for International Development (USAID) and the World Bank, it was told that the lower riparians ought to be consulted and their concerns accommodated. Although Turkey did not solicit the prior-approval of the downstream countries, it responded to the expectations of these agencies by making a unilateral statement in 1966, in which it agreed to release an average of 350 cubic meters per second. Subsequently, the amount was first raised to 450 and eventually in 1976 to 500 cubic meters per second (Turan 2004, 197-198), (Inan 1994, 228), (Giray 1994, 250).[6] These commitments were verbal, but 500 cubic meters per second was inserted into a commercial agreement signed with Syria in 1987, where Turkey agreed to release this average amount monthly until a final agreement was reached.[7]

Syria and Iraq have both been concerned by Turkey's plans to build a network of dams for power generation and irrigation purposes on the Tigris and the Euphrates rivers. These plans are part of a comprehensive regional development project, the South-eastern Anatolia Project, which is best known by its acronym as the GAP. The lower riparians fear that this project will eventually lead to a situation in which Turkey will consume most of the Euphrates waters and thus deprive them of their major lifeline. Therefore, they have pursued policies sometimes individually, sometimes together, to resist Turkey's implementation of the GAP project. The aggressive approach adopted by the lower riparians has played an important role in the shaping of Turkish foreign policy regarding water in recent years. Turkish water policy, the main features of which will be examined later, has tried to achieve two ends. On the one hand, it has tried to establish through persistent efforts that Turkey is entitled to use the waters of the two rivers since most of the waters, particularly of the Euphrates, originate in Turkey; on the other hand, it has tried to develop means

[4]See also DSI 2005, 113.

[5]The plans for the dams were initiated in 1963 but the construction did not start until 1965 and completed in 1975.

[6]Giray (1994) notes that the increase to 500 m took place in 1976 at the time of the construction of the Keban dam; see also Kut (1994, 110).

[7]Article 6 which obliges Turkey to release a monthly average of 500 m/sec until the final allocation of waters (Rezmi Gazete (1987), see also Annex 12 for the Protocol).

whereby water-related conflicts may be channelled into peaceful domains and hopefully resolved through cooperative efforts.

The problematical relationship with Syria and Iraq regarding the waters of the Tigris and the Euphrates rivers has overshadowed the fact that Turkey has other transboundary rivers on which cooperation rather than conflict has characterized the interstate relationships. In the following section, I will examine examples of cooperation between Turkey and its neighbours on transboundary rivers and analyze why such cooperation has proven possible. I will then examine the difficulties encountered in cooperating with Syria and Iraq. Finally, I will briefly mention some projects that have been developed by Turkey to make its unused waters available to the Middle Eastern countries including Turkey's southern neighbours.

3 Cooperating with neighbours: joint development of transboundary rivers

Twenty two percent of the land borders of Turkey, or 615 kilometers, are formed by transboundary rivers (see Table 1). Ozden Bilen, a former director of State Hydraulic Works (Devlet Su Isleri, DSI) notes that a country that has eight neighbours is compelled to employ, as much as possible, cooperative means to achieve security and stability (Bilen 2000, 96-97).

The part of the land borders that are comprised of rivers and dry riverbeds has been measured on 1:25.000 scale maps by Colonels (ret.) A. Ilker Arisoy and Yasar Turkoglu, both formerly members of the General Directorate of Maps of the Turkish Armed Forces, as part of a seminar paper entitled "Evaluation of Turkey's Inland Waters that Form National Boundaries from the Perspective of Political Geography" which they prepared in 1998 for a MSc program in Geography at Gazi University in Ankara.

It appears that both Turkey and its neighbours in its Northeast and Northwest have availed themselves to cooperation as regards the joint utilization of transboundary rivers. One exemplary project for cooperation was developed

Table 1 Length of borders of Turkey (DSI website 2007)

Country	Length of land borders (km)	Length of river forming the border (km)
Syria	911	141
Azerbaijan	18	18
Georgia	276	75
Armenia	328	202
Iran	560	76
Iraq	384	236
Bulgaria	269	126
Greece	203	192
Total	2,949	1,067

between the Soviet Union and Turkey at a time when the Cold War was still prevailing. In 1973, the two countries reached an agreement to jointly construct a dam on the Arpacay (Ahuryan) tributary of the Aras (Araks) which formed the border with Armenia S.S.R. before reaching the main stem of the river. Since further down the Aras also forms a border with Iran, Turkey secured the latter's consent regarding the building of the dam. The dam itself was completed in 1986.[8] Its reservoir with a storage capacity of 510 million cubic meters (MCM) covered land on both sides of the border. It was agreed in advance that the waters would be divided evenly between Turkey and the Armenian S.S.R. During both the construction and later the operation stages, technicians from the two countries have cooperated smoothly within the framework of a JTC to make annual plans for sharing the waters. Turkey uses its share to irrigate the fertile Igdir plain.[9]

Turkey and Greece have worked jointly in another area of water management: flood control. The Meric forms most of the Turkish-Greek border. In 1951, the two countries agreed on the preparation of a master plan for flood control works and contracted a firm to prepare the plan. Each party would pay for its own part of the construction. The plan was completed and some projects recommended in the plan have been implemented. But the plan's ultimate success has been dependent on the upper riparian Bulgaria, from whom cooperation was, for a long time, not forthcoming (Bilen 2000, 98-99).

In the past, Bulgarian authorities have always tried to have their dams full. Therefore, when there were major rains, lacking storage capacity, they had to open the dam gates to release all the incoming water, flooding the downstream areas. Since the major political changes that Bulgaria has undergone in recent years, opportunities for cooperation have improved. For example, during a period of drought in summer 1993, Turkish rice fields along the Meric River were in severe need of water. Reservoirs on the Turkish side were exhausted. Turkey negotiated with Bulgaria to have the latter release water from its reservoirs. Eventually an agreement was reached so that Bulgaria allowed almost 16 MCM to be released downstream. Turkey paid 12 US cents per cubic meter for the water it received (Yavuz 1997, 563). Despite such progress, an early warning system and further cooperation for flood control continue to be needed.

It may be asked whether Turkey's willingness to pay for water from the Bulgarian dams would not constitute recognition that the waters of Meric, a transboundary river, in fact belonged to Bulgaria. In answering the question, it has to be pointed out that Turkey's landscape (no fault of Bulgaria) does not allow it to develop sufficient water storage capacity to meet the shortages created by the drought while Bulgaria had its dams full and could therefore release some of its water downstream. The shortage was temporary and derived from climatic

[8]The dam was financed entirely by the Soviet Union as part of a package of economic assistance which also included the construction of a steel plant in Iskenderun and an aluminum refinery in Seydisehir.

[9]This summary relies heavily on Bilen (2000, 98); similar information is also present in Giray (1994, 246).

conditions specific to that year. There was no overall shortage of water and the question of an equitable division of the river's waters was not being debated. Furthermore, Turkey's willingness to pay for the water was in harmony with its position on the Tigris and the Euphrates that all waters of a transboundary river originating in one country belong to that country.

More recently, cooperation activity has been intensified between Bulgaria and Turkey. Based on an agreement that had been signed between Turkey and the Bulgarian Socialist Republic on 23 October 1968 on the use of the rivers that flow through the territories of both countries, Bulgarian authorities have agreed on a protocol that obliges them to give advance notice should water be released from the dams on the Meric and its tributaries Tunca and Arda (see Annex 3). They have also expressed their willingness to provide technical information about the dams on these rivers in order to be better prepared for floods.[10] Currently, the two governments are working on plans to construct a joint dam on the Tunca River for the purpose of controlling floods and generating hydropower. Planning is at an early stage and many questions such as how much of the cost and how water will be shared will inevitably require elaborate negotiations. It has already been agreed that fifty percent of the power generated will be given to Bulgaria and the other half to Turkey.[11] Technical efforts are progressing though the specific arrangements for the construction of the dam including cost-sharing arrangements are not clear at the time of writing. Though not documented, the joint efforts on the Arpacay River with Armenia S.S.R. may be regarded as a positive example for this prospective endeavour.

Finally, a technical committee has begun to function between Turkey and Georgia since 1998 to deal with water-related issues in the Coruh (Chorokhi) River. Of particular interest is the development of an early notification (advance warning) system in case of emergency discharges coming from dams on the Turkish side.[12]

To conclude, a cooperative mood has characterized Turkey's relations with its neighbours as regards transboundary rivers. Remarkably, even during the Cold War, it was proven possible to develop and implement a joint dam project with the Soviet Union. A question of major concern is which factors encourage and facilitate cooperation between neighbours? It may have already come to our attention that in states like Bulgaria and Georgia, cooperative attitudes followed regime changes which, in the case of Georgia, meant becoming independent. This suggests that political regimes may be an important factor in determining how a society engages in cooperative projects with its neighbours. In analyzing Turkey's

---

[10]A copy of the protocol dated on 23 February 2000 has kindly been provided by the Ministry of Foreign Affairs.

[11]Minutes of the joint meeting, held from 30 January to 1 February 2006. Document kindly provided by the Ministry of Foreign Affairs.

[12]Minutes of the meeting held in Ankara on 7 February 2006. Document is the courtesy of the Turkish Ministry of Foreign Affairs.

relations with its Southern neighbours, the nature of their regimes may be kept in mind as a constraining factor on cooperative endeavours.

In the examples we have looked at so far, none of the projects were of such a magnitude that either party felt that its future economic prosperity was intimately tied to the realization of that project. By becoming a partner in a cooperative endeavour, neither side achieved such a superior position that it could inflict unacceptably high deprivations or harm on its respective neighbour. Rather, both sides felt that by cooperating each would gain something of which it would be deprived if it failed to cooperate. It is interesting that the overall relations between two countries, however problematical they may have been, did not discourage the development and implementation of joint projects. Greece and Turkey were on good terms, for example, when their cooperation on the Meric River started, but relations grew progressively worse during the 1960s and 1970s due to the beginning of the Cyprus conflict (see Kramer and Schellig in this volume). The flood control projects continued throughout the period in question. With regard to the dam on the Arpacay River, the Cold War, particularly as it became a matter of routines over time, did not stand in the way of a cooperative project between members of the opposite camps provided that such projects did not affect the strategic balance between them. In fact, such cooperation was seen as a step in building mutual confidence and as a testimony to the fact that the two parties did not anticipate major developments that would threaten the stability of their relationship.

Finally, two related observations may be made. First, in advance of the commencement of cooperation, each party must have some confidence in its ability to affect the behaviour of the other in order to prevent a totally unacceptable outcome. Second, cooperating parties must possess a pragmatic orientation toward cooperation rather than one that can be characterized as ideological. Cooperative endeavours necessitate both ability and willingness on the part of all actors to transcend ideological differences in order to work together.[13]

4 Cooperating in the difficult Southern Environment

We have already seen that the emergence of water issues, which became a major concern of Turkish foreign policy, was to a large degree the product of Turkey's decision to utilize the waters of the Tigris and the Euphrates rivers for power generation and eventually also for irrigation. The downstream riparians, Syria and Iraq, immediately perceived Turkey's plans to develop these rivers as a threat. They turned to a variety of means to resist these plans. They argued that as the downstream countries, they had acquired rights of prior use, concluding that any plans Turkey had would be subject to their approval since Turkey's utilization would inevitably have an impact on their current use. During this time, a set of principles

[13]For a more comprehensive discussion on the role of ideology as an impediment to cooperation, see Turan and Kut (1997, 139-145).

regarding the utilization of transboundary waters was being discussed and developed by the United Nations under the auspices of the International Law Commission. According to them, the waters of transboundary rivers were to be used equitably and optimally, and further utilization by one riparian should not cause appreciable harm on others. In practice, these principles were interpreted as protecting exclusively downstream countries since upstream countries were in a position to utilize the river waters without being dependent on the goodwill of those downstream.[14] Sensing that the international environment was in their favour, Syria and Iraq both tried to internationalize their problems with Turkey. These efforts commenced initially when Turkey asked the World Bank and USAID for credits to help finance the Keban Dam in 1965. Syrian arguments at the time appear to have impressed these agencies. A Turkish government official who was involved in the negotiations complained: "During the construction of Keban, [International Cooperation Agency] ICA (USAID) confronted us as if it were Syria's attorney…"[15]

In the subsequent years, Syria and Iraq continued their efforts to stop Turkey's projects by utilizing a multiplicity of means. They continued, for example, their efforts to stop international lending agencies from making funds available to Turkey with some success. Turkey was unable to find funding from the international lending agencies for the construction of the Ataturk Dam, and to the surprise of the lenders, proved capable of financing and constructing the dam with domestic sources. Syria and Iraq also launched a campaign against Turkey among the Arab countries, arguing that the Tigris and Euphrates rivers had historically been Arab rivers which Turkey was wresting away from them. For example, when the credit agreements for the construction of the Birecik Dam on the Euphrates River were signed, Syria started lobbying against the construction of the dam in the Arab League and in December 1995 got Egypt, Bahrain, Kuwait, Oman, Qatar and the United Arab Emirates to call on Turkey to agree to equally share the waters of the Euphrates River with Syria (Mazlum 2002). The arguments which were based on international principles regarding the utilization of transboundary waters, did not serve the interests of all Arab countries in the same way, however. Syria, for example, was the upper riparian on the Orontes River but had no interest in guaranteeing rights of prior use to the downstream country Turkey, a privilege it wanted to be observed on the Euphrates. The status of the riparians to the Jordan River was even more complicated. It proved easier to criticize Turkey by mobilizing feelings of enmity against her, derived from centuries of Ottoman rule and Turkey's parting ways with the Arabs to join the West (which was considered to be Arab betrayal)[16].

---

[14]On this point, see also Bagis (1994, 22).

[15]Quoted in Turan (2004, 198, 13).

[16]Gruen (1994) seems to capture the essence of this background when he writes, "Although Turkey and its Arab neighbors share elements of an Islamic culture and religious heritage, these common ties have been overshadowed by the legacy of mutual suspicion and distrust stemming in part from 'Arab resentment of four centuries of Ottoman rule and Turkish feelings of betrayal by the Arab revolt during World War I.", 264.

The most serious challenge to peaceful relations between Turkey and Syria came, however, when Syria's President al-Assad decided to extend support and offer safe haven to a Kurdish ethnic and separatist movement, the PKK. The movement kept part of the Turkish military occupied with an externally supported domestic challenge while the Turkish government devoted substantial resources to the anti-terrorist campaign which might have been allocated otherwise for development purposes. The PKK was also able to set up organizations in Europe and mobilize segments of the Western European public to criticize the anti-terrorist campaign of the Turkish government, as well as its plans to build further dams. The Syrian government had hoped that its support to the PKK might give it a bargaining card for water in its conflict with Turkey. But the Turkish government was adamant in its rejection of linking terror and water because of its belief that making terrorists a part of the negotiating process would only invite further terror.[17]

Although Iraq and Syria had a convergence of interests in receiving a maximum amount of water from Turkey, Turkey's relations with Iraq were on the whole better for three reasons. First, Iraq felt somewhat less pressed for water because in addition to the Euphrates, it had access to the waters of the Tigris whose waters were utilized much less by Turkey. Secondly, it had an ongoing multi-dimensional rivalry with Syria. Only a part of this conflict was related to Iraq's not receiving sufficient water from Syria from the Euphrates River. Much of it came from other sources. For example, the two countries competed with each other for the domination of B'ath socialism, while Syria developed close relations with Iran to arrest the desire of Iraq to be the dominant Arab power in the Levant. Finally, Iraq was constantly involved with external engagements and therefore wanted to preserve good relations with Turkey. Until 1974, Iraq was faced with a Kurdish insurgency that was supported by Iran. The peace accord reached in 1974 and the subsequent withdrawal of the Iranian support ended the insurgency. Soon after the Iranian Revolution, Saddam launched a war against Iran which lasted several years, resulting in a stalemate. In 1991, Iraq launched an invasion against Kuwait, an undertaking that was repelled by an international coalition led by the United States. After that, Iraq was paralyzed in its Northern and Southern regions by the US designed and implemented Operation Provide Comfort, which banned the Iraqi government from flying war planes over these regions. Finally, it got invaded by the United States in 2003.

The relations between Turkey and its southern neighbours, particularly with Syria, were generally driven by a lack of mutual trust, a condition that made it difficult to develop mutually beneficial cooperation including areas involving the utilization of water. The fact that the Syrian and Iraqi regimes were insecure confounded their inability to collaborate with the outside world even though such cooperation might be rewarding. Some examples might illustrate the nature of the difficulties. The lack of trust in the other, for example, brought with it a concern about a politically motivated food embargo which, in turn, produced an obsessive

[17]Syrian backing of the PKK and the problems it has generated in Turkish-Syrian relations are summarized in Beschorner (1992, 37) and passim; see also Kut (Fall 1993, 9).

occupation with food security (Turan and Kut 1997, 140). Governments mobilized their citizens to support efforts to achieve food security, but then got trapped into exaggerating their water needs. Such a position was hardly conducive to cooperation, which would require optimal utilization as its guiding concept.

Moreover, during the late 1980s, it became known that Syria would begin the construction of the Tishreen Dam near the Turkish border. The maximum depth of the dam reservoir was limited to 20 meters. On the Turkish side of the Euphrates River, a final dam, the Karkamis, had been planned as the last dam before the river entered Syria. In 1998, Turkey suggested to Syria that these projects should be combined, to build one dam together, rather than building two dams separately - one on each side of the border. The new dam would be 40 meters in depth and generate more power than the two independently built dams. The power generated would be divided between Turkey and Syria. Syria responded that the feasibility studies and the completion of construction plans would cause delays whereas it wanted to commence with construction works immediately. Turkey suggested that the two sides could form a technical committee and produce the needed studies in a short time, given Turkey's experience in dam design and construction. Syria chose to continue with its own project, however. Similarly, Turkey later proposed to Syria that a joint dam should be built in the region where the Tigris forms the border between the two countries for a short distance. The idea was similar to that of the Arpacay Dam: power and water for irrigation would be shared among the two riparians. However, on the Syrian side positive interest failed to materialize.[18]

In an atmosphere of distrust, sometimes even an honest mistake may be perceived as a hostile act. In 1997 during a religious holiday, the General Directorate of State Hydraulic Works, in anticipation of reduced demand for electricity, closed off some turbines in the Ataturk Dam on the Euphrates River. It failed to inform the Foreign Ministry to give the Syrians advance notice for what was a brief interlude of reduced flow. Vigilant protests were to follow immediately. Turkey released extra water after the holiday to demonstrate that this was an honest case of oversight and that no other interpretation should be conceived (Gruen 1994, 265).

Distrust leads the political leaders of a country to perceive ill intentions on the part of other countries with which they have to interact, making it difficult for parties to perceive mutual benefits. Turkey's construction of dams on the Euphrates and Tigris rivers has a strong hydropower production component. This meant that at any given time, Turkey's own interest would dictate the release of a considerable amount of water downstream in order to produce electricity. This reality should assure the downstream riparians that the upstream riparian would suffer significant losses itself if it ever tried to impound most of the waters of a river. For example, a deputy director of the DSI explained in a written note to a British scholar that there were eight turbines in the Ataturk Dam, each of which needed 225 cubic meters per second to operate. Assuming that about six turbines would be operating at a given

---

[18]The paragraph summarizes Bilen (2000, 100). It should be noted that Bilen who, at the time of these discussions, was Deputy Director of State Hydraulics Works, became its General Director later.

time, no less than 1,350 cubic meters per second would have to be released downstream (Beaumont 1994, 206). This is in far excess of the 500 cubic meters per second guaranteed in the 1987 Turkish-Syrian Protocol (see Annex 12). Yet, even with this and other similar information, constant fears of damming the waters so none are left for the downstream countries are ever present.

Not surprisingly, Syrian actions have stimulated similar responses on the Turkish side. The then Prime Minister, Suleyman Demirel, asserted on one occasion: "I do not believe in worrying about threats of war resulting from development projects in Turkey. If there is threat, we will repel it. Turkey has deterrence. It will have more deterrence in the coming period. Turkey will build more such works. The more it builds, the fewer threats it will be faced with."[19]

A turning point in Turkish-Syrian relations came in 1998 in an unexpected fashion. Having failed to convince the Syrians that they should stop supporting a terrorist movement, Turkey decided that it might consider military action. The commander of the Turkish Third Army responsible for the defence of the Southern border issued an ultimatum that unless Syria stopped hosting the leadership of the PKK within two days, military action would follow. Earlier, the Syrians had denied that they were doing anything of the sort even after Turkish authorities had provided them with the address and the telephone numbers of the PKK headquarters in Damascus. Recognizing that the two countries were on the verge of war, the Hafez al-Assad Government felt obliged to close down the PKK headquarters in Damascus and sent its leader away. Since that time, relations have improved including questions concerning water. The improvement process was initiated by a meeting of the officials of the two governments in Adana on October 20, 1998 (Mazlum 2002, 191) and has continued smoothly since that time.

Although the relations between Turkey and her southern neighbours, especially Syria, have in the past often been characterized by disagreement regarding the waters of the Euphrates River, efforts to keep channels of communication open in order to become familiar with each others' viewpoints, to examine whether changes were taking place, to explore ways of cooperating and to demonstrate that the parties are committed to resolving their differences by peaceful means, have never been lacking.[20] As early as in 1964, Turkey had initiated separate and joint meetings with Syria and Iraq to offer them up to date information on the construction of the Keban Dam. These contacts culminated in a trilateral meeting in Baghdad in 1965 during which Turkey proposed the establishment of a Joint Technical Committee (JTC). Although the idea was accepted in principle, immediate disagreements between Iraq and Syria regarding whether the meeting should focus only on the Euphrates or also include the Tigris River, produced a deadlock. The idea was laid to rest (Bilen 1997, 112-113).

The idea of a JTC was revived in 1982 when Iraq and Turkey decided to re-establish it. Syria joined the following year (Beschnorner 1992, 40). Since that time

[19]Quoted in Beschnorner (1992, 42).

[20]A concise and excellent summary of the history of cooperation between Turkey, Syria and Iraq can be found in Kibaroglu and Unver (2006, 34-38).

the committee has met with some regularity. While representatives on the committee have exchanged some technical information, the basic outstanding issues between Turkey and her Southern neighbours cannot be dealt with by technical committees because they derive from major differences in policy positions. If one were to summarize these, two major points stand out. First, Turkey argues that the Euphrates and Tigris rivers constitute a single basin and river system while Syria and Iraq want to treat the two rivers separately, and concentrate particularly on the waters of the Euphrates River. Second, since the waters of the two rivers, especially the Euphrates River (93 percent) emanate from Turkey, Turkey argues that it is willing to negotiate only the *allocation* of the water. The lower riparians, on the other hand, consider the waters as belonging to all countries in the basin and therefore talk about *sharing* or *dividing* the Euphrates' waters.

The most comprehensive offer Turkey has put forth in order to establish a basis on which Turkey, Syria and Iraq could cooperate in managing the waters of the Euphrates and the Tigris rivers has been the Three Stage Plan. This plan was initially presented at a tripartite ministerial meeting in Ankara on June 26, 1990. It was coolly received and then rejected by Syria and Iraq (Kut 1993, 12). The plan envisaged a process of three consecutive stages which would result in the preparation of a master plan for *allocating* the water among the three countries. During the first stage, an inventory of water resources would be taken in the Euphrates-Tigris Basin. The associated activities would include collecting meteorological data and measuring river flows using the same standard instruments, and exchanging this information among the riparians. The second stage would comprise of the classification of soils and the determination of irrigation needs using the same standards and procedures. This would lead to the third stage where a master plan would be made for water allocation in the basin (Bilen 1997, 112-113).

Why were the downstream riparians not enthusiastic? We have already discussed two principles that they did not accept. To begin with, the plan treated the two rivers as forming part of the same basin while Syria and Iraq were interested talking only on the Euphrates, each for different reasons. Iraq is already assured of sufficient water from the Tigris and would therefore like to keep it out of the discussion. Syria, on the other hand, is interested only in the Euphrates since the Tigris does not flow through its territory. It considers the Euphrates as its lifeline. Furthermore, the plan did not change the Turkish position that the waters might be *allocated* but not *divided or shared.* Yet the plan contained elements of a breakthrough that were ignored if not unnoticed by the downstream countries. As Kut observes:

> "First, Turkey was proposing joint action to determine the common utilization of the waters by all three countries, thereby indirectly disclaiming full sovereignty over the rivers without calling it as such. Second, the plan was suggesting the application of high and appropriate technologies in order to minimize water requirements for agriculture, thus injecting the element of conservation as an essential solution to what is basically a problem of scarce resources. Together, the plan also proposed to regulate the flow of the Euphrates River according to seasonal needs of the downstream riparian countries - instead of keeping a steady flow which may not match the seasonal agricultural demand. . ." (Kut 1993, 13).

The plan necessitated a change in the mind set with which water problems are taken up among the three countries. The interpretation of territorial sovereignty and territorial integrity would have to be moderated to accommodate the making of joint studies and the collection and sharing of standardized information. Further, the implementation of the principles of "equitable and reasonable utilization" of water resources and "not causing appreciable harm to other riparians" would constrain inflexible interpretations of sovereignty (Bilen 1997, 113). These were points that the lower riparians were not willing to concede. Rather, they chose to insist that Turkey should increase the amount of water made available to Syria to 700 cubic meters per second (58 percent of which would then be released to Iraq). It is not clear whether either of the countries is in immediate need of this water or would be able to utilize it fully if the water were made available. But, in anticipation of a future state of affairs when they might need more water and fearing that close international cooperation might undermine domestic authoritarian structures which are to a large extent sealed from international influences, and preferring to maintain a state of emergency to bring about national integration, governments have shunned cooperation and limited their position to insisting on increasing the amount of water released downstream.

The changing international environment (the end of bipolarity, the US invasion of Iraq) and the shift in the foreign policy of Syria toward Turkey that came about in 1998 have reduced the tensions between Syria and Turkey. For example, during a visit in August 2001, the Syrian Minister of Irrigation called for a revival of the JTC. The two countries issued a joint communiqué that they would cooperate in the area of agricultural water research and the participation of Syrian experts in the international training programs of the GAP administration (Mazlum 2002, 201). This last point has been realized. At the time of writing, the Iraqi government had asked that the question of the waters of the Tigris and the Euphrates be taken up at the next Joint Economic Committee meeting between Turkey and Iraq, but the Turkish government had declined the request as an inappropriate item for the agenda of such a committee (Hurriyet Newspaper, 30 October 2006, 8).

5 Water as a saleable commodity

Holding a substantially different viewpoint than her Southern neighbours regarding the waters of the Tigris and the Euphrates, Turkey has come forth with offers of shipping water that it does not use to other Middle Eastern countries where water needs are more keenly felt. The offers of shipping water, while intending to demonstrate that Turkey is sensitive to the water needs of its southerly neighbours, also reflects the evolution of a recent line of thinking: The waters that come from national sources constitute a commercial commodity and as such, can be sold to interested parties.

The various proposals that have been put forth, including some that have been realized, constitute the topics of some other chapters of this book. The list includes the construction of pipelines to send national waters to various countries of the Middle East[21] (Kolars and Mitchell 1991, 289); (Dellapenna 1996, 234), the construction of suspended underwater pipelines to send water to Northern Cyprus and on to Israel and Palestine (Yavuz 1997, 562-563) after the failure of an experiment to ship the water in balloons (Nachmani 2000, 84-85), the sale of water from a dam on the Manavgat River to be shipped to buyers by tankers[22].

As the water needs of the neighbouring countries intensify, it may be possible to develop the unused national waters into a commodity which neighbouring countries may purchase. Currently, however, countries in the region experiencing water shortages seem not to be disposed favourably toward either coming to rely on other countries for their water supply or toward treating water as a commodity for which a price has to be paid. Future projects will be constrained by the same factors as exist now: the projects must be technologically feasible, economically viable and politically acceptable (Gruen 1994, 282).

6 Conclusion

Water has been rising in importance as a concern of Turkish foreign policy. As a latecomer to irrigated farming, Turkey has been trying to develop its water resources during recent years. This has provoked concerns among Turkey's southerly neighbours that gradually Turkey will use more and more water, leaving less and less for them to use. Conflict bound relations have been characteristic. Nevertheless, Turkey has worked to find peaceful ways through which the fears of its neighbours could be assuaged and bases for cooperation could be found. In its water-based relations with other neighbours, Turkey has demonstrated that it is possible to develop and implement projects to the mutual benefit and satisfaction of both sides. There is no reason that similar cooperation may not be achieved with her southerly neighbours. This calls for a patient process of confidence building and relying on the accumulation of successful steps for further success.[23] Turkey's moving toward

---

[21]Gruen mentions a mini-pipeline project that provides city water for Syrian cities, Amman, and the West Bank and Gaza which he defends as being much less expensive and more practical (1992, 42); see also Giray (1994, 243) and Gruen (1994, 281).

[22]In an agreement signed in 2004, Turkey agreed to sell and Israel to buy 50 million m.per annum of treated water for a period of fifty years. See article 1 of the Agreement between the Republic of Turkey and the State of Israel Concerning the Purchase of Water, March 4, 2004, Jerusalem. Text at the courtesy of the Ministry of Foreign Affairs of Turkey.

[23]Kolars and Mitchell express a similar opinion when they argue that "... failures should not be considered as indications of an impossibility of cooperation in international basins. The Turkish experience gained on its international rivers with the USSR and Greece suggests that, by carefully planned sequential approach to the problem, constructive and successful cooperation can be achieved." (1991, 291)

membership in the European Union (EU) should only facilitate this process since this is one of the intentions of the EU Water Framework Directive (WFD).[24]

It is not, however, realistic to expect Turkey to develop unilateral proposals to alleviate regional water problems when its neighbours have so far not gone beyond asking for larger amounts of water to be let downstream. It is equally unrealistic to expect Turkey to subscribe to the EU WFD, prepared without Turkey's input[25], until Turkey joins the EU and begins to take part in the EU decision making processes. Current EU documents tend to emphasize the needs of the lower riparians and demonstrate little sensitivity to Turkey's needs and rights (Kibaroglu and Unver 2006, 33). It is to be hoped that improved relations with Syria and with an Iraq, whose political status remains undetermined at the time of writing, will achieve an atmosphere of mutual confidence among them so that they may plan the uses of the waters of the Euphrates and Tigris rivers among them.

Difficult as it may sometimes seem, progress is possible. The warming of relations between Turkey and Syria, as already indicated, has produced a training programme for Syrian water officials in Turkey. Improved relations have most recently made it possible for the two countries to talk about the possibility of building a joint dam on the Orontes. Syria's withdrawal of support from the PKK and its recent willingness to conduct joint operations against the terrorist organization will only make it easier to pursue further cooperative endeavours regarding the mutual utilization of the waters of the Euphrates. The change in Turkish-Syrian relations and the positive impact of these changes on water related issues suggest that it is useful to integrate questions of water into broader and more inclusive programs of regional political and economic cooperation and development. Concentrating exclusively on water can easily lead to perceptions of a zero-sum game whereas a more integrated approach may multiply the areas of give and take, thereby enhancing the potential of generating positive sum games. In this regard, it may be useful to refer to the work of Euphrates-Tigris Initiative for Cooperation (ETIC), a community of scholars from the region working to develop a common frame of reference as a prelude to offering creative ideas to tackle the water related issues (Kibaroglu and Unver 2006, 39-40).

The promise of comprehensive approaches to water problems does not necessarily make them easy to develop and implement. There is much to be done until such time as integrated approaches are adopted. The three countries can both work individually and cooperate in achieving increased utilization of water conservation technologies, in adopting the cultivation of crops that require less water, in using methods to recycle water and in preserving the quality of their waters. The potential for selling unused national waters also needs to be developed. Some international support may be available for such undertakings.

---

[24]See Directive 2000/60/EC of the European Parliament and the Council of October 23, 2000, Official Journal of the European Communities, 22.12.2000, L.327/1.

[25]Aksoy (2006, 23) notes that much of the documentation prepared by the EU Commission is based on insufficient and sometimes inaccurate information as well as prejudgments bearing little relationship to facts.

As can be seen, the conclusion has gravitated toward Turkey's southerly neighbours. This is where water related questions are more likely to constitute a concern for Turkish foreign policy-makers than in those in other parts of the country. This is a challenge rendered more formidable by the alleged effects of global warming that appear to be reducing the precipitation in the region.

References

Aksoy O (2006) Avrupa Birligi'ne Katilim Surecinde Turk Sulari. Stratejik Analiz, no 80

Bagis AI (1994) Water in the region: Potential and prospects. In: Bagis AI (ed) Water as an element of cooperation and development in the Middle East. Hacettepe University and Friedrich Naumann Foundation, Ankara

Beaumont P (1994) The growing pressures on water resources in the Middle East and the need for new approaches to provide solutions. In: Bagis AI (ed) Water as an element of cooperation and development in the Middle East. Hacettepe University and Friedrich Naumann Foundation, Ankara

Beschorner N (1992) Water and instability in the Middle East. International Institute of Strategic Studies, London

Bilen O (1997) Water issues in the Middle East. South-eastern Anatolia Project Regional Development Administration, Ankara

Bilen O (2000) Orta Dogu Su Sorunlari ve Turkiye, 2nd edition TESAV, Ankara

Dellapenna J (1996) The two rivers and the land between: Mesopotamia and the international law of transboundary waters. BYU Journal of Public Law, 10(2)

Devlet Su Isleri (2005) DSI in Brief: 1954-2005. General Directorate of State Hydraulics Works, Ankara

DSI website (2007) Devlet Su Isleri (DSI). www.dsi.gov.tr/topraksu.htm Accessed 1 March 2007

Giray S (1994) Su Sorununda Turkiye'nin Siyasal Boyutlari. In: Bagis AI (ed) Water as an element of cooperation and development in the Middle East. Hacettepe University and Friedrich Naumann Foundation, Ankara

Gruen GF (1992) The water crisis, Rev. ed. A Simon Wiesenthal Center Report, Los Angeles

Gruen GF (1994) International Regional Cooperation: Preconditions and Limits. In: Bagis AI (ed) Water as an element of cooperation and development in the Middle East. Hacettepe University and Friedrich Naumann Foundation, Ankara

Hurriyet Newspaper, 30 October 2006

Inan Y (1994) Legal Dimensions of International Watercourses. In: Bagis AI (ed) Water as an element of cooperation and development in the Middle East. Hacettepe University and Friedrich Naumann Foundation, Ankara

Kibaroglu A, Unver O (2006) Frat-Dicle Havzasinda Isbirligi Icin Yeni Acılımlar. Stratejik Analiz, no 80

Kolars JF, Mitchell WA (1991) The Euphrates river and the Southeast Anatolia Development Project. University of Southern Illinois Press, Carbondale

Kut, G (1993) Burning Waters: Hydropolitics of the Euphrates and the Tigris. In New Perspectives on Turkey, no 9 (Fall)

Kut G (1994) Ortadogu'da Su Sorunu ve Turkiye. In: Ulman H (ed) Ortadogu ve Turkiye. TUSES, Istanbul

Mazlum I (2002) The hydropolitics of Turkey and Syria on the Euphrates-Tigris Basin: From the tense past to a cooperative future. Proceedings, IIIeme Congres International Du Dialogue Turco-Arabe, Istanbul, 22-26 May 2002. Institute for the Study of the Balkans and the Middle East, Istanbul

Nachmani A (2000) Scant resources: The problem of water in Cyprus. Mediterranean Politics, 5(3)

Official Journal of the European Communities (2000) Directive 2000/60/EC of the European Parliament and the Council of October 23, 2000. 22.12.2000, L.327/1
Resmi Gazete (1987) 10 Aralik 1987, No. 19660
Turan I, Kut G (1997) Political-ideological constraints on intra-basin cooperation on transboundary water. Natural Resources Forum, 11(2)
Turan I (2004) Water and Turkish foreign policy. In: Martin LG, Keridis D (eds) The future of Turkish foreign policy. MIT Press, Cambridge, Mass.
Yavuz H (1997) The Manavgat Project of Turkey: Water, an economic good. Water Resources Development, 13(4)

Turkey's Water Diplomacy: A Theoretical Discussion

Paul A. Williams

1 Introduction

The term 'water diplomacy' connotes explicit and purposeful communication between representatives of different states charged with negotiating a resolution to contentious issues related to the mutual use of common rivers. In fact, though, communication is seldom confined to the formal exchange of official views, as these issues often encourage harder forms of bargaining. As the entrenchment of opposing legal positions tends to prevail at the formal diplomatic level, states are inclined to employ more tacit exercises of influence using means ranging from positive inducements to coercion.

It is unsurprising that upstream-downstream constellations protract riverine disputes. Indeed, Realist International Relations theory, depicting international politics as an anarchic realm that compels states, the main actors, to be rationally preoccupied with survival (i.e., territorial sovereignty and integrity), to mistrust each other, to minimize interdependence, and to maximize relative gains in a zero-sum setting, all of which make relations competitive and war-prone, aptly applies to riverine conflicts. Basin states in geographically asymmetrical positions tend to advance legal principles that correspondingly oppose each other, with upstreamers espousing territorial sovereignty and equitable utilization, and downstreamers advocating territorial integrity and no appreciable harm to (their) prior uses. Moreover, because upstream developments often instil fear of material losses in downstream states, the latter employ coercive linkage strategies to forestall these events, exacerbating conflict.

Despite their conflictive tendency, however, upstream-downstream interactions rarely, if ever, end in 'water war'. Basin states also negotiate solutions, even if environmental and human-rights concerns are rarely incorporated, to problems of water scarcity associated with flow-altering hydrological installations. Most of

P.A. Williams (✉)
Department of International Relations, Bilkent University, Ankara, Turkey
e-mail: paulw@bilkent.edu.tr

A. Kibaroglu et al. (eds.), *Turkey's Water Policy*,
DOI 10.1007/978-3-642-19636-2_11,

Turkey's (trans-)boundary rivers feature agreements, albeit generally bilateral and sometimes informal, to divide water volumes, convene joint technical organs, establish joint construction projects and exchange data. According to Liberal International Relations theory, trade ties and regimes can shift state preferences in a more peaceful direction. That is, states may choose to address scarcity issues, and the uncertainty that surrounds them, by demanding new information necessary to craft solutions, and negotiations based on data exchange may reinforce cooperative behaviour. Furthermore, positive linkage strategies, employed especially by outside parties who can supply material inducements or enmesh disputants into integrative relationships that lessen the saliency of the zero-sum quality of the immediate conflict, can also play a valuable role.

Realism and Liberalism are rationalist approaches, assuming interests to be fixed in interactions, with variation only in strategies states use to attain their preferences. These theories suggest that cooperation is conditioned to different extents by objective material parameters, discounting the potential for change residing in social aspects of international relations. The Constructivist approach to International Relations inverts this rationalist-materialist logic, suggesting that it is the broader social context of relations that imbues material objects with their most essential meaning, the one that informs actors in a relationship how they should act towards the objects in question. Participants in an interaction engage in non-verbal as well as discursive actions, which are guided by shared understandings of identities, interests and appropriate behavioural norms and in turn either reproduce these understandings or change them in ways that improve cooperation or exacerbate conflict. For instance, while the issuance of *fait accomplis* reflects its relatively advantageous material position, even a salient upstream country like Turkey (i.e., vis-à-vis the Tigris-Euphrates system) has exhibited concern over its 'water rich' image, while implicitly accepting that its preferred legal principle of 'equitable and reasonable utilization' might be transformed from an instrumental negotiating stance into a more principled mode of reaching a flexible multilateral solution. Moreover, crucial shifts in the larger social context of basin-state relations have also influenced the probability of resolving specific water disputes.

The three aforementioned theoretical frameworks pertaining to international conflict and cooperation are surveyed below. Each successive section delineates the main assumptions and arguments of the respective approaches—Realism, Liberalism and Constructivism—before turning to discuss how each paradigm accounts for different aspects of Turkey's transboundary water relations.

2 Realism

Realism is highly applicable to analysis of transboundary water relations. This theory of International Relations depicts international politics as an anarchic realm where the main actors, i.e. states, assumed to be strategically rational, pursue the essential goal of survival, specifically territorial sovereignty and integrity. Because states can never fully ascertain whether others are defensively motivated—that is, attempting to maximize their respective levels of security to protect what they have already attained, or offensively driven to attain more than what they have, they inherently mistrust each other and prefer to be as self-sufficient as possible (Waltz 1979). States in this 'self-help' system thus experience irremediable degrees of insecurity that compel them to maximize relative gains and capabilities (Grieco 1993), thereby transforming most of their mutual interactions into zero-sum games. Consequently, international politics is competitive and war-prone.

Realism is not monolithic. One controversy within this school of thought centres on whether states are security seekers or power maximizes. Defensive Realists hold that states generally require only a certain minimum level of capability to be secure, with pursuit of additional increments signalling aggressive motives and consequently triggering counter-balancing behaviour by others. This presupposes that defensive capabilities have an advantage over, and can be differentiated from, offensive ones, making it possible for a state to obtain security without threatening others. Conversely, Offensive Realists argue not only that the difficulty of ascertaining motives is compounded by a fundamental similarity between defensive and offensive means, but also that anarchy and uncertainty about intentions compel states to ensure their survival by maximizing power, thereby fuelling the 'security dilemma' (Mearsheimer 2001, 20-21, 35-36). For example, upstream-downstream disputes have the potential to escalate because, in an atmosphere of mistrust featuring a perceived lack of full disclosure of plans and data (typical of competition between rival powers), the same hydrological installations that relevant upstream-state authorities explicitly support, as designed to enhance an array of national socio-economic goals tend, to be criticized in downstream states as threatening to their own economic security. Therefore, downstream-state officials typically try to deter or slow the completion of upstream projects and may in turn incur greater hostility, and heighten the resolve to finish the projects, on the part of upstream-state actors.

Another issue revolves around whether unipolar orders are more stabilizing for the international political system than multipolar (or bipolar) ones. Defensive Realism embodies 'balance-of-threat' theory, which holds that counter-balanced states have to possess aggressive *intentions* in addition to aggregate power, offensive capabilities and geographical proximity (Walt 1987, 21-26). Consequently, even hegemons can choose *not* to adopt capabilities, postures and strategies that make others insecure. Hegemonic Stability Theory goes so far as to assert that unipolar distributions of power even favour the provision of cooperative regimes and public goods. Arising out of a different intellectual tradition, but reaching a somewhat analogous conclusion to that of Hegemonic Stability Theory, the neo-Gramscian concept of 'hydrohegemony' suggests that an upstream state can achieve a type of 'positive hegemony' (Warner 2008, 280) combining the power inherent in its geographically dominant position with some degree of legitimate authority vis-à-vis the downstream countries. Offensive Realism assumes that all states seek hegemony to guarantee their survival, even as the quest for hegemony *per se* suffices to threaten others (Layne 2004, 109-115). This would imply that risks of war increase more when one state is ascending to hegemony than after it has indisputably established itself in that position. Analogously, efforts to build upstream projects may be more destabilizing to upstream-downstream relations than the successful operation of finished projects, as those downstream opponents who view these projects as conferring additional power upon others to curtail their access to a crucial flow of economic benefits also see the construction stage as the critical window of opportunity to prevent this power imbalance from materializing; afterwards, operational projects become key sources of leverage for upstream actors.

A third division among Realists concerns whether the anarchic structure of the international system suffices to explain balancing and war in the absence of other substantive goals. While Waltz (1959, 230-238) argued that anarchy was merely a 'permissive' cause of war, not an 'efficient' motive, as a Neorealist, Waltz (1979) argued that anarchy, assumed to be a constant 'ordering principle,' and the variable distribution of material capabilities (primarily military), were enough to account for the general conflictive quality of inter-state relations. In criticizing Waltz (1979) for having an under-determined theory of war, Mearsheimer (2001, 20) addresses this perceived lacuna by arguing that the structural properties of the system itself create pressures for balancing and war by inducing states to seek both hegemony and the overwhelming offensive capabilities that accompany it.

Others question Waltz's (1979) suggestion that precise motives for *specific* conflicts, which Neorealism relegates to the domain of foreign policy, are less important because they originate at the *unit* (domestic or intra-state) level of analysis. Neoclassical Realists, by contrast, directly address 'efficient' motives for state action stemming from domestic political and economic systems (Layne 2006, 7-11). Thayer (2000) even revives Morgenthau's argument that human nature causes war, but rests it on the findings of evolutionary biology. Lying at the core of this thesis is the impetus of material scarcity. Postulation of a causal nexus between scarcity and conflict extends from Thomas Malthus to more recent arguments attributing war to 'environmental scarcity' (Homer-Dixon 1999, Matthew 1999, Klare 2002). Indeed, any discussion of transboundary water relations would seem patently deficient without recognition of the role of scarcity as a key motivation behind conflict and its escalation.

2.1 Realism in transboundary water relations

Realist thought, developed to explain great-power politics, does not map precisely onto the contours of river basin relations revolving around available supplies of a localized natural resource. Interestingly, though, the symmetrical array of littoral states ringing a common lake or flanking a boundary river exemplify 'common pool resources'. For these natural goods, relevant consumers cannot be excluded, or at least at high cost, from enjoying the available benefits, but consumption is rival, in the sense that each state's use subtracts from the total benefits available to everyone else (LeMarquand 1977, 9). Thus, common pool resources' politics bear some analogical resemblance to great powers' zero-sum security competition. The substantive distinction between the two domains lies in the fact that, unless rivals can agree on an optimal assignment of usage rights, restraints and responsibilities, they are more likely to destroy the *resource* than each other, thus illustrating the renowned 'tragedy of the commons' (United Nations 2006, 209).

In contrast to common pool resources' configurations, private goods are both rivalrous and excludable. Common pool resources might even be 'privatized' through legal or technological means of exclusion. While Waterbury (2002, 10) asserts that upstream storage does not obliterate the commons' dimensions of a transboundary watercourse, as water must eventually be released, salient upstream actions asymmetrically and unreciprocally diminish the quantity and quality of water available downriver, to the practical extent of excluding downstreamers from access (LeMarquand 1977, 10, Matthew 1999, Williams 2003). Indeed, Homer-Dixon (1999, 137-141) describes 'simple scarcity conflicts' as those occurring over scarce, vital and physically capturable resources, such as water.

Thus, salient upstream-downstream constellations unsurprisingly lend themselves to protracted riverine rivalry. It is not infrequently remarked upon that the 'English word *rival* comes from the Latin *rivalis*, meaning one using the same river as another' (United Nations 2006, 215). This rivalry should intensify to block unilateral construction of upstream hydrological installations, which pose the potential for excluding downstreamers from enjoying the myriad benefits stemming from the freer flow of water (Naff and Matson 1984, 193). Therefore, users in downstream areas, where historically prior development tends to occur (McCaffrey 1990, 235), have obvious incentives to try to forestall upstream actions via linkage strategies that draw on power resources from outside the water-related issue-area (Keohane and Nye 2001). Moreover, political leaders also harbour fears of losing social prestige and power at home lest their traditional reputations for delivering water to citizenries come under the sort of serious challenge posed by another state's hydrological developments (Allan 2002, 215-217).

As noted above, Offensive Realism underscores that antagonism between great-power rivals is likely to resemble a 'security dilemma' that escalates to war because the two sides cannot necessarily differentiate between defensive and offensive capabilities, let alone motives for using these instruments. Apart from the obvious fact that upstream-downstream antagonism is based on more pronounced geographical asymmetries of position and interest, which tends to reinforce extant degrees of inter-state mistrust, especially in situations where shared hydrological data is in scarce supply, downstream-state actors are correspondingly ill-equipped to distinguish key intrinsic economic operations of upstream installations, especially where the latter permit substantial water diversions, from potential instrumental uses of these mechanisms to accomplish larger political goals (as suggested in Naff and Matson 1984, 184, Gleick 1993, 86-89).

2.2 *Realism in Turkey's water diplomacy*

Turkey's water diplomacy fits the Realist paradigm in a number of respects. Disputes over the Euphrates and Tigris rivers occupy a prominent position on Turkey's water-related foreign-policy agenda because of these rivers' national prominence. While Turkey's total mean annual run-off amounts to 186 billion cubic meters (BCM), only 112 BCM is estimated to be cost effective to develop, and the Euphrates-Tigris accounts for nearly 47 percent of this latter figure. The country's water-use demands are also rising. The supply being utilized is expected to increase due to: population growth, which could lower Turkey's annual per-capita water supply from 1,500 cubic meters to 1,000 cubic meters between 2000 and 2030; the need to improve upon respective 55 and 36 percent municipal access rates to sewage and wastewater treatment services; and efforts to augment domestic hydropower to meet more of Turkey's increasing energy needs, estimated to reach 285 billion kilowatt hours (kWh) in 2010. By then, the Southeast Anatolia Project (GAP) is projected to harness Euphrates-Tigris rivers' water to supply nearly 10 percent of Turkey's energy and one-fifth of its irrigable land (Kibaroglu et al. 2005, 5-9).

At the same time, Turkey's upstream position on both branches of this river basin offers commensurably advantageous access. While Turkish territory comprises small percentages of the Euphrates and Tigris basin areas, it supplies 90 and 40 percent of respective stream flows (according to Kolars 1994, 46, tributary flows affected by Turkey's groundwater use boost respective percentages to 98 and 49). Syrian and Iraq users had been exploiting the flow of water prior to Turkey's 1974 construction of the Keban Dam on the Euphrates River, with Iraqi withdrawals of Euphrates water reaching an estimated 45 percent of average annual flow in the 1960s (Naff and Matson 1984, 90). Even though Turkish officials could cite drought as a cause of lower volume, the concurrent filling of both Keban Dam, a hydropower-only installation, and Syria's Tabqa Dam, compounded crucial short-term shortages in Iraq, which, in the context of hostile relations, nearly went to war with Syria ostensibly for curtailing water releases (Naff and Matson 1984, 93-95, Kibaroglu et al. 2005, 59), instead of joining Syria, as in 1990, to protest Turkey's dam filling.

Downstream concerns stem from the capacity of Turkey's five Euphrates dams to impound nearly three years' worth of annual average water supply. Fear surrounded even the filling of reservoirs behind the four facilities permitting only non-consumptive uses, one (Keban) constructed separately and the other three (Karakaya, Birecik and Karkamis) built within the GAP framework, but intensified as work proceeded in the 1980s on the multi-purpose Ataturk Dam, the largest of the five, and its affiliated Urfa (diversion) tunnels. The Turkish government ran into increasing difficulty securing foreign financing for dam construction even before initiation of work on the Ataturk Dam in 1983, and outside funds came only after Turkey's assurance of 350 cubic meters per second for Iraq in 1966 vis-à-vis the Keban and 500 cubic meters per second for Syria in 1980 vis-à-vis the

Karakaya (Bilen 1997, 114-115). Following Ataturk Dam's completion, downstream-state objections dissuaded all but private consortia from funding the Birecik and Karkamis dams, and objections from non-governmental organizations caused the withdrawal by 2001 of all potential external financiers of the Ilisu Dam on the Tigris (Kibaroglu et al. 2005, 63).

Though lacking in sufficient military capabilities, downstream-state linkage strategies were not restricted to formal protests and moves to block external channels of monetary support, but included backing of terrorist activities. Syria began permitting PKK incursions into Southeast Turkey in 1984, one year after the Ataturk Dam underwent construction, and in 1987, agreed, in close proximity to another security accord, on a temporary protocol again committing Turkey to provide 500 cubic meters per second on average, measured over bimonthly periods, at the Syrian border (see Kibaroglu and Scheumann in this volume). This rough sequence of events reinforced perceptions of the politicized nature of the water allocation (Beschorner 1992, 37).

The filling of Ataturk Dam in 1990 actually increased Turkey's potential capacity to condition its willingness to accommodate downstream water needs on reciprocal security assurances. The probable impact of then President Turgut Ozal's pre-1990 threats to cut off water and private UN discussions on using Turkey's Euphrates dams to deprive Iraq of water during the first Gulf War (Gleick 1993, 89) increased after Ataturk Dam's completion. As expressed by former Prime Minister Suleyman Demirel in 1991, 'I do not believe in worrying about threats of war resulting from development projects in Turkey.... Turkey has deterrence. It will have more deterrence in the coming period. Turkey will build more such works. The more it builds, the fewer threats it will be faced with' (quoted in Beschorner 1992, 42). Nonetheless, Ankara eventually resorted to the actual threat of massive use of force in late 1998 to obtain Syria's expulsion of now-imprisoned PKK leader Abdullah Ocalan and the related Adana Accord. Moreover, while terrorist activity, which could affect dispersed irrigation-related works more easily from a logistical standpoint than spatially concentrated storage installations, may have directly retarded GAP's overall progress, especially in meeting irrigation and Tigris development goals, the security situation probably had greater indirect effects by slowing land reform and on-farm training (Kolars 1994, 67-68, 70-75).

The slowing of the GAP does not imply Turkey's concession of sovereignty over Euphrates-Tigris water. While Turkish governments abjure the principle of 'absolute territorial sovereignty' in favour of 'limited sovereignty', calling the Euphrates a 'transboundary' river imputes an exclusive right to exploit it until it flows into Syria. Former Prime Minister Demirel even stated that 'water is an upstream resource and downstream users cannot tell us how to use our resources', with others in authoritative positions echoing that the 'rivers concerned are not subject to international rules' (quoted in Bagis 1997, 577). The prevailing position is usually articulated more subtly. Since 1984, Turkish experts have consistently advocated the 'Three-Staged Plan for Optimal, Equitable and Reasonable Utilization of the Transboundary Watercourses of the Euphrates-Tigris Basin', which proposes to inventory all riparians' water and land resources before evaluating both sets jointly in order to determine actual water needs and to convene a joint body to assemble and exchange data on temporal fluctuations in the level of these needs (Kibaroglu et al. 2005, 61). While founded on principles of scientific rationality, the likely result of this plan's acceptance as a basis for tripartite negotiations would be to reveal the lesser viability of Syrian and Iraqi irrigation expansion plans (Scheumann 1998, 128). Moreover, adamant upstream adherence to this preferred mode of resolving water-allocation issues could be perceived as another fait accompli to gain more time in which to complete projects, the adverse effects of which cannot be ascertained until construction finishes. This is also an issue that has complicated Turkish-Georgian relations on the Coruh River (Kibaroglu et al. 2005, 52-53, 89).

3 Liberalism

Despite their predominant propensity towards conflict, upstream-downstream interactions virtually never escalate to the point of inter-state 'water war' and cooperation is prevalent to varying degrees (Deudney 1999, Homer-Dixon 1999, Allan 2002, Williams 2003, Kibaroglu et al. 2005, Selby 2005, Conca 2006, United Nations 2006). Liberal International Relations theory, especially Neoliberal Institutionalism, which encompasses regime theory, argues that anarchy makes cooperation difficult but not impossible where states stand to obtain larger absolute gains than each could otherwise get from unilateral action (Keohane 1989, Baldwin 1993). In the prisoner's dilemma game that exemplifies the cooperation-under-anarchy problematique, uncertainty centres on whether potential partners will honour an agreement. Cooperation is allowed by the fact that most states interact over an indefinite time period and can thereby respond to each other's behaviour in tit-for-tat fashion if they can also agree on how to define, detect and punish cheating and other forms of non-compliant behaviour (Axelrod and Keohane 1993).

Liberalism variously emphasizes the role of democracy, trade and institutions in strengthening cooperation. Indeed, Immanuel Kant's notion of 'perpetual peace' roots cooperative trends in transboundary flows of ideas, goods and people (Doyle 1986). Based on Krasner's (1983, 1) famous definition of international regimes as 'principles, norms, rules and decision-making procedures around which actor expectations converge in a given issue area', Neoliberal Institutionalism (Keohane 1989) views institutions as essential for reaching both primary agreements on substantive issues and secondary cooperation on enforcing commitments. The fact that regimes are products of state cooperation in the first instance begs the question of whether regimes can independently impel more cooperation (Hasenclever et al. 1997, 43-44). While not necessarily rejecting the 'hegemonic stability thesis' that only a powerful actor expecting commensurable gains can surmount collective action dilemmas by supplying the public goods of regime creation and enforcement itself, Liberals nonetheless hold that regimes can outlast hegemonic decline because they are costly to replace and may continue to yield greater benefits than unilateral action (Keohane 1984).

Neoliberal Institutionalists expect regime effectiveness to be higher in certain issue-areas. To the extent that they feature relative symmetry of interests and power, lesser proximity to vital national interests and a smaller probability of generating absolute gains from cooperation that can be converted into military power (Keohane and Nye 2001, Powell 1993), economic issue-areas are often more conducive to cooperation than military security (Jervis 1983). Otherwise, interactions over economic issues that impinge on vital security matters may exhibit the escalatory dynamics of these more highly competitive matters. Then, engaged third-party leadership may play a crucial role in catalyzing regimes (Young 1991) by providing opportunities for integrative linkages or package deals at higher levels of interaction (Benvenisti 1996, 401), thus altering the strategic calculi of actors in a positive direction.

3.1 Liberalism in transboundary water relations

Certain problems related to international watercourses lend themselves more easily to cooperative resolution than others. Salient geographical and hydrological characteristics facilitate, to a greater or lesser extent, the formation of cooperative international regimes to manage economic uses of different multi-state systems of freshwater (LeMarquand 1977, 7-11). Situations where the relevant actors enjoy a relatively more symmetrical relationship to a common supply of water and non-consumptive uses of water (i.e., those, such as flood control or hydropower, that return withdrawn water, in the same quantity and quality, to the river channel) prevail should be the most conducive to cooperative arrangements, as indicated by numerous agreements regulating joint uses of 'international rivers' that demarcate boundaries. Cooperation confronts slightly more formidable obstacles when upstream and downstream states are negotiating to apportion the costs of 'integrated development opportunities' involving the production of new public goods, such as flood control or hydropower, but becomes even more difficult in relation to the question of how to induce individual and collective restraint on the part of symmetrically arrayed consumptive users of a common pool resource before it is depleted. However, the hardest test for Liberal IR theory involves the resolution of 'upstream-downstream conflicts', where asymmetry of territorial position is compounded by the predominance of consumptive uses of water. In these cases, incentives for cooperation, assuming that upstream states posses a preponderance of military deterrence capabilities, may come from the upstream states' countervailing cooperative interests, for instance, in protecting economic and social values attached to occupation of downstream position in other disputed river basins or in bolstering a 'good neighbourly' image for the purposes of promoting larger economic and political goals (Ibid, pp 10-12).

Territorial proximity raises the potential value of positive relations with downstream states. Upstream states now espouse 'limited sovereignty', which entails liability for 'significant' harm to downstream uses (Wapner 1998, 280), rather than 'absolute territorial sovereignty', and often engage in prior notification and consultation related to project construction. Embrace of the 'equitable and reasonable utilization' principle, reflecting recognition of the historically laggard pattern of development upstream, may also imply the need for ancillary data gathering and exchange efforts, which can mitigate the severity of competition and nurture habits of cooperation (Benvenisti 1996, 401). Indeed, Neoliberal Institutionalism, the direct successor to Regime theory, identifies the need for information as a core part of regime demand (Hasenclever et al., 33-34). In the larger context of trade ties and institutional integration, which lessen the economic and political exigencies of the perceived need for self-sufficiency or territorial conquest (Deudney 1999, 205-207), upstream states may be more predisposed to limit their own allocatively inefficient consumption of water in agriculture. 'Win-win' forms of cooperation identify opportunities for assembling 'baskets of benefits' that enhance each party's absolute welfare gains by transcending issues of volumetric allocation to encompass financial flows, hydropower trade, data sharing and peace agreements, all of which nonetheless require institutional capacity-building with the assistance of such key outside actors as the EU, World Bank and the UN Development Program (United Nations 2006, 224-231).

3.2 Liberalism in Turkey's water diplomacy

Despite the high profile of its riverine water controversies, Turkey has signed accords to allocate streamflow in common watercourses, establish joint technical bodies, fund construction projects, exchange data and provide prior notification and consultation in relation to its infrastructure. As expected, Turkey faces stronger issue-specific incentives to cooperate on freshwater usage in shared river basins where it is downstream, but has also committed itself to joint action with respect to the use of river water that originates on Turkish territory but later demarcates one of its international boundaries. Conversely, arrangements where Turkey occupies a purely upstream position seem less committal, but bear some similarity in certain substantive terms to other agreements. Moreover, as in similar cases elsewhere in the world, Turkey's water-related interactions with various neighbours in particular basins are influenced not only by the territorial configuration of the watercourse in question but also by the quality of its overall bilateral relations with those neighbours (Brunée and Toope 1997, 47; United Nations 2006, 220).

Downstream position provides a logical incentive to collaborate. In the Meric river basin, Turkey has worked with upstream Bulgaria, with which it shares a 15-km boundary section of the Tundja tributary, to alleviate flooding and drought. Their 1968 agreement committed them to 'good neighbourly' limits on mutual harm, data exchange and dispute settlement by a joint commission, and after the

Cold War, Bulgarian economic decline even bolstered the prospects for further cooperation by reducing negative externalities (e.g., upstream pollution) and paving the way for Turkey to purchase water and commit to building infrastructure on Bulgaria's stretch of the Arda tributary and to importing the generated hydropower, thus enmeshing water cooperation in complementary trade ties. Sharing the 187-km downstream boundary section of the Meric that reaches the Aegean Sea, Turkey and Greece have agreed on permitted infrastructure, data exchange, prior notification and dispute settlement. Poor environmental and water-quality protections have opened political space for institution-building by outside parties, especially the European Union, the German government, the IUCN and UNESCO (Kibaroglu et al. 2005, 26–37). Conversely, Turkey's downstream vulnerability on the smaller Orontes River, which also demarcates a 31-km section of the international boundary via a 1939 Franco-Turkish treaty, was until recently aggravated by tense relations with Syria and a long-standing dispute over Hatay province. However, the 1998 Adana Accord gave a new impetus to cooperation, manifested in a 2004 trade agreement implying Syrian recognition of Turkey's possession of Hatay and an accord creating a bilateral technical commission to study building a joint dam on the Orontes (see Scheumann / Tereci / Sagsen in this volume; Kibaroglu et al. 2005, 71). An important part of efforts to build this 'friendship dam' involves positive potential linkages to this area of energy cooperation (Turkish Daily News 2007).

Turkey has engaged in cooperation even where it enjoys upstream status. In the Kura-Aras basin, upstream Turkey has numerous agreements with Georgia, Armenia and Iran. Water relations feature border delimitation, technical cooperation and joint dam construction, but, unlike the Meric and Orontes basins, Turkey has agreed to allocate various streamflows to these fellow riparians on a 50-50 basis and even to provide Iran a minimum flow of Aras (Sarisu tributary) water. Because the Arpacay tributary (Aras) delineates a long Turco-Armenian boundary, a natural locus for cooperation prior to rising Turco-Armenian tensions, the two countries had created a bilateral commission, divided water equally and agreed on joint diversion-dam construction (Kibaroglu et al. 2005, 38-46). Turkey's 1927 agreements with Georgia also cover the Coruh Basin, but, while these could not anticipate post-Cold War concerns over coastal erosion in downstream Georgia, Turkey, which values the energy transit trade with its neighbour (i.e., flow of crude oil and gas via the Baku-Tbilisi-Ceyhan and Baku-Tbilisi-Erzurum pipelines), has demonstrated an interest in assisting the study of its planned dams' ecological impact (Kibaroglu et al. 2005, 47-54).

Even the highly competitive Tigris-Euphrates rivers feature tenuous cooperation. While signed in close proximity to a security accord, the 1987 Turco-Syrian Protocol nonetheless informally committed Ankara to a 50-50 allocation of Euphrates water to Syria. Moreover, before Ataturk Dam's 1990 filling, Turkish authorities engaged, though not to downstreamers' satisfaction, in prior notification, consultation and streamflow releases from other dams (Gleick 1993, 107-108). Some contend that Turkey's dams even provide the public good of flow regulation (Bilen 1997, 68-73, Allan 2002, 255-266). Information sharing has taken place, and, while it cannot overcome larger political tensions, the 1983-1993 Joint

Technical Committee (JTC) meetings, where Turkey advocated its Three-Stage Plan, expressed commitment to finding a cooperative solution. In low-profile settings, water scarcity may propel conservation and innovation efforts, such as those to reduce the three-quarters share of Turkey's water use in irrigated agriculture (Kibaroglu et al. 2005, 8).

4 Constructivism

The previous rationalist approaches assume that state goals are relatively fixed, with variation only on strategies selected to maximize given values in an interaction (Hasenclever et al. 1997, 23-24, 136). States act rationally in pursuit of material interests. International conflict represents the initial default outcome, although the liberal argument that absolute gains from continued cooperation should diminish problems of cheating opposes the Realists' scepticism that states can surmount relative-gains concerns (Grieco 1993). Conflict and cooperation basically follow from the structure of material incentives, but even scholars who emphasize rationality of state action sometimes concede that outcomes are influenced by the social context of international relations (Wendt 1995).

The Constructivist approach to International Relations opposes rationalist-materialist accounts. It places more emphasis on the independent influence of ideas, including identities and the constituent elements of Krasner's (1983) seminal definition of regimes, than material forces, which are argued to be significant only within larger social structures, which also consist of shared knowledge and agent or actor practices that (re)produce these structures (Wendt 1995, 73). More importantly, not only the interests, but also the very identities, of actors can change as a result of social interchange. In processes of interaction, actors can learn new identities, but one actor may also be teaching another to adopt preferred roles (e.g., enemy, rival or friend) by symbolizing the latter '*as if* it already had that identity' (Wendt 1992, 421). However, Wendt (1999, 346-347) also recognizes that humans engage in 'ideological labour', discursive activity that generates social knowledge of interdependence and builds the necessary trust to avoid collective action dilemmas. Others explicitly emphasize linguistic practices. Habermas's theory of 'communicative action' stresses the capacity of practical reasoning, centred on speech acts and reasoned argumentation (Muller 2001). This can even lead to the 'argumentative self-entrapment' of those who find themselves accountable for fulfilling the normative implications of statements initially put forward to advance instrumental bargaining positions (Risse 2000, 23). Simply put, actors can find their own words unintentionally moving them towards cooperation.

International Relation's theoretical Constructivism tends to reveal an underlying substantive concern for how social interaction shifts actors' identity conceptions and interests in a cooperative direction, as implied in the phrase 'Anarchy is what states make of it' (Wendt 1992). Constructivist accounts imply that anarchy and states are co-constituted, as states enjoy sovereign rights and responsibilities, including respect for the meta-principle of *pacta sunt servanda* ('agreements shall be served'), upon which issue-specific regimes can be founded (Hasenclever et al., 169-175). Yet, Constructivism is not a theory of cooperation, as interaction can exacerbate hostilities, and it posits that change is possible, not certain (Wendt 1995), as anarchy and states' predominantly rivalrous identities are shaped by powerful logics of reproduction (Wendt 1999).

4.1 Constructivism in transboundary water relations

Perhaps innately aware of the aforementioned risk of rhetorical entrapment, the representatives even of powerful upstream states purposefully and carefully craft discourse suiting their interests in regime formation and collaborative action. As international water law presents a logical and salient discursive template by which regimes may be shaped in certain directions, and not others, for specific river basins, reaching a consensus agreement on the textual content of this law becomes as politicized a task as that of negotiating particular basin accords. The 1997 UN (framework) Convention on the Law of the Non-navigational Uses of International Watercourses passed only after arduous negotiation and never obtained enough ratifications to become effective. The document provided no practical guidelines on reconciling its competing master principles, 'equitable and reasonable utilization' (Article 5) and the 'obligation not to cause significant harm' (Article 7), or on resolving competing claims, and fewer basin accords embody either or both of these principles, as well as that of prior notification (Conca 2006, 110-118; United Nations 2006, 218).

In its earliest gestation, international water law barely transcended crude codification of asymmetrical positions, with upstream states favouring 'absolute territorial sovereignty' and downstream states embracing 'absolute territorial integrity'. The current prevailing theory of 'limited territorial sovereignty' (Kibaroglu et al 2005, 20) signals progressive movement towards reasoned compromise, since it entails that the predominantly upstream development of 'equitable uses', which does not mean equal divisions of water (Gleick 1993, 106-107), needs to take downstream 'acquired rights' into account (Waterbury 2002, 28). The implied limitation on states' rights to exploit transboundary resources signals emergence of a shared understanding that individual states may be held liable for 'damaging other states ecosystems' (Wapner 1998, 280). Moreover, the fact that the Convention even contains an 'obligation not to cause significant harm' puts downstream interests on a level legal footing (Conca 2006, 100-101). Watercourse agreements that proscribe 'significant harm' also tend to provide for prior notification, data exchange, and peaceful dispute resolution, the latter two appearing more frequently in general, while accords enshrining 'equitable use' often include corollary water allocations (Conca 2006, 114-118).

Providing a weak foundation for articulating shared conceptions of a 'common' resource, international legal developments offer an even more precarious basis for substantively more complex collaboration within the holistic framework of 'integrated watershed management'. Most agreements are dyadic, even for multi-riparian basins. Fulfilling an 'obligation not to cause significant harm' can be consistent with preserving ecological values and human needs (Waterbury 2002, 29), but most of the few basin accords upholding this principle have not 'created significant entry points for environmental protection' (Conca 2006, 118). Some even argue that, precisely by enticing disputants to advance flexible negotiating positions, contrary to fixed rules that create deadlock, the vague standard of 'equitable use' may allow for periodic water reallocations that respond to newly emergent needs (Benvenisti 1996). Others similarly emphasize the value of non-binding 'contextual regimes' as the basis for 'international ecosystem law', which emphasizes 'future-oriented, flexible and adaptable frameworks that nonetheless allow for the elaboration of specific norms on individual issues' (Brunée and Toope 1997, 42). Here, third parties might supply not only material assistance, but also shared knowledge of 'best practices' and non-state-centric modes of discourse.

4.2 Constructivism in Turkey's water diplomacy

Understood in an International Relations Constructivist mode, Turkey has exerted itself in two key ways vis-à-vis its transboundary waters. In relation to the Tigris-Euphrates rivers, it has tried to eschew a 'water rich' image, while implicitly accepting the risk, albeit low, that its advocacy of 'equitable and reasonable utilization', if accepted by Syria and Iraq as a basis for negotiating a permanent tripartite treaty, could transform an instrumental negotiating stance into a principled basis for committing Turkey to provide a certain flow of water downstream. In this vein of logic, Turkey's ongoing EU accession process may entail outside introduction of 'green' discourse at the expense of the technocratic language that constitutes the dominant frame within which its transboundary water uses have been developed and negotiated.

Several factors fostered Turkey's politicized 'water rich' profile. In proximity to announcement of the Three-Stage Plan, a 1984 government study overestimated excess runoff in Turkey's 26 drainage basins at two thirds of the total (Kolars and Mitchell 1991, 291, 294-295). Ozal's 'Peace Pipeline' proposal, a component part of the 1987 Protocol that also pledged to deliver 500 cubic meters per second of Euphrates water at Syria's border, also implied that Turkey could compensate for any permanent shortage there by supplying water from its wholly domestic Seyhan and Ceyhan rivers. Moreover, the 1989 GAP Master Plan underestimated future maximum Euphrates water consumption at 10.4 BCM, equivalent to 335 cubic meters per second (Kolars 1994, 75), thus dovetailing with the 1990 Syro-Iraqi demand of 700 cubic meters per second (Gruen 1993, 103-104). Finally, in 1991, eminent hydrologist Malin Falkenmark placed Turkish per capita water supply at 4,600 cubic meters per year, much higher than Syria's 1,300 (Gruen 2000, 3). Ankara later moved to undermine the language of plenty, as when the foreign minister stated in 1992 that, 'We are ready to cooperate in this respect [water], but I must point out that. . .. Turkey is not a country which has abundant water resources. We may soon face problems in meeting our own needs' (quoted in Beschorner 1992, 44). This discursive turn suggests that even powerful upstream states remain conscious of the need to attend to the normative legitimacy of their bargaining positions. While Turkey represents one of three states that voted against the 1997 UN Convention, its complaints about what it claims to be the undue importance attached to the 'obligation not to cause significant harm' and inclusion of mandatory dispute settlement have some broader intellectual resonance (Benvenisti 1996, Brunée and Toope 1997, Waterbury 2002).

Conversely, discourse may also impel Turkey towards more intensive cooperation. Turkish experts' consistent advocacy of the Three-Stage Plan, embodying their preferred technocratic language, is premised on a basic understanding that water scarcity is non-negligible and requires curtailing those downstream claims, based on non-scientifically formulated mathematical quotas, that Turkey can 'reasonably' expect to fulfil within the spirit of 'equitable utilization' (Kibaroglu et al. 2005, 61-62). If multilateral relations improve and basin societies become more receptive to freer and more transparent exchanges of ideas and data, downstream actors may become willing to invoke the principle of 'equitable and reasonable utilization' to defend values associated with environmental restoration of endangered ecosystems, like the Mesopotamian Marshlands in southern Iraq and their inhabitants' unique traditions, to the extent that an argument can be credibly made for how and which upstream activities are as damaging now as actions by Saddam Hussein's Baathist regime were in pre-2003 history. The Meric River, where two riparians belong to the EU and the third, Turkey, is a candidate member, indicates how concerted external involvement may deepen local actors' cooperative attitudes by directing them to take on 'win-win' tasks of environmental protection (Kibaroglu et al. 2005, 32-36, 86-90). Cooperative tendencies are additionally supported by the language of the European Water Framework Directive of 2000, which stipulates that EU members should coordinate with non-members (United Nations 2006, 221).

5 Conclusions

At first glance, Realist theories provide a useful analytical framework for understanding Turkey's water diplomacy, the tenor of which predominantly reflects the mixture of material incentives for competition suggested by its respective geographical positions on various transboundary watercourse systems. For instance, the country's governments have displayed their greatest reluctance to relax their bargaining stance on the Tigris-Euphrates system, both because of the nature of issues specific to that basin and to compensate for a weaker position on the Orontes River. While Turkey's greater desire for basin-centric collaboration in the Meric and Orontes might reflect its downstream position, and its partial boundary position considerably mitigates incentives for competitive unilateralism on the Kura-Aras tributaries, these factors do not explain the existence, albeit highly imperfect and tenuous, of cooperative actions vis-à-vis the Coruh and Tigris-Euphrates rivers. As Liberalism would anticipate, in the latter cases, extraneous economic inducements, especially in the form of trade ties, exist to some degree.

However, the overall quality of Turkish cooperation on this issue remains sub-optimal. As Constructivist analysis might suggest, this is in part the result of implicit discursive efforts to avoid having to play a role of basin-wide provider of Tigris-Euphrates water, which would be more normatively appropriate for a country with a 'water rich' identity. On the other hand, the discourse of 'equitable and reasonable utilization' could, albeit with low probability at present due to downstream-state actors' reluctance to embrace it (reflecting what seems to be a gaping deficit of shared social knowledge of the problem at hand), form a principled basis for conducting ongoing dialogue leading to agreement on more favourable apportionments of Euphrates-Tigris water to downstream states, but in explicit service of inchoate ecosystem values, the benefits of which might diffuse more widely to current and future populations at large.

References

Allan JA (2002) The Middle East water question: hydropolitics and the global economy. I.B. Tauris, London and New York

Axelrod R, Keohane RO (1993) Achieving cooperation under anarchy: strategies and institutions. In: Baldwin DA (ed) Neorealism and neoliberalism in world politics. University Press, New York Columbia, pp 85-115

Bagis AI (1997) Turkey's hydropolitics of the Euphrates-Tigris basin. Water Resources Development 13:567-581

Baldwin DA (1993) (ed) Neorealism and neoliberalism in world politics. University Press, New York Columbia

Benvenisti E (1996) Collective action in the utilization of shared freshwater: the challenges of international water resources law. The American journal of international law 90:384-415

Beschorner N (1992) Water and instability in the Middle East. Adelphi Paper 273. International Institute for Strategic Studies, London

Bilen O (1997) Turkey and water issues in the Middle East. Southeastern Anatolia Project (GAP) Regional Development Administration, Ankara

Brunée J, Toope SJ (1997) Environmental security and freshwater resources: ecosystem regime building. The American journal of international law 91:26-59

Conca K (2006) Governing water: contentious transboundary politics and global institution building. The MIT Press, Cambridge (MA) and London

Deudney DH (1999) Environmental security: a critique. In: Deudney DA, Matthew RA (eds) Contested grounds: security and conflict in the new environmental politics. SUNY Press, Albany, pp 187-219

Doyle MW (1986) Liberalism and world politics. American Political Science Review 80:1151-1169

Gleick PH (1993) Fresh water resources and international security. International Security 18:79-112

Grieco JM (1993) Anarchy and the limits of cooperation: a realist critique of the newest liberal institutionalism. In: Baldwin DA (ed) Neorealism and neoliberalism in world politics. Columbia University Press, New York, pp 116-140

Gruen GE (1993) Recent negotiations over the waters of the Euphrates and Tigris. In: International Water Resources Association (ed) Proceedings of the international symposium on water resources in the Middle East, Urbana-Champaign, pp. 100-108
Gruen GE (2000) The politics of water and Middle East peace. American Foreign Policy Interests 22:1-21
Hasenclever A, Mayer P, Rittberger V (1997) Theories of international regimes. Cambridge University Press, Cambridge, UK
Homer-Dixon T (1999) Environment, scarcity, and violence. Princeton University Press, Princeton
Jervis R (1983) Security regimes. In: Krasner S (ed) International regimes. Cornell University Press, Ithaca, London, pp 173-194
Keohane RO (1984) After hegemony: cooperation and discord in the world political economy. Princeton University Press, Princeton
Keohane RO (1989) Neoliberal institutionalism: a perspective on world politics. In: Keohane RO (ed) International institutions and states power: essays in international relations theory. Westview Press, Boulder, pp 1-20
Keohane RO, Nye JS (2001) Power and interdependence: world politics in transition (3rd edn). Longman, New York
Kibaroglu A, Klaphake A, Kramer A, Scheumann W, Carius A (2005) Cooperation on Turkey's transboundary waters. Research report, German Federal Ministry for Environment, Nature Conservation and Nuclear Safety, Berlin
Klare MT (2001) Resource wars: the new landscape of global conflict. Henry Holt and Company, New York
Kolars J (1994) Problems of international river management: the case of the Euphrates. In: Biswas A (ed) International waters of the Middle East from Euphrates-Tigris to Nile. Oxford University Press, Oxford, pp 44-94
Kolars JF, Mitchell WA (1991) The Euphrates river and the Southeast Anatolia development project. Southern Illinois University, Carbondale, Edwardsville
Krasner S (1983) Regimes and the limits of realism: regimes as autonomous variables. In: Krasner (ed) International regimes. Cornell University Press, Ithaca, London, pp 355-368
Layne C (2004) The war on terrorism and the balance of power: the paradoxes of American hegemony. In Paul TV, Wirtz JJ, Fortmann M (eds) Balance of power: theory and practice in the 21st century. Stanford University Press, Stanford, pp 103-126
Layne C (2006) The peace of illusions: American grand strategy from 1940 to the present. Cornell University Press, Ithaca, London
LeMarquand D (1977) International rivers: the politics of cooperation. Westwater Research Centre, Vancouver
Matthew RA (1999) Scarcity and security: a common-pool resource perspective. In: Barkin JS, Shambaugh GE (eds) Anarchy and the environment: the international relations of common pool resources. SUNY Press, Albany, pp 155-175
McCaffrey S (1990) The non-navigational uses of international watercourses. In: American Society of International Law (ed) Proceedings, Eighty-Fourth Annual Meeting, pp 228-235
Mearsheimer JJ (2001) The tragedy of great power politics. WW Norton and Company, New York and London
Muller H (2001) International Relations as Communicative Action. In: Fierke KM, Jørgensen KE (eds) Constructing international relations: the next generation. ME Sharpe, Armonk, London, pp 160-178
Naff and Matson (1984) Water in the Middle East: Conflict or Cooperation. Westview Press, Boulder
Powell R (1993) Absolute and relative gains in international relations theory. In: Baldwin DA (ed) Neorealism and neoliberalism in world politics. Columbia University Press, New York, pp 209-233
Risse T (2000) 'Let's argue!': communicative action in world politics. International Organization 54:1-39

Scheumann W (1998) Conflict on the Euphrates: An analysis of water and non-water issues. In: Scheumann W, Schiffler M (eds) Water in the Middle East: potential for conflicts and prospects for cooperation. Springer, Berlin, pp 113-135

Selby J (2005) Oil and water: the contrasting anatomies of resource conflicts. Government and opposition 40:220-224

Thayer B (2000) Bringing in Darwin: evolutionary theory, realism, and international politics. International Security 25:124-151

Turkish Daily News (2007) A friendship dam to be built on Orontes River by Turkey and Syria. 5 April 2007

United Nations (2006) Human development report 2006: Beyond scarcity: Power, poverty and the global water crisis. United Nations Development Program, New York

Walt SM (1987) The origins of alliances. Cornell University Press, Ithaca, London

Waltz KN (1959) Man, the state, and war: a theoretical analysis. Columbia University Press, New York

Waltz KN (1979) The theory of international politics. Random House, New York

Wapner P (1998) Reorienting states sovereignty: Rights and responsibilities in the environmental age. In: Litfin K (ed) The greening of sovereignty in world politics. The MIT Press, Cambridge (MA) and London, pp 275-297

Warner J (2008) Contested hydrohegemony: hydraulic control and security in Turkey. Water Alternatives 1(2):271-288

Waterbury J (2002) The Nile basin: national determinants of collective action. Yale University Press, New Haven, London

Wendt A (1992) Anarchy is what states make of it: the social construction of power politics. International Organization 46:391-425

Wendt A (1995) Constructing world politics. International Security 20:71-81

Wendt A (1999) Social theory of international politics. Cambridge University Press, Cambridge, UK

Williams PA (2003) The common and uncommon political economies of water and oil "wars." ASAM Review of International Affairs 3:13-28

Young O (1991) Public leadership and regime formation: on the development of institutions in international society. International Organization 45:281-308

Turkey's Position towards International Water Law

Annika Kramer and Aysegul Kibaroglu

1 Introduction

Turkey's position towards transboundary water cooperation is widely perceived as being reluctant. This view mainly originates from disputes that arose over the Euphrates and Tigris rivers between Turkey, being the upstream country, and the downstream riparians, Syria and Iraq. In addition, Turkey's vote against the 1997 United Nations Convention on the Law of the Non-navigational Uses of International Watercourses (UN Watercourses Convention) as well as Turkey's refusal to discuss transboundary water issues within the context of the 2002 Organisation for Security and Co-operation in Europe (OSCE) Economic Forum[1] explains the view of Turkey's critics.

However, as will be seen in the following chapters and the Annex of this book, Turkey has entered a number of bilateral agreements on transboundary water resources management with its neighbours. Turkey is also a signatory to multilateral agreements on the protection of marine environments and freshwater ecosystems, such as the Convention on Wetlands of International Importance, called the Ramsar Convention, the Convention for the Protection of the Mediterranean Sea against Pollution and the Convention on the Protection of the Black Sea against Pollution.

---

[1]For further details refer to the summaries of the OSCE 10th Economic Forum and of its respective preparatory meetings, available at http://www.osce.org/eea/documents.html.

A. Kramer (✉)
Adelphi, Berlin, Germany
e-mail: kramer@adelphi.de

A. Kibaroglu
International Relations Department, Okan University, Istanbul, Turkey
e-mail: Aysegul.Kibaroglu@okan.edu.tr

A. Kibaroglu et al. (eds.), *Turkey's Water Policy*,
DOI 10.1007/978-3-642-19636-2_12, © Springer-Verlag Berlin Heidelberg 2011

2 General principles of international water law

International water law, just as international law, is based on general legal principles such as, *inter alia*, the principle of equality of states, good neighbourliness, peaceful settlement of disputes, achieving international cooperation in solving international problems. Two principles, in particular, are considered to be cornerstones of international water law (compare e.g. Brunnee and Toope 1997):

- the principle of equitable and reasonable utilisation and participation in the development of the watercourse by all riparian states
- the obligation not to cause significant harm to another watercourse state by using the international watercourse

The principles of international water law were first formulated by the International Law Association (ILA) when it published its Helsinki Rules in 1966. Since then, however, little progress has been made in finding international consensus over their application and no binding international treaty has come into force.

The conflicting approaches to allocation and management of transboundary waters can be categorised with the help of three theories:

- The theory of absolute territorial sovereignty states that every nation can utilise the waters of an international river flowing on its territory as it likes, regardless of the consequences in other countries and without the duty to consult.
- The theory of absolute territorial integrity regards an international river as the common property of its co-riparians, which means that no state is allowed to deprive the others of the benefits of the waters in question. Consequently, the lower riparian has the right to demand the continued, uninterrupted flow of water from the territory of the upper riparian.
- The theory of limited territorial sovereignty reflects the general legal principle of 'sic utere tuo ut alienum non laedas' (one should use his own property in such a manner as not to injure that of another) and is based on the assertion that every state is free to use the waters of shared rivers flowing through its territory as long as such utilisation does not prejudice the rights and interests of the co-riparians.

Turkey does not acknowledge that downstream countries have the right of co-sovereignty on waters of upstream countries or vice versa (Kibaroglu 2002). In the past, Turkey has referred to the Helsinki Rules (1966) by highlighting that the ILA considered the principle of equitable and reasonable utilisation to be the guiding rule while the no harm rule was one among a series of elements to be considered in determining whether a given use was 'equitable and reasonable'. Traditionally Turkey has also stressed the principle of "Good Neighbourliness" which considers other riparians' interests in dealing with 'transboundary' and 'international' rivers.

In 2003 The Turkish Ministry of Foreign Affairs formulated a set of principles which delineates Turkey's official policy regarding the use of transboundary rivers (Turkish Ministry of Foreign Affairs 2003). These have been reiterated in Turkey's Water Report 2009 (DSI 2009):

- "Each riparian state in a transboundary river has the sovereign right to make use of the water in its territory.
- Riparian states must make sure that their utilisation of such waters does not inflict 'significant harm' on others.
- Transboundary rivers should be used in an equitable, reasonable and optimum manner. ('Equitable use' does not mean the equal distribution of waters of a transboundary river among riparian states."

Whereas Turkey explicitly distinguishes between the terms "international rivers" and "transboundary rivers" and considers international rivers only to be those that constitute a border between two or more countries such as the Meric River which forms the border between Turkey and Greece and the Arpacay River (Aras basin) where it forms the border between Turkey and Armenia. While such boundary rivers are to be shared equally between the riparian countries, the water of transboundary rivers should be allocated equitably.

However, Turkey has been reluctant in signing multilateral agreements that lay down the principles of international water law - especially when they include compulsory mechanisms for dispute settlement and the procedures for prior notification. For this reason, Turkey, as one of only three countries, has voted against the UN Watercourses Convention. The arguments Turkey put forward during negotiations of the UN Watercourses Convention sheds further light on this position.

Equally, Turkey has not signed four of the five environmental conventions[2] under the auspices of the United Nations Economic Commission for Europe (UNECE), namely the

- Convention on the Protection and Use of Transboundary Watercourses and International Lakes (Helsinki Convention)
- Convention on Access to Information, Public Participation in Decision-making and Access to Justice in Environmental Matters (Aarhus Convention)
- Convention on Environmental Impact Assessment in a Transboundary Context (EIA/Espoo Convention)
- Convention on the Transboundary Effects of Industrial Accidents.

3 Turkey's position pertaining to the multilateral agreements on freshwater resources

3.1 UN Watercourses Convention (1997)

The UN Watercourses Convention (1997) is intended to establish general principles for the use and management of international watercourses and to assist the state parties in the resolution of disputes. It encourages states to enter into specific

---

[2]Turkey is party to the UNECE Convention on Long-range Transboundary Air Pollution since 1983.

agreements concerning the watercourses they share. It contains 37 articles dealing with the obligations of riparian states to utilize international watercourses in an equitable and reasonable manner while taking all appropriate measures to prevent the causing of significant harm; to consult with each other; to protect the environment and to resolve disputes.

To date only 22 states have become parties to the convention whereas 35 ratifications, acceptances, approvals or accessions are needed for it to enter into force. While Turkey states that the Convention has lost its credibility (Turkish Ministry of Foreign Affairs 2003), other authors argue that the large number (103) of votes for the adoption of the convention[3] in the United Nations General Assembly indicates general acceptance among the other member states (Wouters 1999). Stephen McCaffrey argues that even if the UN Watercourses Convention never comes into force, it is likely to be the starting point for negotiations for agreements on transboundary waters (McCaffrey 2001). On the other hand some analysts denied that the Convention would contribute to the resolution of disputes over the uses of the Danube, Ganges, Jordan, Nile, Mekong or the Euphrates. Asit Biswas asserts that while the Convention offers general guidance to co-riparian states, its vague language can result in varied, even conflicting interpretations of the principles contained therein (Biswas 2000). While Ellen Hey explains this with the Convention's provisions, namely that it neither furthers sustainable water use nor regulates the discretionary powers of watercourse states (Hey 1998).

Turkey's position

During the negotiation process for the UN Watercourses Convention Turkey played a leading role among the nations opposing the draft articles that had been prepared by the International Law Commission (ILC). When explaining their refusal, Turkish officials stated (among other things) that "the Convention goes beyond the scope of a framework" (Kibaroglu 2002). Turkey specifically raised the following points during negotiations.[4]

- The main purpose of the Convention should be to achieve an equitable and reasonable arrangement regulating water utilisation among the watercourse states: 'any other approach turns the Convention into a document which unilaterally restricts, in terms of both quantity and quality the utilisation rights of states in which watercourses originate; due attention should also be paid to establishing an equitable balance of rights and obligations among all watercourse

[3]The UN Convention on the Non-Navigational Uses of International Watercourses was adopted by 103 votes in favour to 3 against (Turkey, China, and Burundi) with 27 abstentions on May 21st, 1997 in New York.

[4]Compare UN General Assembly, 51st session: Summary Records of the 12th to 25th and 52nd to 62nd meetings: 6th Committee held at headquarters, New York between 17 September 1996 and 4 April 1997, U.N. Doc A/C.6/51/SR.12 and following.

states'.[5] Yet, according to the Turkish Government, while these requirements were taken into account to a certain extent in the general principles set forth in Part II of the Convention, the same cannot be said of Part III and Part IV of the document. In fact, as regards Part III (Articles 11-19) of the Convention, Turkish authorities delivered severe criticism mainly on the point that Part III comprises detailed procedural arrangements such as determining the period for reply to notification, ignoring the basic fact that each international watercourse possesses different and specific characteristics. Accordingly, the Turkish proposal was to reduce the dispositions of Part III to a minimum and to set forth certain general principles regarding planned measures.[6]

- Turkish authorities had a series of criticisms about Article 5 (equitable and reasonable utilization and participation) as well. They were eager to make it clear that the principle of equitable and reasonable utilisation should be understood in light of the fundamental principle of the sovereign rights of states over their territory. It should also be applied by taking fully into account all the particularities of the watercourses, including the distinction of whether they are transboundary by nature or international (forming a boundary) between states. In relation to the first paragraph of Article 5, Turkish officials suggested that the principle of 'optimal utilisation' should aim at both protecting the watercourse and at optimising the interests of riparian states in a way that avoids water waste. Yet, the first paragraph of Article 5 does not clearly state that 'optimal utilisation' should not be restricted to the protection only but should be seen also as comprising the concept of 'efficient use'.[7]
- Turkey also asserted that the obligation not to cause significant harm should be subsidiary to the principle of equitable and reasonable utilisation. In other words, if a state made use of a watercourse in conformity with the principle of equitable and reasonable utilization, the exercise of that right should not be limited by a second criterion.
- Turkey claimed that the draft convention was broader than had been intended in the General Assembly resolution 51/206; it should have merely established general principles, the application of which would be determined by means of specific agreements taking account of the particular characteristics of each watercourse. Contrary to what should be the case with a framework convention, the draft convention established a mechanism for prior notification on planned measures which had no basis in general and customary international law, and which created an obvious imbalance among states by setting up an obligation to obtain prior approval on planned measures from other riparians.
- A framework convention was not the appropriate place for setting out obligatory dispute settlement rules; the latter should be left to the discretion of the States concerned.

---

[5] See The Statement of Turkey: UN GAOR, 51st Session, 99th Plenary Meeting, U.N. Doc. A/51/PV.99 (1997), 5.

[6] See U.N. Doc A/C.6/51/SR.62 (1997), 8.

[7] See U.N. Doc. A/C.6/51/SR.61 (1997), 3.

- Parties to existing agreements should be free to choose whether or not to accept the principles set forth in the draft articles. As for future agreements, there again the parties must be free to conclude both, agreements that took account of the framework convention's provisions and agreements that diverged there from, even to a substantial extent.

Turkey's cautious attitude towards the ILC 1994 Draft and its refusal to endorse the 1997 UN Watercourses Convention originates from the deep concern about the fact that certain articles of the Convention might restrain its official stance in future negotiations for the allocation and management of the Euphrates-Tigris rivers system. Turkish officials are very keen on the point that the Convention should keep the original aim of constituting a framework document. Therefore, they go on asserting that the Convention clearly comprises provisions, which go far beyond that scope (Kibaroglu 2002, 256-7)

3.2 UNECE environmental conventions

The UNECE was established in 1947 to encourage economic cooperation among its member States. It is one of five regional commissions under the administrative direction of UN headquarters. It has 56 member States, and reports to the UN Economic and Social Council. The broad aim of UNECE's environment activities is to safeguard the environment and human health, and to promote sustainable development in its member countries in line with Agenda 21. UNECE has negotiated five environmental treaties, all of which are now in force. Of all these Conventions Turkey has only been party to the "Convention on long-range transboundary air pollution" since 1983.

3.2.1 UNECE Helsinki Convention (1992)

The Convention on the Protection and Use of Transboundary Watercourses and International Lakes (UNECE Helsinki Convention) was adopted shortly before the Rio Conference in 1992 in Helsinki. It entered into force in 1996 and up to date 37 UNECE countries and the European Community have ratified the Convention, including most of the countries in Eastern Europe and some of the Caucasus and Central Asia (EECCA) countries. The Convention provides a legal framework for transboundary cooperation on water bodies, which is highly relevant in the UNECE region. It has helped countries in the region to conclude river basin agreements and establish joint monitoring and assessment programs; and it has provided a basis for consultations on problematic issues, the exchange of information between riparian countries and public information. In contrast, the 1997 UN Watercourses Convention has had less impact in the region and has, for example, not been signed or ratified by any of the EECCA countries, apart from Uzbekistan which acceded in 2007.

Most of the agreements negotiated since the beginning of the 1990s, with the creation of new transboundary waters due to the collapse of the Union of Soviet Socialist Republics (USSR), are modelled on the UNECE Water Convention. Another strength of the UNECE Helsinki Convention is that it is embedded in a broader set of environmental policies dealing with transboundary cooperation.

3.2.2 UNECE Aarhus Convention (1998)

The UNECE Convention on Access to Information, Public Participation in Decision-making and Access to Justice in Environmental Matters, also named "Aarhus Convention" was adopted on 25th June 1998. It has been signed by 35 countries and the EU at the pan European Conference of Environment Ministers in the Danish City Aarhus. The Convention establishes several rights of the public with regards to the environment.

The first goal of the Convention is the access to environmental information and thus the right for everyone to receive information that is held by public authorities. This includes information on the state of the environment, policies or measures taken and on the state of human health and safety. The second goal is to support the right of public participation in environmental decision-making. That means that such arrangements have to be made by public authorities so that the public affected and environmental non-governmental organizations (NGOs) are informed. They could thus comment on proposals for projects affecting the environment, or plans and programmes relating to the environment. These comments have to be considered in decision-making and information should be provided on the final decisions. The third important goal of the Convention is the implementation of the right for everyone to have "access to justice" and thus the possibility to go to law with problems concerning environmental matters.

The Parties to the Convention are required to make the necessary provisions so that public authorities will aid these rights to become effective. The Convention became effective in October 2001 when the sixteenth state signed the agreement. Currently 44 states including the EU are parties to the Convention.

3.2.3 UNECE EIA/Espoo Convention (1991)

The Convention on "Environmental Impact Assessment in a Transboundary Context" is a UNECE convention signed in Espoo, Finland in 1991. It entered into force in 1997. It consists of 20 articles, defining the legal parameters for the signing parties. Its general goal is to expand the EIAs across national boundaries, in the case that national projects will likely have adverse impacts on adjoining countries. The Convention stipulates the commitments of its signatories to evaluate the environmental impact of certain activities at an early stage of planning. A general obligation of participating states will be to notify and consult each other on all major projects. Currently the convention is signed by 44 parties including the EU.

The Convention has been supplemented with the Protocol on Strategic Environmental Assessment - the SEA Protocol which is a result of the second meeting of the parties to the Convention and which was adopted in January 2003. Shortly after the adoption the protocol was opened for signature at the Ministerial "Environment for Europe" Conference in Kiev, where 35 parties signed it. When the protocol comes into force it requires its parties to evaluate the environmental consequences of their plans and programmes. A goal of the protocol is to support public participation in numerous sectors, where the public will also have the right to comment and be informed about the reasons for decisions. Above all, the protocol puts emphasis on health questions, involving the Worlds Health Organization and the obligations made at the London Ministerial Conference on Environment and Health in 1999.

3.2.4 The Convention on the Transboundary Effects of Industrial Accidents (1992)

The 1992 Convention on the Transboundary Effects of Industrial Accidents is designed to protect human beings and the environment against industrial accidents by preventing them as far as possible, by reducing their frequency and severity and by mitigating their effects. It promotes active international cooperation between the contracting Parties, before, during and after an industrial accident.

The aim of the Convention is to help its Parties to prevent industrial accidents that can have transboundary effects, to prepare for them and to respond to them. The Convention also encourages its Parties to help each other in the event of such an accident, to cooperate on research and development, and to share information and technology.

The Convention was adopted in Helsinki on 17 March 1992 and came into force on 19 April 2000. The Conference of the Parties was constituted as the Convention's governing body at its first meeting in Brussels on 22-24 November 2000. Currently, the convention has been signed by 40 parties including the European Union (EU).

The Protocol on Civil Liability for Damage and Compensation for Damage Caused by Transboundary Effects of Industrial Accidents on Transboundary Waters, adopted in Kiev on 21 May 2003, is a joint instrument to the Convention on the Transboundary Effects of Industrial Accidents and to the Convention on the Protection and Use of Transboundary Watercourses and International lakes.

3.2.5 Turkey's position vis-à-vis the UNECE Conventions

Little information is available on Turkey's position during the UNECE Helsinki Convention negotiations. Turkey pledged that the scope of the convention be restricted to questions of water quality and pushed for the phrase "The parties shall take all appropriate measures to prevent, control and reduce any transboundary impact" (art.2, para.1) to be replaced with the narrower term "transboundary pollution". Turkey's efforts were, however, unsuccessful (Demeter 2001). Furthermore, it can be assumed that one of Turkey's major concerns about the UNECE Water Convention was the dispute resolution mechanism, and not just the focus on environmental regulations.

On the other hand, Turkey has been consistently advised by the European Commission, through its regular progress reports (e.g. European Commission 2005), to become party to those international conventions (UNECE environmental conventions, in particular) to which the EU is a party (see Sumer and Muluk in this volume). That communication between the EU and Turkey constitutes, in fact, an obligation which Turkey needs to fulfil before the membership to the Union (Saner 2006). The concerned institutions in Turkey, namely the Ministry of Foreign Affairs, Secretariat General for EU Affairs, Ministry of Environment and Forestry and the State Planning Organization have been studying and analyzing the UNECE conventions particularly by comparing and contrasting them with the national legislation. In this respect, for instance, Act. 4982 on Right of Information Acquirement which was adopted in 2003 at the Turkish parliament may be evaluated as being in the scope of the Aarhus Convention to the extent that it lays down the guidelines and procedures for individuals to exercise their right of information acquirement in accordance with the principles of equality, neutrality and openness which are the fundamentals of democratic and transparent administration.

The conventions have also been under review particularly with respect to their possible impacts or pressures on Turkey's official stance vis à vis transboundary waters. Moreover, the fact that the amendments were adopted by the UNECE Helsinki Convention (2003) and the EIA/Espoo Convention (2001) which allow accession by countries outside the UNECE region that are not members of the UNECE also seemed to be another complicating factor for the Turkish authorities in their deliberations of membership to these conventions: while these amendments have not yet entered into force, they would invite other countries of the world to use the Conventions' legal frameworks and to benefit from its experience. The possibility that Syria and Iraq may become parties to these Conventions in the future adds further thorny questions in balancing the transboundary water strategy with the stipulations of these conventions (Ministry of Foreign Affairs 2010).

4 Participation in other regional agreements

4.1 Black Sea

Turkey's Black Sea coastline is 1,400km long, making it the country with the second longest Black Sea coastline and Turkey's most important fishing region. However, catches have been declining due to over-fishing and the Sea's changing ecosystem. Turkey contributes to Black Sea pollution and suffers from the degradation of Black Sea ecosystems. The Black Sea receives large quantities of mostly untreated domestic wastewater from Turkey, mainly from the Kizilirmak, Sakarya, and Yesilirmak rivers. In addition, pollution from the Danube, Dnieper and other streams and sources affect the water quality on Turkey's Black Sea coast.

Regional cooperation in protecting the Black Sea is manifested in the Convention on the Protection of the Black Sea against Pollution (Bucharest Convention), signed in Bucharest in April 1992. It was ratified by all six countries bordering on the Black Sea (Bulgaria, Georgia, Romania, Russian Federation, Turkey and Ukraine) at the beginning of 1994. The basic objective of the Convention was to substantiate the general obligations of the contracting parties to prevent, reduce and control pollution in the Black Sea in order to protect and preserve the marine environment and to provide a legal framework for co-operation and concerted action to fulfil this obligation.

The Bucharest Convention includes a basic framework of agreement and three specific Protocols, these are:

1. the control of land-based sources of pollution;
2. dumping of waste; and
3. joint action in the case of accidents (such as oil spills).

Implementation of the Convention is managed by the Commission for the Protection of the Black Sea Against Pollution (also sometimes referred to as the Istanbul Commission), and its Permanent Secretariat in Istanbul, Turkey. The support provided to governments for developing and implementing the Black Sea Strategic Action Plan, took the form of a series of Global Environmental Facility (GEF), Technical Assistance to the Commonwealth of Independent States (TACIS) and PHARE projects (Programme of the EU to assist the applicant countries of Central and Eastern Europe), and smaller donor initiatives, coordinated within a loosely defined programmatic framework described as the Black Sea Environmental Programme (BSEP). The BSEP established its headquarters in Istanbul with the support of the Government of Turkey. The most important achievements of BSEP were the Transboundary Diagnostic Analyses and the regional Strategic Action Plan for the Rehabilitation and Protection of the Black Sea (UNDP 2004).

4.2 Mediterranean Sea

Several coastal zones and near shore areas in Turkey are critically affected by pollution of the Mediterranean. These include areas of importance to tourism, such as the coast from Kemer to Alanya, as well as areas of biological importance such as the Goksu Delta (protected area for waterfowls) and the Bay of Iskenderun. In the north-east Mediterranean, most of the land based pollution loads originate from Turkey. The pollution loads consist of agricultural run-off, domestic and industrial wastewater discharges and organic pollution carried by rivers and streams (Samsunlu et al. 2002).

Regional environmental cooperation in the Mediterranean Sea is laid down in the Barcelona Convention for the Protection of the Mediterranean Sea against Pollution.[8] The convention is complemented by six protocols[9]. Today, 21 countries and the EU are party to the convention: Albania, Algeria, Bosnia and Herzegovina, Croatia, Cyprus, Egypt, France, Greece, Israel, Italy, Lebanon, Libya, Malta, Monaco, Montenegro, Morocco, Slovenia, Spain, Syria, Tunisia, Turkey.

The United Nations Environment Programme (UNEP) is responsible for the secretariat of the Barcelona Convention and its Action Plan, through the Mediterranean Regional Coordination Unit in Athens, Greece. The Mediterranean Action Plan (MAP) was the first Regional Seas Programme of UNEP established within the framework of the Barcelona Convention.

Within this framework, Turkey declared fourteen Specially Protected Areas according to the addendum protocol of the Barcelona Convention. Several Turkish NGOs and the Ministry of Environment and Forestry (MoEF) have collaborated successfully with MAP's Specially Protected Areas Regional Activity Centre, SPA/RAC (Tunis, Tunisia) in addressing various issues pertinent to the protection of endangered species such as sea turtles and monk seals, and their habitats. Furthermore, Turkey contributed to the MAP Coastal Area Management Programme (CAMP) with the "The Bay of Izmir" project, which was officially launched in June 1990 following an agreement signed between the Turkish Government and MAP. A total of 11 activities were envisaged by the agreement, but only half of them were implemented. However, major positive changes could be observed in the management and the actual development of the Izmir Metropolitan Area, partly attributable to the CAMP initiative (UNEP/MAP/PAP 2005). Achievements of the CAMP project include:

- An EIA of the Izmir Sewage Treatment Project,
- database on environmental / development issues and environmental zoning of the Izmir area,
- An Integrated Management Study for the Area of Izmir,
- Improved water quality in the bay, and
- Improved institutional capacity for integrated coastal area management (including application of tools and techniques such as Geographical Information System - GIS - and EIA).

---

[8]Adopted in Barcelona, Spain, 16 February 1976; entered into force: 12 February 1978; amended: Barcelona, Spain, 9-10 June 1995. New Title: Convention for the Protection of the Marine Environment and the Coastal Region of the Mediterranean (entered into force on 9 July 2004).

[9]The Protocol for the Prevention of Pollution in the Mediterranean Sea by Dumping from Ships and Aircraft; the Protocol Concerning Cooperation in Preventing Pollution from Ships and, in Cases of Emergency, Combating Pollution of the Mediterranean Sea; the Protocol for the Protection of the Mediterranean Sea against Pollution from Land-Based Sources; the Protocol Concerning Specially Protected Areas and Biological Diversity in the Mediterranean; (in force) as well as the Protocol for the Protection of the Mediterranean Sea against Pollution Resulting from Exploration and Exploitation of the Continental Shelf and the Seabed and its Subsoil; and the Protocol on the Prevention of Pollution of the Mediterranean Sea by Transboundary Movements of Hazardous Wastes and their Disposal (pending).

4.3 Conventions on wetlands and biodiversity

Turkey has shown its commitment to protecting valuable wetland ecosystems by signing, in 1994, the Convention on Wetlands of International Importance Especially as Waterfowl Habitat (Ramsar Convention), adopted in the Iranian city of Ramsar in 1971. With the intention of preventing further loss of wetland areas, the Ramsar Convention established a list of wetland areas of international importance. Designation by a member state of a site in its territory for the Ramsar list awards the wetland site international protection. All parties to the convention are obliged to consider the conservation of listed wetlands in the planning of land and water resource use. Further, they are requested the "wise use" promotion of all non-listed wetlands in their territory (Ramsar Convention, Article 3, No. 1). Until today, 13 Turkish wetlands were included in the List of Ramsar Sites. On January 30, 2002, the Turkish Regulation on Conservation of Wetlands was put into force. The "National Wetlands Committee" which was established through this regulation became an important tool in developing coordination and collaboration between relevant institutions.

The Convention on Biological Diversity (CBD) (1992) assigns its contracting parties broad duties aimed at the protection of the long term productivity and diversity of ecosystems and habitats. It covers the conservation of biological diversity, the sustainable use of its components, as well as the fair and equitable sharing of the benefits arising from the utilisation of genetic resources. The Government of Turkey ratified the CBD in 1996 and has participated in several European activities related to the implementation of the convention in forestry and forest biodiversity (Arancli 2002). Awareness is increasing in Turkey about the conservation of its rich and diverse biological resources and the need for their sustainable management. The National Environmental Action Plan and the National Biodiversity Strategy and Action Plan emphasise the importance of integrating biodiversity needs into the development and implementation of relevant sectored policies, and also envisages the involvement of all stakeholders.

5 New concepts gaining prominence in Turkey's transboundary water policy: benefit-sharing and confidence building measures

It is evident that Turkey upholds the principle of equitable, reasonable (and optimum) utilization of transboundary rivers as the backbone of customary water law. The principle "not to cause significant harm" is also accepted as the subsidiary norm. Yet, in addition to these legal norms some new concepts have been entering into the rhetoric of Turkish water diplomacy notwithstanding the fact that these concepts did not gain legal prominence. In this context, a concerned Turkish diplomat has expressed that "Turkey's transboundary water policy is aimed to efficiently utilize and 'share the benefits' of transboundary water resources through cooperation among riparian states. The riparian states should adopt a comprehensive approach to the matter. Such an approach calls for determination to discuss all water-related issues in a transboundary context. It entails sharing of responsibility as well as the benefits" (Rende 2004).

Moreover, 'confidence building measures' are suggested as a means to dispel mistrust and create the appropriate environment for meaningful cooperation. In this respect, information and data-sharing at basin level is emphasized. It has been recommended that confidence building efforts could be initiated through informal exchange of information and data at experts' level. Such a dialogue can be further developed by discussions on general principles to be applied in the utilization of the water resources of the basin. In the Turkish official contention, "an effective form of confidence building measures in this context could be to share the benefits of water through joint projects having to do with agriculture, irrigation, hydropower generation, provision of safe drinking water and water for sanitation and last but not least projects designed to protect the environment" (Rende 2007, 167).

The Turkish diplomacy underlines the importance of settling disputes over transboundary waters in a peaceful manner by adopting diplomatic procedures, namely the method of negotiations. Here, it is strongly emphasized that transboundary water issues should be negotiated between the riparian states only; the third party intervention such as mediation attempts cannot be acceptable (Ministry of Foreign Affairs 2010). Within the framework of negotiations, riparian states should have the political will to engage in genuine cooperation. Such cooperation could lead to reaching a common understanding on the utilization of the transboundary waters in the interest of all (Rende 2007, 169).

Furthermore, Turkey has started harmonizing its domestic legislation with that of the EU in the field of environment and water resources. Hence, the on-going EU accession process has become one of the main determining factors of Turkey's water resources policies along with contemplations on recent developments at global and regional levels, the contractual obligations, as well as further considerations of present and future water needs of the fast growing population.

References

Arancli S (2002) Biodiversity and natural resource management in Turkey. Paper presented at Environmental Connectivity: Protected Areas in the Mediterranean Context, Malaga, Spain, 26-28 September 2002

Biswas A (2000) Presentation given at the CDG Conference on International Water Courses. Berlin 27th October 2000

Brunnée J, Toope SJ (1997) Environmental security and freshwater resources: Ecosystem regime building. The American Journal of International Law, vol 91: 26-59

Demeter B (2001) Schutz und Nutzung internationaler Binnengewässer – Zwei Rahmenkonventionen im Vergleich. Frankfurt am Main, Peter Lang

DSI (2009) Turkey water report 2009. DSI General Directorate of State Hydraulic Works, Ankara. http://www.dsi.gov.tr/english/pdf_files/TurkeyWaterReport.pdf. Accessed 25 May 2010
European Commission (2005) Regular report on Turkey's progress towards accession, Brussels
Hey E (1998) The Watercourses Convention: To what extent does it provide a basis for regulating uses of international watercourses? Review of European Community and International Environmental Law, 7 (3): 291-300
Kibaroglu A (2002) Building a regime for the waters of the Euphrates-Tigris river system. Kluwer Law International, London The Hague New York
McCaffrey S (2001) The contribution of the UN Convention on the law of the non-navigational uses of international watercourses. Int. J. Global Environmental Issues, vol. 1, no 3/4: 250-263
Ministry of Foreign Affairs (2010) Personal communication with the officials, Ankara
Rende M (2004) Turkey's water resources management and transboundary water policy. Speech delivered at the 4th Biennal Rosenberg International Forum, Ankara, 2004
Rende M (2007) Water transfer from Turkey to water stressed countries in the Middle East. In: Shuval H, Dweik H (eds) Water resources in the Middle East: Israeli Palestinian water issues — From conflict to cooperation., Springer, Berlin, pp 167-169
Samsunlu A, Maktav D, Kapdaslii S (1996) Transboundary water issues between Greece-Bulgaria and Turkey: the case of Meric / Evros River. In: Bogardi I (ed) Transboundary water resources management. Institutional and engineering approaches. Springer, Berlin
Saner E (2006) Turkiye, Su ve Uluslararasi Boyutu. Paper presented at the Union of Chambers of Turkish Engineers and Architects- Proceedings of Congress of Water Policies, Ankara
Turkish Ministry of Foreign Affairs (2003) The looming water shortgage and Turkey's water management in a transboundary context. NATO Parliament Assembly, 48th session. Istanbul, 15-19 November 2002
UNDP (2004) Black Sea Ecosystem Recovery Project - Phase 2. Project Document. http://www.gefweb.org/Documents/Council_Documents/GEF_C23/IW_-_Regional_-_Black_Sea_Ecosystem_-_ Project_Document.pdf. Accessed 03 June 2005
Wouters P (1999) The legal response to international water conflicts: the UN Watercourses Convention and beyond. German Yearbook of International Law 42: 292-336

Meric River Basin: Transboundary Water Cooperation at the Border between the EU and Turkey

Annika Kramer and Alina Schellig

1 Introduction

The Meric basin, one of the major river systems of the eastern Balkans, is shared by Bulgaria, Greece and Turkey. Water needs for irrigation and flood control are the main disputed issues in the basin, particularly between Turkey and Bulgaria. In the past, political distrust between the three countries hampered co-operation. However, recent rapprochement between Turkey and Greece, Bulgaria's joining of the European Union (EU) and the prospect of EU membership for Turkey are expected to have positive effects on transboundary water management.

2 Geographical and hydrological setting

The Meric River system rises in Bulgaria and flows along the Turkish Greek border into the Aegean Sea. The basin's main river is called Meric in Turkish, Maritsa in Bulgarian, and Evros in Greek.[1] The Meric basin, including its main tributaries - the rivers Arda and Tunca[2] that mainly flow in Bulgaria and the River Ergene that flows entirely in Turkish territory - has a drainage area of about 52,600 square kilometers of which 65 percent lie in Bulgaria, 87 percent in Greece and 28 percent in Turkey (UNECE 2007). Its main river, the Meric, is about 500km long and has its source in

[1]Other spellings include Merich, Maritza, Marica, Hebros.

[2]Other spellings include Ardas, Tunca, Tundzha.

A. Kramer (✉)
adelphi, Berlin, Germany
e-mail: kramer@adelphi.de

A. Schellig
Alina Schellig, Berlin, Germany
e-mail: alina@schellig.net

A. Kibaroglu et al. (eds.), *Turkey's Water Policy*,
DOI 10.1007/978-3-642-19636-2_13, © Springer-Verlag Berlin Heidelberg 2011

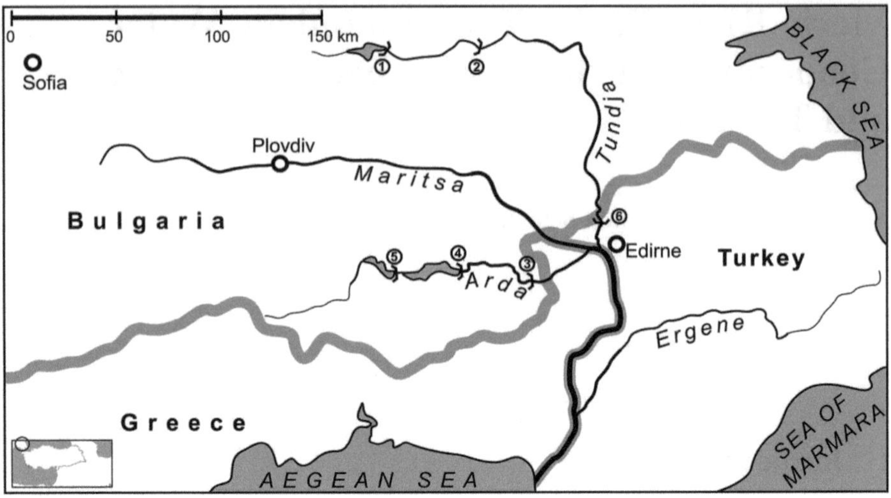

(1) Georgi Dimitrov Dam (3) Ivailovgrad Dam (5) Kardjali Dam
(2) Jdrebchevo Dam (4) Studen Kladnetz Dam (6) Suakacagi Dam (planned)

Fig. 1 Meric River, its main tributaries and selected dams

the Rila mountains near the summit of Musala. From its spring, it flows eastwards through tectonic valleys between the Balkan and Rhodope Mountains, passing the city of Plovdiv and the Eastern Rumelia plain. At the point where the Meric is close to the three-way border between Bulgaria, Greece, and Turkey, it first forms a natural boundary between Bulgaria and Greece for about 15km, then, for about 187km, it forms the border between Turkey and Greece in the Thrace Region before finally entering the Aegean Sea.[3] Figure 1 shows a map of the Meric River, its main tributaries and selected dams, and Table 1 provides an overview of the context for cooperation on the river.

Shortly after the three-way border, close to the Turkish city of Edirne, the Meric is joined by the Arda River from the south and the Tunca River from the north. The Arda River springs from the eastern Rhodope Mountains in Southern Bulgaria from where it flows eastwards. After 240km on Bulgarian territory, it flows for 30km in Greece before it discharges into the Meric River at the Turkish border. The Tunca River has its source in the Stara Planina Mountains (Balkan Mountains) in the centre of Bulgaria. From there it flows eastwards along the Balkan Mountains towards the Turkish border. For about 15km, Tunca River forms the border between Turkey and Bulgaria. Then, it flows about 30km through Turkey to enter the Meric. The Ergene River springs in the Istranca Mountains in Turkey and joins the Meric about 30km before the mouth of the river.

[3]As the Meric river forms the border between Greece and Turkey, Turkish authorities consider it to be an "international river" between Turkey and Greece and a "transboundary river" where it crosses the border between Bulgaria and Turkey (Yanik 1997).

Table 1 Cooperational context in the Meric basin

Meric Basin area: 52,600 km²; mean annual discharge 8 BCM		
Riparian position	Basin area (percent of total) Contribution to annual discharge	Main water uses
Bulgaria upstream	34,067 km^2 (65 percent) 5.7 BCM (71 percent)	irrigation, hydropower
Greece downstream	3,685 km^2 (7 percent) 0.5 BCM (6 percent)	irrigation, conservation area
Turkey downstream	14,850 km^2 (28 percent) 1.8 BCM (23 percent)	irrigation
Main agreements and covered issues		
Turkey - Bulgaria	1968 – flood protection, data exchange, joint studies, no harm principle, dispute settlement (see Annex 3) 1998 – joint infrastructure projects (see Annex 6) 2002 – exchange of data on water quantity (see Annex 7)	
Turkey - Greece	1934 – specifications for infrastructure, data exchange, prior notification, dispute settlement (see Annex 1) 1955 – joint construction of flood control 2001 – general MoU on cooperation on environmental protection	
Bulgaria – Greece	1964 – no harm principle, exchange information on floods, joint studies on infrastructure 1971 – cooperation in the field of electric energy and the use of waters of transboundary rivers since 1992 – cooperation in EU INTERREG programme, establishment of water quality monitoring network 2002 – joint efforts to monitor water pollution; water protection and use in the spirit of EU WFD	
Unsettled issues		
Quantity	No agreement on water quantity allocation. Conflicting interests mainly between Turkey and Bulgaria	
Quality	No agreement on water quality standards, exchange of water quality data insufficient, lack of wastewater treatment plants upstream, water quality of concern in Turkey and Greece	
Flooding	Early warning and flood protection unsatisfactory for Greece and Turkey	

Where the Meric River enters the Aegean Sea near the Gulf of Saroz, it forms a delta of about 188 square kilometers, of which 150 square kilometers lie on Greek territory. The river delta is a typical Mediterranean delta formed by alluvial deposits and shaped by interaction with the sea. The delta is one of the most important Mediterranean wintering areas for about three-hundred species of birds, including the last 15 surviving pairs of Royal Eagle (Zardava 2004). 100 square kilometers of the Greek part of the delta (Evros Delta) is protected as a "Wetland of International Importance" under the Ramsar Convention.[4] It also enjoys the status of Special

[4]In addition, the area enjoys the status of a Special Protection Area (Directive 79/409/EEC for the conservation of birds and their habitats) and as a Site of Community Importance (Directive 92/43/EEC for the conservation of natural habitats as well as the wild fauna and flora - NATURA 2000 Network).

Protected Area and NATURA 2000 site (UNECE 2009). However, the natural delta ecosystem has been modified by human activity since 1950 (Ramsar Convention 1999).

The climate in the basin is influenced by continental conditions in the northern and Mediterranean conditions in the southern part. Generally the climate in the basin is characterised by cold rainy winters and dry and hot summers. Precipitation ranges between 900 to 1,100 mm in the mountainous regions in Bulgaria and about 600 mm in the lower regions of the basin. Maximum precipitation on upper Meric is usually experienced in May and June, while the precipitation minimum can be observed in July and August. Climate conditions in the sub-basins of the southern tributaries vary according to the Mediterranean influence and maximum precipitation occurs, in accordance with the Mediterranean climatic influence in winter, during December and January (Arda).[5]

According to the variation of local climate, different flow patterns can be observed for the Meric and its tributaries. Flow patterns of the Meric and Tunca have shown great seasonal and annual deviations. During summer, particularly in dry periods, the Tunca flow rate reduces drastically, partly caused by the operation of dams upstream in Bulgaria. Based on data gathered by the Bulgarian Ministry of Environment and Water at flow monitoring stations close to the Bulgarian-Greek border for the period 1961-1998, the average annual discharge rate of the Meric is 108 cubic meters per second (UNECE 2007). However, there is high seasonal variety, and discharge rates observed at flow monitoring stations in west Edirne fluctuate between 1,679 cubic meters per second (max) to 10.4 cubic meters per second (min). The Ergene River also shows huge variations in flow between summer and winter/spring. Flow rates of the Ergene during summer months fall to nearly zero, whereas intermittent flooding has been seen in the region during the rainy (spring) seasons (Yanik 1997). The total annual water discharge rate of the Meric basin is 8 Billion cubic meters (BCM). In the drainage area in Turkey, the Ergene contributes 1.2 BCM, Tunca 0.4 BCM and Meric East Bank 0.2 BCM per year. In Bulgaria, the Tunca River contributes 0.6 BCM, the Meric and the Arda add 5.1 BCM per year to the overall water potential. The Meric River basin tributaries in Greece contribute about 0.5 BCM/year (Ozis et al. 2002).

Water quality in the basin suffers from agricultural run-off and the discharge of untreated wastewater. Nitrate loads up to 50 mg/l have been measured in the Meric and up to 25 mg/l in the Arda and Tunca rivers (Regional Environment Accession Project 2002). Furthermore, large amounts of sediment are carried by the Meric River which is made up of sand because soils in the basin are mainly covered by sand and loam layers (Yildiz 1999). Near the Edirne Bridge, 1,200,000 cubic meters of sediment per year has been measured.

[5]Artinyan et al. 2007; personal communication with Roumen Arsov, Faculty of Hydrotechnics, University of Architecture, Civil Engineering & Geodesy, 20 June 2006, and Dobri Dimitrov, National Institute of Meteorology and Hydrology (NIMH), 26 June 2006.

3 Current and future water uses

The main water use in the basin is for irrigation. In-stream use for hydropower generation takes place in Bulgaria. In addition, the delta region provides good fishing grounds for Turkey and Greece.

Water use in Turkey

The Turkish part of the Meric basin includes highly productive agricultural lands where rice is grown; other crops are sugar beet, sunflowers, corn, vegetables and fruits. As a consequence, the agricultural sector uses 82 percent of the total water withdrawal which reached 1,352 Million cubic meters (MCM) per year in 2009, whereas industry accounted for 13 percent and domestic water use for 4 percent. Total water withdrawal is expected to increase by almost 50 percent by 2015 (UNECE 2009).

Seven main dams and one regulator operate in the Turkish part of the basin, in addition to several small dams on minor tributaries (UNECE 2009), providing irrigation water for about 60,000 hectares, flood control and supplying drinking water. About 95 percent of the basin drainage area, i.e. 1,239,102 hectares of land is arable and 395,194 hectares, is irrigable, yet, only 328,039 hectares are technically and economically categorised as irrigable land (DSI XIth Regional Directorate, no year). As of 2003, the total irrigated area with surface and groundwater resources was 144,639 hectares. Plans are in place to increase irrigation agriculture; irrigation systems are under construction on a further 54,879 hectares, whereas 328,879 hectares are at the project and planning stage. Projections for 2015 foresee that 1,560 MCM/year of water will be used in the agricultural sector (UNECE 2009).

The basin area in Turkey covers the most developed parts of the country with the urban centres of Edirne and Kirklareli. Both urban wastewater and solid waste volumes have increased due to population growth. Untreated domestic wastewater has been identified as one of the main sources of pollution in the Turkish part of the basin. Leakage from official as well as illegal waste disposal sites poses another threat to water quality (UNECE 2009). Industrial pollution occurs mainly in the vicinity of the cities of Luleburgaz, Corlu and Cerkezkoy originating from organised industrial sites as well as illegal wastewater discharges. Furthermore, nitrogen, phosphorus and pesticides are washed into the river from agricultural land. As a result of these diverse pollution sources, the water of the Ergene River sub-basin has been classified as Class IV ("very polluted") and the Meric River sub-basin as Class III ("polluted") (UNECE 2009).

Water use in Bulgaria

In Bulgaria, the rivers are used for agriculture and hydropower production as well as for domestic and industrial water supplies. There are 21 main dams in operation with a total storage capacity of more than 3,000 MCM (Arsov 2004). The total number of reservoirs in the Bulgarian part is as high as 722, and most of the small dams are used for irrigation purposes and fish-breeding (UNECE 2009). The basin area (about 34,000 square kilometers) is home to 2.3 million people (INWEB 2010). The main cities are Plovdiv, Stara Zagora, Haskovo, Sliven, and Yambol. The area suffers from water stress because of drought and deterioration in water quality (Regional Environment Accession Project 2002).

Approximately 55 percent of total water withdrawals are used in agriculture, 39 percent for non-consumptive energy production, whereas the industrial and the domestic sectors only account for 3 percent and 1 percent respectively (UNECE 2009). The Meric Plain includes some of Bulgaria's most fertile agricultural land (Regional Environment Accession Project 2002). Areas of relatively intensive agricultural production can be found here, particularly in the vicinity of Plovdiv (Penov et al. 2003). Main crops are cereals, vegetables, cotton and tobacco; orchards are cultivated in the lower valley areas, whereas the hilly regions serve as pastures, vineyards and potato fields (Artinyan et al. 2007). In general, it can be observed that water use for irrigation experienced a drastic reduction in Bulgaria during the 1990s.[6] Water use efficiency is low with average water losses in irrigation systems of 57 percent, reaching as much as 75 percent in some regions (ECSSD 2003).

Pollution sources are run-off from agricultural production and stock-breeding, and industrial and urban effluents (Centre for European Constitutional Law 2001). Only about 67 percent of the population is connected to sewerage systems, and only 30 percent of wastewater is treated (UNECE 2009). However, several wastewater treatment plants are currently under construction or planned.[7] Diffuse sources of pollution from agricultural runoff are the second biggest factor impacting on water quality (UNECE 2009). Furthermore, industrial activities might cause heavy metal pollution: the lead and zinc industry in Bulgaria is based on mining and processing operations near Plovdiv. Although most of the mines have been shut down and pollution declined remarkably since 1990, a recent study done by the Bulgarian Academy of Science shows that upper parts of the Arda and Meric Rivers are still significantly polluted by heavy metal and copper. However, it was found that its impact on water quality at the border is rather marginal.[8] The river receives industrial waste from various areas, but the quantity of pollutants was markedly reduced due to the economic decline of the country in the 1990s.

Water use in Greece

The Greek part of the basin area only amounts to 3,700 square kilometers and is home to about 85,000 people (Skias and Kallioras 2007). Apart from Alexandroupoli, there are no major cities, and industrial activity is also very low.

---

[6]From 1.2 BCM in 1991 to less than 0.1 BCM in 1997 (ECSSD 2003).

[7]For more detailed information see "The Study on Integrated Water Management in the Republic of Bulgaria. Presentation for the Basin Council Meeting of EABD. December 11th, 2007". http://www.bd-ibr.org/details_file_download.php?a_id=347&PHPSESSID=3f7aba02355508783449df362e0a6f19. Accessed 17 July 2010.

[8]Personal communication with Mariyana Nikolova, Institute of Geography, Bulgarian Academy of Science, 27 June 2006.

The principal source of pollution is domestic wastewater, especially from the towns of Orestiada and Didimoticho (Centre for European Constitutional Law 2001).

No comprehensive data on water use in Greece has been accessible, but it is assumed that freshwater is increasingly used for irrigation. Other water users are commercial fishery and tourism in the delta area (Centre for European Constitutional Law 2001). A number of dams store water for irrigation (UNECE 2009) while i.e. the dam on the Arda River, close to the Bulgarian boarder, regulates water discharge from the power plant of the Bulgarian Ivailovgrad dam. Water from this reservoir is used to irrigate 30,000 hectares (Tzovaridis et al. 1996). Close to the river delta, about 15,000 hectares are used for agriculture (cotton, sugar beet, sunflowers, tomatoes, asparagus). In the delta area, agriculture is restricted due to nature conservation and unfavourable soil conditions (Ramsar Convention 1999).

4 Potential impacts on downstream riparians and the Mediterranean Sea

Diversion and storage of water for irrigation purposes, mainly in Bulgaria, result in reduced river flows downstream. Because of water shortage, Turkey has, on some occasions, been deprived of irrigation for paddy fields (Ozis et al. 2002). Moreover, low groundwater levels (Mylopoulus et al. 2008) are reported and low inflow of freshwater increases salt water intrusion into the river delta and upstream. Increased salt loads can be observed up to 35km upstream of the mouth of the river, making the water inappropriate for irrigation. Furthermore, low freshwater inflows cause siltation problems in the delta (Samsunlu et al. 1996).

In fact, flooding is a major problem in the Meric basin for the downstream riparians Turkey and Greece. In the 2000s a series of severe floods occurred, indeed the most disastrous floods happened in spring 2005[9] and 2006, and in November 2007. In 2005, Bulgaria was affected by five major floods, of which two directly affected the Meric River basin. The Turkish town of Edirne experienced serious flooding caused by heavy rain and the overflow of the Ivailovgrad reservoir into the Arda River in February 2005. It inundated houses and farmland in Greece and Turkey and caused damage, estimated at 50 million US Dollars in the area around Edirne (DSI 2005).[10] In August 2005, the cyclone *Ihtiman* was followed by heavy rainfall and flooding on the Meric.[11] In March 2006, the lower part of the river basin

[9]Estimated recurrence interval of this flood event was 1,000 years (UNECE 2009b).

[10]12,000 hectares of agricultural land were damaged and two bridges collapsed (personal email correspondence with officials from the Devlet Su Isleri (DSI) Edirne Regional Office, 03 June 2005).

[11]See Dimitrov (no year) http://www.balwois.com/balwois/administration/full_paper/ffp-873.pdf. Accessed 17 July 2010.

was again heavily flooded when rain and snow melt pushed the river over its banks.[12] Extreme water levels were reached in Turkey, with peak flood discharges of the Meric River at Ipsala reaching 2,632 cubic meters per second on 17 March 2006 (Darama 2009).

Although experts from Bulgaria, Greece and Turkey agree upon the fact that the major floods were primarily caused by the coincidence of unfavourable flow patterns and rare extreme meteorological conditions,[13] they disagree upon the influence of water management. According to Turkish and Greek experts, the management of Bulgarian reservoirs is not appropriate: maintaining high water levels in reservoirs close to the border (e.g. Ivailovgrad) increases the risk of uncontrolled flooding, they argue, because in cases of heavy rainfall and snow melt, Bulgaria ought to flood agricultural land downstream in order to prevent dams from breaking.[14] On the other hand, experts from downstream countries also mention that inappropriate floodplain management in Greece and Turkey, such as changing land use patterns and intensification of agriculture as occurred in the 1970s and 1980s, also increased risk and vulnerability of the downstream riparians.[15] Furthermore, the impact of floods is exacerbated by the lack of an appropriate early warning system between Bulgaria and the downstream riparians.[16]

Low water quality in the basin is another concern in Turkey and poses a threat for the protected delta. According to the measurements taken at the Kapikule border quality monitoring stations between 1985 and 2001, Turkish authorities claim that the Meric River enters Turkey as polluted (class III) and the Tunca River enters as heavily polluted and has to be characterized as a modified water body (class IV) (UNECE 2009).

High sediment loads of the river cause sand accumulation in the riverbed particularly near Edirne, and forms sand islets on both sides of the river. Several trees rise on these sand islets, and grow into forests. This situation causes severe coastal erosion as a result of the increase in the roughness coefficient in the riverbed (Yildiz 1999).

[12]According to REUTERS the worst floods in 15 years (cited from Visible Earth / http://visibleearth.nasa.gov/view_rec.php?id=20541 accessed 17 July 2010).

[13]Personal communication with Dobri Dimitrov, NIMH, 26 June 2006, and Thymio Papayannis, WWF Greece, 20 June 2006; see also Darama 2009.

[14]Personal communication with Thymio Papayannis, World Wide Fund for Nature (WWF) Greece, 20 June 2006, Stelios Skias, Democritus University of Thrace, 29 June 2006, and Spyros Tasoglou, Central Water Agency, Greek Ministry of Environment, Physical Planning and Public Works, 20 June 2006; see also Darama 2009 and Skias and Kallioras 2007.

[15]Personal communication with Stelios Skias, Democritus University of Thrace, 29 June 2006.

[16]Personal communication with Stelios Skias, Democritus University of Thrace, 29 June 2006, and Dobri Dimitrov, NIMH, 26 June 2006.

5 Status of cooperation

Any cooperative initiative in the Meric basin needs to be considered within the broader context of political relations between the riparian countries. Relations between Greece and Turkey, in particular, have been far from friendly over the years. After the second Greco-Turkish war (1919-1922)[17] the main issues have been the Cyprus dispute and conflicting territorial claims in the Aegean Sea, with the 1996 *Kardak Crisis* over a deserted island in the Eastern Aegean, which brought serious diplomatic confrontation between both countries. In addition, the Meric basin is situated in Thrace, an area that is home to diverse communities with Turkish minorities living in both the Greek and the Bulgarian part of Thrace as well as Greek minorities that used to live in Turkey. The minority conflict is the oldest such issue between Turkey and Greece and has been the main problem affecting Bulgarian-Turkish relations since the end of World War II (Petkova 2002).

Since 1999, however, Turkish-Greek relations have entered a new rapprochement era, which is largely due to close cooperation between the Foreign Ministers of the two countries. Turkish-Greek joint committees have been established and several agreements on promoting cooperation in e.g. environmental protection or combating terrorism have been reached. Meanwhile, 45 rounds of exploratory talks have been held (the latest took place in Istanbul, in July 2010) to sound out each other's positions; progress, however, has been limited (European Commission 2009).

Membership in the EU provides a general framework for cooperation between Greece and Bulgaria. The Accession Partnership between the EU and Turkey further provides opportunities for cooperation between Turkey and its European neighbours. With regards to potential water cooperation, it can be observed that the water management system in each of the three riparians is undergoing a transition period moving towards the European aquis communautaire, though different in each case (Skias and Kallioras 2007). Similar water management systems and approaches in line with the EU framework directives may facilitate transboundary water cooperation among the three riparians in the future. Furthermore, the Water Framework Directive (WFD) explicitly requires Member States to assign transboundary river basin districts for basins lying within the EU's territory and to coordinate in particular all programmes to achieve the WFD's environmental objectives. Where international river basins extend beyond the territories of the Union, the WFD requires Member States only to "endeavour" to establish appropriate coordination and to produce a single river basin management plan with the relevant non-Member States, with the aim of achieving the objectives of the WFD throughout the river basin district. However, as explained below, no such coordination is yet taking place between the Meric's riparians.

[17]Known in Turkey as the Turkish War of Liberation.

5.1 Turkey and Greece

Earlier agreements between Turkey and Greece on the Meric River mainly covered the construction of facilities for flood protection, erosion control and water diversion. Recently, the two countries entered a Memorandum of Understanding (MoU) on environmental issues.

The first agreement on the Meric between Greece and Turkey was signed in 1934. The "Agreement pertaining to the construction of hydraulic facilities on both banks of the Meric-Ebros River" (see Annex 1), mainly covered specifications for infrastructure that both parties were allowed to build for flood protection and erosion control. It also included provisions for exchanging topographic data, notification to the other party prior to construction, and for dispute settlement between the two parties.

Another agreement relating to the construction of flood control measures on the Meric River was signed between Turkey and Greece in 1955. The text of the agreement was, however, not published. According to Bilen (2000), the agreement provided for the construction of flood control measures in accordance with a master plan, drawn up by the Harza Engineering Company. Each government would undertake the construction and financing of the work on its own territory. However, only some of the facilities envisaged by the master plan have been realized (Bilen 2000). In order to iron out disputes arising from the master plan and to carry out hydraulic works on both sides of the Meric River, Turkish-Greek technical delegations convened in 1963 and agreed on the "Protocol on the improvements of the River Meric watercourse that constitutes a significant portion of the Turkish-Greek Thracian Border" (see Annex 2). This protocol encompassed articles on the modification of the border, as exchange of land was necessary to build infrastructure on the river. Any disputes over this matter would have been assigned to a General Engineer, appointed by the French Ministry of Agriculture. Other articles included stipulations on specific technical issues of water infrastructure construction.

A "Memorandum of Understanding Concerning Cooperation on Environmental Protection" was concluded between Turkey and Greece in 2000. It stipulated that the two parties "shall exchange scientific, technical and legal information among governmental bodies and shall encourage such exchange among academic institutions" (Article 2). "Coordination of co-operation in the different fields of activities shall be managed by a Joint Committee comprising five representatives from each of the two countries" (Article 8). The possible fields of cooperation named do not, however, include river management. Yet, some of the areas mentioned, such as "combating marine pollution", "Environmental Impact Assessment", "land-based sources of pollution", provide options for cooperation relating to the management of the Meric River.[18]

[18]According to the former Greece Minister of Foreign Affairs, George A. Papandreou, the most important achievement attained up to this point concerns the establishment of communication between appropriate carriers. http://www.papandreou.gr/papandreou/content/Document.aspx?d=6&rd=7739474&f=1380&rf=2038276612&m=4565&rm=9378861&l=1. Accessed 17 July 2010.

There have also been several attempts at cooperating on technical projects. For example, during the 3rd meeting of the Greek-Turkish Joint Committee on environmental cooperation,[19] the two ministerial delegations discussed a proposal to cooperate on pollution prevention of the Meric River. In June 2006, Greece and Turkey signed a protocol to work together on flood mitigation in a project funded under the EU Cross Border Cooperation programme for which a technical committee was set up. However, the planned flood mitigation measures have never been implemented, because they involved points of high military/national importance and touched questions about the positioning of the borderline that both sides could not agree on (Skias and Kallioras 2007).

5.2 *Turkey and Bulgaria*

In 1968, Turkey and Bulgaria signed the "Agreement between the Republic of Turkey and the People's Republic of Bulgaria on the Cooperation of the Utilization of the Waters of the Rivers Flowing in the Territories of the Two Countries" (see Annex 3), which refers to the principles of international law and good neighbourly relations. The main objective was to regularise beneficial use of boundary and transboundary rivers and to provide for flood protection. The parties committed themselves to cooperate in research and the study of ventures which would be beneficial to both of them, to not inflict serious damages on each other by constructing and operating facilities on the rivers, to exchange information on floods and icing as rapidly as possible, and to exchange hydrological and meteorological data. A Turkish-Bulgarian Joint Commission composed of equal numbers of experts from both countries was authorised with settling disputes which may have arisen during the implementation of the agreement.

The 1975 "Agreement on Long Term Economic, Technical, Industrial and Scientific Cooperation" (see Annex 4) between the Government of the Republic of Turkey and the Government of the People's Republic of Bulgaria states that cooperation between the concerned Turkish and Bulgarian enterprises and organisations shall be simplified in all the fields of economy including "energy production and irrigation, including the joint use of the waters whose shores are on both countries, for energy production and irrigation purposes" (Article 5).

Recognising the need for cooperation to alleviate the severe consequences suffered by both parties due to drought, the "Agreement on Assistance and Cooperation in the Field of Water for Reducing the Negative Effects of the Drought" of 1993 (see Annex 5) was signed. It states that Bulgaria, on a one-off basis and limited to 1993, should provide additional water to Turkey from the river Tunca. In turn, Turkey should allocate US Dollars 0.12 per cubic meter of water provided by Bulgaria. Accordingly,

[19]Agreed Minutes of the third meeting of the Greek-Turkish Joint Committee on environmental cooperation. Available at http://www.ecolex.org/server2.php/libcat/docs/bi-82300.pdf. Accessed 16 March 2010.

Turkey purchased 15,866,000 cubic meters of irrigation water from Bulgaria at 1,903,904 US Dollars cost (Turkish Parliament Research Commission 2002).

In 1998, Bulgaria and Turkey signed an agreement on co-operation in the energy and infrastructure sectors (see Annex 6), in which Bulgaria agreed to contract Turkish companies for two major infrastructure projects: the Gorna Arda hydropower project and construction of a stretch of the Meric highway. In return, Turkey was to purchase a certain amount of electricity at a fixed price from Bulgaria. The Gorna Arda hydropower project was launched in 1999. It included rehabilitation of existing dams as well as construction and operation of three new dams on the Arda River near the Turkish border. In 2000, however, the Turkish Ceylan Holding Company, which won the bid, experienced financial difficulties. No alternative contractor was commissioned and the projects did not get beyond their planning phases. Turkey stated that this was non-fulfilment of the 1998 agreement and stopped purchasing electricity from Bulgaria in 2003 (Buechsenschuetz 2003). The project was delayed further and it was not until September 2009 that the Bulgarian government sealed a letter of approval for the construction of the project by an Austrian-Bulgarian Consortium.[20] The hydropower project with a capacity of 175 MW is expected to produce 440 GWh per year and to cost 500 million Euros which shall be covered by the consortium.[21]

The Turkish-Bulgarian Joint Committee for Economic and Technical Cooperation signed the "Agreement on the Approval of the 15th Term Protocol" in 2002.[22] Under the subheading "Environment", both parties agree to further environmental cooperation for the protection of surface and groundwater resources and water related environments. Under "Energy and Environment", the Turkish side repeated the request to establish a joint technical working group to investigate the conditions for building the Suakacagi Dam on the Tunca River.[23] The Bulgarian side confirmed that this issue would be addressed promptly. Both sides agreed to continue hydrological data exchange in order to prevent flooding and to exchange data regarding water levels and releases from dams on the Meric, Arda and Tunca. They further agreed that the Technical Working Group which was created under the 1968 Agreement should continue its regular activities.

A protocol was signed between the Devlet Su Isleri (DSI) and the National Institute of Meteorology & Hydrology (NIMH) of Bulgaria in 2002 which covers the installation, operation, and maintenance of a flow observation telemetry station

[20]http://www.novinite.com/view_news.php?id=110374. Accessed 17 July 2010.

[21]http://noe.orf.at/stories/387176/, Accessed 17 July 2010.

[22]The Protocol of the Fifteenth Session of the Turkish-Bulgarian Joint Committee for Economic and Technical Cooperation contains provisions on trade and economic relations: bilateral trade relations, trade promotion activities, standardisation, industry and transport (inter alia, road transport and maritime transport), telecommunications and postal services, agriculture, and environment (Sofia, 22-23 January 2002, Resmi Gazete, 2002-07-03, No. 24804, 3-36; see Annex 7).

[23]Joint construction of the Suakacagi Dam has been planned with Bulgaria starting in 1968 but has not been completed until to date. The dam would irrigate 50,000 hectares, protect 2,000 hectares from being flooded and operate three turbines, two on the Bulgarian and one on the Turkish side to generate 100 Gigawatt per hour (GWh)/year energy.

on the Meric River in Svilengrad to improve monitoring and to generate hydrometeorological data in periods of flood. In addition, to support future collaboration between NIMH and DSI in joint projects related to Mediterranean Hydrological Cycle Observing System (MED-HYCOS), North Atlantic Treaty Organisation (NATO), European Commission (EC), Water Observation and Information System for Balkan Countries and others, a framework agreement was signed in November 2002 between the two institutions.[24]

Since then, the two institutions have cooperated in projects on flood risk management funded under the EC PHARE Cross-Border Cooperation (CBC) programme which involved the creation of a hydrometeorological database and the installation of flood warning and water information systems (Dagdas 2007, Sezen et al. 2007).

5.3 Greece and Bulgaria

The main transboundary water agreement between Greece and Bulgaria is the 1964 agreement on common use and management of joint surface water resources. It stipulates that parties must avoid causing damage to the other party by the construction of infrastructure, that parties exchange hydrometeorological data and information on floods, and that they will carry out feasibility studies on joint infrastructure projects (Tzovaridis et al. 1996). In 1971, an agreement was concluded between Greece and Bulgaria for the establishment of a Greek-Bulgarian Committee that would deal with hydropower and with the use of shared water resources (INWEB 2004).

Bulgaria and Greece are both parties to the United Nations Economic Commission for Europe (UNECE) Helsinki Convention, which provides a legal framework for cooperation in transboundary water management. Moreover, as members of the EU, both countries have to comply with the relevant European directives. As a new member of the EU, Bulgaria has already implemented all relevant EU environmental norms in the water sector into national legislation - including the WFD (Arsov 2004). Greece has only recently incorporated the WFD in its national legislation. While both countries are in the process of developing basin management plans for their respective parts of the Meric basin, a transboundary river basin district has not been assigned and the two riparians are far from producing a single international river basin management plan for it.

An agreement on cooperation in the field of environmental protection[25] concluded by Bulgarian and Greek authorities in 2002 explicitly refers to the

[24]Framework agreement between the NIMH and the DSI, signed on 27 November 2002.

[25]Agreement between The Ministry for the Environment, Physical Planning and Public Works of the Hellenic Republic And The Ministry of Environment and Water of the Republic of Bulgaria On Cooperation in the Field of Environmental Protection, 2002. The main objective of the agreement is to promote cooperation in various areas of environmental protection, e.g. water, coastal zones, bioderversity, spatial planning and environmental impact assessment.

WFD when stipulating to "Promote water protection and use, in the spirit of the EU WFD, and cooperate for sustainable management of transboundary water sources, in accordance with the international conventions to which they are Contracting Parties." (Article 2, (4)). In addition, the parties agreed to conduct joint research efforts in the area of environmental monitoring (namely air, water and soil pollution) and thus to establish monitoring systems and provide bilateral linkage between them, or use existing monitoring systems provided within the framework of the relevant EU legislation (Article 2, (3)).

Cooperation in scientific and technical fields is well established among the two riparians. Within the framework of the INTERREG programme of the European Commission, Bulgaria and Greece have been cooperating on research projects on transboundary waters since the early 1990s. Special emphasis was put on aspects of the management of shared waters. For example, pollution measuring stations were installed and equipped on the Bulgarian section of the rivers Nestos/Mesta, Strymonas/Struma and Evros/Meric. A recent project funded under the PHARE-CBC programme focussed on early warning in case of floods and accidental pollution.[26] The project involved setting up automatic hydrometeorological stations and harmonizing data as well as capacity building for local and regional stakeholders.

6 Outstanding issues

Up to now, only bilateral agreements exist which cover issues of flood protection and joint infrastructure projects as well as general environmental cooperation including conservation of protected areas. Issues of water allocation, on the other hand, remain unsettled. Conflicting interests in water resource development of the Meric basin mainly exist between Bulgaria and Turkey. Turkey's plans to increase irrigated areas in the Meric basin would aggravate the situation. In order to make more water available for irrigation in Turkey, it was proposed that Turkey should consider the possibility, despite the additional cost, to unilaterally construct off-stream storage facilities which may collect excess winter outflow from Bulgarian and Greek dams (Ozis et al. 2002). In addition, Turkey has proposed joint dam projects with Bulgaria, which would also serve as a means for flood control.

Furthermore, existing agreements do not cover legal provisions on water quality standards. Likewise, arrangements on the exchange of data and information mainly focus on information on floods, while cross-border availability of data on water quality is reported to be a problem (Mylopoulos et al. 2004). In addition, no agreement exists that would provide for a minimum inflow of freshwater into the

[26]Capacity Improvement for Flood Forecasting of Meric/Evros River Basin in the BG-GR CBC Region, Project Fiche available at http://ec.europa.eu/enlargement/fiche_projet/document/2006_018-387.03.02%20Capacity%20Improvement%20for%20Flood%20Forecasting.pdf. Accessed 23 March 2010.

delta, satisfying the water needs of the ecosystem as well as preventing salt water intrusion and siltation.

While there are several on-going technical projects between Bulgaria and Turkey as well as between Bulgaria and Greece, none of these water-related initiatives involve all three riparians. However, there are relations between local governments that aim at promoting cross-border cooperation among the three countries (Skias and Kallioras 2007).[27] Currently, these networks deal mainly with issues such as tourism, labour, health and culture, but they also facilitate cooperation on humanitarian assistance e.g. during flooding. While they have thus no mandate to take actions with regards to water management, they might provide an entry point for trilateral projects with regards to managing the water resources of the Meric basin. Papayannis (2004) states that while the population in the border region is open to cross-border cooperation, there is reservation at the government level, especially from the military authorities. The role of local level networks could thus be to put pressure on national government to facilitate transboundary cooperation on issues of importance for the local population (see also Skias and Kallioras 2007).

Flood mitigation and risk management

One of the most urgent fields of action in the basin is flood protection. Even though agreements exist, adherence to them has not been satisfactory in the past. After the severe floods of March 2005, Turkey is reported to have sent Bulgaria a note of protest because of her alleged failure to abide by the bilateral agreement. Likewise, Greece also blamed Bulgaria for flood incidents (Andonova and Velinova 2005, The Sofia Echo, 11 March 2005, see also Skias and Kallioras 2007). Since then, several gauging stations were installed on Bulgarian territory along the Meric, Arda and Tunca rivers. These allowed warning downstream riparians 15 hours in advance of the occurrence of the 2006 spring flooding. Likewise, the Prime Ministers of Greece and Bulgaria signed an official declaration in April 2006 that outlined a new framework for bilateral partnership regarding, among others, “policy measures and actions to avert floods in river Evros basin” and decided that the first step shall be the construction of monitoring and early warning systems (cited from Skias and Kallioras 2007). Bilateral technical projects such as those funded under the PHARE-CBC programme have worked to improve early warning capacities of the basin countries. However, the monitoring system is still considered as insufficient for good flood risk management (Darama 2009). Furthermore, trilateral cooperation is necessary to improve the early warning system (UNECE 2009). Skias and Kallioras (2007) report on a Greek initiative to establish a trilateral

[27]The network Euroregion Evros-Maritza-Meric for example, was established in 2001 through an agreement signed by representatives of the Border Line Unions and the Prefectures of Evros (Greece), Edirne (Turkey), and the area of Haskovo (Bulgaria). For more information see http://www.evroregion.gr/profile.html. Accessed 17 July 2010.

working group to discuss the flooding topic. However, the working group only met once in October 2006.

Operation of dams, management of floodplains and early warning have been identified as crucial in averting disasters. The adopted coping strategies in the Meric basin, however, remain rather technocratic than management-oriented: the construction of additional dams upstream is often considered as the favourable solution for addressing the flooding issue. This becomes obvious, for example, in the planning of another extensive dam project, the Suakacagi Dam, on the Tunca River close to the Bulgarian-Turkish border: the Suakacagi Dam project has been planned since 1968. In the aftermath of the 2005 floods, Ankara and Sofia agreed to advance the project in order to mitigate flood problems in Turkey. The two sides agreed on appointing experts to develop the project and to establish a Turkish-Bulgarian joint technical commission for the implementation of the project. The dam is expected to not only serve as flood protection means but also to further benefits to the two countries. It could, for example, also provide irrigation water for the area around Edirne and Kirklareli in Turkey (The Sofia Echo, 03 June 2005). In April 2005, a technical delegation from Bulgaria paid a visit to the DSI Regional Directorate in Edirne. Consensus was reached regarding the dam site and a protocol was signed. However, to date, the project has not passed the planning stage due to disagreement between Turkey and Bulgaria over land rights.[28] Joint dam projects that provide benefits for both riparians - such as the Suakacagi Dam or the 1998 electricity-for-infrastructure bilateral deal that included the building of dams on the Arda River - may serve as a first step in transboundary cooperation if based on a fair agreement to share benefits arising from the dam project.

In the future, not only infrastructure should be considered when addressing problems of flood control and water allocation, but also the development of an integrated basin wide management approach. While transboundary cooperation on the flooding issue has proven to be complicated on the trilateral as well as on bilateral level, existing bilateral work on flood forecasting should still be considered as an entry point for broader cooperation in water resources management and existing initiatives further be pursued.

With regards to a more integrated approach to flood risk management, the European Floods Directive[29] provides orientation for long-term developments in the basin. It explicitly states that Member States shall coordinate their flood risk management practices in shared river basins, including third counties, and shall in solidarity not undertake measures that would increase the flood risk in neighbouring

[28]See Hürriyet "A city sinks for want of international cooperation," 16 March 2006, http://www.hurriyet.com.tr/english/4090775_p.asp. Accessed 17 July 2010, and Darama 2009.

[29]Directive 2007/60/EC on the assessment and management of flood risks that entered into force on 26 November 2007, and aims to reduce and manage the risks of floods. The proposed management approach shows parallels to the Water Framework Directive (WFD), particularly with regard to the management by river basin districts according to WFD Art. 3 and the coordination mechanisms (http://ec.europa.eu/environment/water/flood_risk/index.htm. Accessed 17 July 2010).

countries. Furthermore, the Model Provisions on Transboundary Flood Management adopted by the fourth meeting of the parties to the UNECE Helsinki Convention (Bonn, 20-22 November 2006) can provide guidance for the riparian states of the Meric River basin.

Water quality and environment

A second field of action that could bring mutual benefit to all riparian countries is nature protection and conservation in the basin area, especially the protection of the Ramsar Site, i.e. Evros Delta. Projects for the Lower Meric Valley Flood Plain as a biosphere reserve also exist on the Turkish side. Several initiatives have been taken to foster transboundary conservation activities:

- The Turkish Ministry of Environment and Forestry with support from United Nations Educational, Scientific and Cultural Organisation (UNESCO) Regional Bureau for Science in Europe has organised an international conference on "Biosphere Reserves and Transboundary Cooperation between Bulgaria, Greece and Turkey" which took place in Edirne in July 2005.
- In 2001, the Mediterranean Wetlands Initiative (MedWet) developed a project to foster transboundary collaboration to manage and protect the Meric River and its wetlands. Unfortunately for internal reasons, funding for the project was not approved.[30]
- The European Green Belt initiative under the leadership of the German Federal Agency for Nature Conservation and International Union for Conservation of Nature and Natural Resources (IUCN) aims to transform the former Iron Curtain area along the east-west divide of Europe into a protected corridor, thus acting as a symbol of unity between East and West. One stretch of the European Green Belt route follows Bulgaria's borders with Turkey and Greece. Within this context, the Meric basin has been identified as one of the priority sites for transboundary cooperation (IUCN 2004).

However, no concrete results have yet been achieved.

Water quality remains an unsolved issue. Any solution to this problem, arising, to a significant degree, from insufficient wastewater treatment, would need large investments in infrastructure. Ongoing EU-cooperation programmes with accession countries may lead to some alleviation of the problem by supporting the construction of wastewater treatment plants upstream. The Instrument for Pre-Accession Assistance for Turkey, for example, that replaces earlier programmes such as PHARE, includes a priority axis on improved water supply, sewerage and wastewater treatment services with an EC contribution of 134,298,000 Euros.[31] However, the EU member countries and the accession country Turkey are obliged to implement the

[30]Personal email correspondence with Thymio Papayannis, MedWet Senior Advisor, 12 April 2005.

[31]For further information see http://ec.europa.eu/regional_policy/funds/ipa/turkey_environment_en.htm. Accessed 17 July 2010.

EU Urban Wastewater Directive which would result in a significant improvement of the rivers' water quality.

High sediment loads – a consequence of erosion in the catchment areas – causes siltation problems in the river delta and forming of sand islets. Turkey has started a programme to clean the sand islets in order to maintain a regular flow. However, technical cooperation by the other riparians is deemed necessary to fully address this issue (Yildiz 1999). Darama (2009) reports on a cooperation project between Bulgaria and Turkey with regards to the issue that is waiting to be implemented.

Data exchange and scientific cooperation

Exchange of hydrological and water use data seems to be insufficient between the riparian countries. It has been reported that no information is available from the Bulgarian side about any wastewater discharge into the river or any damming of water (Mylopoulos et al. 2004). Establishing scientific exchange between riparian countries can be an invitation for further collaboration.

In the Meric basin, exchange takes place, to some extent, through the International Network of Water-Environment Centres for the Balkans (INWEB) established in 2000.[32] Funded by the UNESCO Regional Bureau for Science in Europe (UNESCO-ROSTE), INWEB organised an international workshop in 2004 with the objective of sharing available data on transboundary watercourses in south-eastern Europe. Participants from all three riparian countries contributed and presented data and information, though incomplete, on the Meric basin. Furthermore, the World Hydrological Cycle Observing System (WHYCOS) offers a network for data exchange. WHYCOS established a global network of national observatories with the objective of creating a relatively transparent data base. The Mediterranean division (MED-HYCOS) has set up five Data Collection Platforms in the Meric basin, four in Bulgaria and one in Turkey.

References

Andonova M, Velinova S (2005) Bulgarian foreign relations flooded. Capital Weekly, Issue 10, 12-18 Mar 2005. http://www.capital.bg/weekly. Accessed 3 June 2005

Arsov R (2004) Bulgarian transboundary rivers. Paper presented at the workshop Development of an Inventory of Internationally Shared Surface Waters in South-Eastern Europe. Thessaloniki, Greece, 18-20 Oct. 2004. http://www.inweb.gr/workshops/papers_surface/BULGARIAN%20TRANSBOUNDARY%20RIVERS.pdf. Accessed 13 March 2010

Artinyan E, Habets F, Noilhan J, Ledoux E, Dimitrov D, Martin E, Le Moigne P (2007) Modelling the water budget and the riverflows of the Meric basin in Bulgaria. Hydrology and Earth

[32]INWEB promotes joint training projects and the sharing of expertise, e.g. through web-based water databases and inventories of transboundary surface and groundwaters in the Balkans and the Mediterranean region. For more information on this initiative see http://www.inweb.gr.

System Sciences Discussions, vol 4: 475-521. http://www.hydrol-earth-syst-sci.net/12/21/2008/hess-12-21-2008.pdf. Accessed 13 March 2010

Bilen O (2000) Turkey and water issues in the Middle East: An examination of the Indus, Colorado, Danube and Jordan-Israel Water treaties and the water agenda of 21st century. Southeastern Anatolia Project (GAP) Regional Development Administration, Ankara, Turkey

Buechsenschuetz U (2003) Turkey's new energy policy hits Bulgaria. Radio Free Europe/Radio Liberty Newsline 19 June 2003. http://www.rferl.org/content/article/1142941.html. Accessed 13 March 2010

Centre for European Constitutional Law (2001) Hellenic-Bulgarian environmental cooperation for Environmental Impact Assessment. Project commissioned by the Hellenic Ministry for the Environment, Physical Planning and Public Works. Project Report http://www.cecl.gr/ceia/index.htm. Accessed 03 June 2005

Dagdas S (2007) Priorities for preventing natural calamities and protection of ecosystems to be studied in border regions related to the cross-border cooperation programme between Bulgaria and Turkiye. Paper presented at International Congress on river basin flood management, Antalya, March 22-24, 2007. www.dsi.gov.tr/english/congress2007/chapter_4/131.pdf. Accessed 13 March 2010

Darama Y (2009) Flood risk management of Maritza river and early warning system. Presentation held at Workshop on transboundary water resources management in South- Eastern Europe 18-20 May 2009, Sarajevo. http://www.unece.org/env/water/meetings/Sarajevo_workshop/presentations/session4/Aegean_Sea/MARITZA.pdf. Accessed 13 March 2010

Dimitrov D (no year) Recent floods in Bulgaria – Information and forecasting services. http://balwois.com/balwois/administration/full_paper/ffp-873.pdf. Accessed 13 March 2010

DSI (2005) Turk Heyeti Bulgaristan Daydi. Basin Bultenleri, DSI Genel Mudurlugu, Basin Musavirligi, 1 April 2005. www.dsi.gov.tr/basin/bulgaristan.htm. Accessed 03 June 2005

DSI XIth Regional Directorate, no year: http://www.dsi.gov.tr/bolge/dsi11/eng/sag.htm. Accessed 03 June 2005

ECSSD (Environmentally and Socially Sustainable Development Department) (2003) Water resources management in South Eastern Europe, Vol 2. The World Bank, Washington DC

European Commission (2009) Turkey 2009 progress report. Commission staff working document {COM(2009) 533}. http://ec.europa.eu/enlargement/pdf/key_documents/2009/tr_rapport_2009_en.pdf. Accessed 13 March 2010

INWEB (International Network of Water Environment Centres for the Balkans) (2004) Inventory of international surface waters. http://www.inweb.gr/workshops/balkan_map.html. Accessed 13 March 2010

IUCN (2004) Conservation without frontiers - Towards a new image for the Balkans. A strategic plan for the IUCN South-Eastern European Programme. IUCN Regional Office for Europe. http://www.lcie.org/Docs/Regions/Balkan/ConservationWithoutFrontiers-TowardsANewImageForTheBalkans.pdf. Accessed 13 March 2010

Mylopoulos YA, Kolokytha E, Vagiona D, Kampragou E, Eleftheriadou E (2008) Hydrodiplomacy in practice: Transboundary water management in northern Greece. Global NEST Journal 10 (3) (Nov 2008): 287-294. http://www.gnest.org/journal/Vol10_No3/287-294_451_MYLOPOULOS_10-3.pdf. Accessed 13 March 2010

Mylopoulos YA, Eleftheriadou E, Kampragou E (2004) The transboundary catchment of River Nestos and the bilateral agreement between Greece and Bulgaria. Conference on Integrated Water Management of Transboundary Catchments: A Contribution from TRANSCAT. Venice, Italy: 24-26 March 2004

Nenkova S (2005) Turkey still wants Gorna Arda, Meric Projects. Bulgarian Economic Review, Issue 6, March 28, 2005. http://www.pari.bg/cgi-bin/ber.home.cgi. Accessed 03 June 2005

Ozis U, Ozdemir Y, Baran T, Turkman F, Fistikoglu O, Dalkilic Y (2002) Turkiye'nin sinir asan sularinin su hukuku ve su siyaseti acisindan durumu. In: Sen Z, Ozis U, Avci I, Bilen O, Zehir C, Birpinar ME (eds) Sinir Asan Sularimiz. Su Vakfi Yayini, Istanbul, pp 36-37

Papayannis T (2004) Shared catchments and wetlands – increasing transboundary cooperation. Paper presented at 5th European Regional Meeting on the implementation and effectiveness of the Ramsar Convention, 4-8 December 2004, Yerevan, Armenia. http://www.ramsar.org/pdf/mtg/mtg_reg_europe2004_docs1d1.pdf. Accessed 13 March 2010

Penov I, Theesfeld I, Gatzweiler F (2003) Irrigation and water regulation systems in transition: The case of Bulgaria in comparison with Latvia, East Germany and Romania. In: Gatzweiler F, Hagedorn K (eds) Institutional change in Central and Eastern European agriculture and environment, vol 3. Food and Agriculture Organisation of the United Nations (FAO) and Humboldt University of Berlin

Petkova L (2002) The ethnic Turks in Bulgaria: Social integration and impact on Bulgarian – Turkish relations 1947-2000. The Global Review of Ethnopolitics, vol 1, no 4: 42-59

Ramsar Convention (1999) Expression of opinion on Greek Ramsar wetlands and possible removal from the Montreux Record, Annexes IV-3. http://www.ramsar.org/cda/ramsar/display/main/main.jsp?zn=ramsar&cp=1-30-168^20976_4000_0__ Accessed 13 March 2010

Regional Environment Accession Project (2002) Protection of waters against pollution caused by Nitrates from agricultural sources, Directive 91/676/EEC, Bulgaria. Final report. Project funded by the EC through the PHARE programme. http://europa.eu.int/comm/environment/enlarg/reap_en.htm. Accessed 03 June 2005

Samsunlu A, Maktav D, Kapdaslii S (1996) Transboundary water issues between Greece-Bulgaria and Turkey: the case of Meric / Evros River. In: Bogardi I (ed) Transboundary water resources management. Institutional and engineering approaches. Springer, Berlin

Sezen N, Gunduz N, Malkarali S (2007) Meric river floods and Turkish-Bulgarian cooperation. Paper presented at the International congress on river basin flood management, Antalya, March 22-24, 2007. www.dsi.gov.tr/english/congress2007/chapter_4/125.pdf. Accessed 13 March 2010

Skias S, Kallioras A (2007) Cross-border cooperation on the flooding problem in the Evros Delta. In: Verwijmeren J, Wiering M (eds) Many rivers to cross: Cross-border cooperation in river management. Eburon Academic Publishers, Delft, The Netherlands

The Sofia Echo, 03 June 2005: New runners for Upper Arda. http://www.sofiaecho.com. Accessed 03 July 2005

The Sofia Echo, 11 Mar 2005: Flood of complaints. http://www.sofiaecho.com. Accessed 11 May 2005

Turkish Parliament Research Commission (2002) Yeralti ve Yerustu Su Kaynaklarimizin Daha Etkin Kullanimini Saglamak ve Komsu Ulkelerle Olan Su Sorununa Cozum Bulmak Amaciyla Kurulan (10/13) Esas Numarali Meclis Arastirmasi Komisyonu Raporu, 6 Mart 2002

Tzovaridis SN, Moustafis NJ, Cavadias GS (1996) Management issues of transboundary rivers between Bulgaria and Greece. In: Ganoulis J, Duckstein L, Litherathy P, Bogardi J (eds) Transboundary water resources management: Institutional and engineering approaches. Springer, Berlin

UNECE (2007) Our waters: Joining hands across borders. First Assessment of Transboundary Rivers, Lakes and Groundwaters. Economic Commission for Europe Convention on the Protection and Use of Transboundary Watercourses and International Lakes. New York and Geneva: United Nations. http://www.unece.org/env/water/publications/pub76.htm. Accessed 13 March 2010

UNECE (2009) Assessment of transboundary rivers, lakes and groundwaters in South-Eastern Europe discharging in the Aegean Sea. Note by the secretariat for the Meeting of the Parties to the Convention in the Protection and Use of Transboundary Watercourses and International Lakes, Fifth Session, Geneva, 10-12 Nov.2009. ECE/MP.WAT/2009/11. http://www.unece.org/env/documents/2009/Wat/mp_wat/ECE_MP.WAT_2009_11_E.pdf. Accessed 13 March 2010

Yanik B (1997) Turkiye'deki Sinirasan ve Sinir Olusturan Su Kaynaklari. Yayinlanmamis yuksek lisans tezi. Istanbul: Teknik Universitesi-Yon: I. Avci.

Yildiz D (1999) Sinir Olusturan ve Sinir Asan Su Kaynaklarimiz ve Kiyidas Ulkeler Arasinda Teknik Isbirligi Gereksinimi. Cevre ve Muhendis TMMOB Cevre Muhendisleri Odasi Yayini, no 18: 28-35

Zardava K (2004) The transboundary River Evros. Presentation held at the workshop Development of an Inventory of Internationally Shared Surface Waters in South-Eastern Europe. Thessaloniki, Greece, 18-20 Oct. 2004. http://www.inweb.gr/workshops/presentations_pdf/surface/Zardava.pdf. Accessed 13 March 2010

Coruh River Basin: Hydropower Development and Transboundary Cooperation

Axel Klaphake and Waltina Scheumann

1 Geographical and hydrological setting

The Coruh[1] River is the longest river of the East Black Sea region and is of high economic importance for Turkey because of its largely undeveloped but economically exploitable hydropower potential. However, the operating and planned dams could also cause environmental damage in Turkey and downstream Georgia, in particular on the Black Sea coast in the Adjaria province.

The Coruh River is located in north-east Turkey and shared by two countries: Turkey and Georgia. Approximately 91 percent of the basin's drainage area (21,100 square kilometers) is, however, in Turkey while Georgia's share amounts to 9 percent only. The principal tributaries of the Coruh River are the Tortum and Oltu rivers in Turkey, and Adzharis and Tsakali rivers in Georgia. In total, the Coruh River is 426km long, ca. 400km of which lies within Turkey's borders. The river also forms a short border (3km) between Turkey and Georgia, and flows for 24km through Georgia. See Figure 1 for a map of the Coruh River, its main tributaries and selected dams, and Table 1 for an overview of the context for cooperation on the river.

The river originates in the western part of the Mescit mountains at a height of over 3,000m, north-west of the Erzurum-Kars Plateau. From these mountains the river first flows west, then turns east with a sharp bend at the Bayburt Plain and afterwards follows a tectonic hollow which separates the East Black Sea coastal mountain chain

[1]Internationally, "Coruh" is the most frequently used name for the river. Other given names are Chorokin, Coroch and Chorokhi.

A. Klaphake (✉)
Deutsche Gesellschaft fuer Internationale Zusammenarbeit, GIZ, Kampala, Uganda
e-mail: axel.klaphake@giz.de

W. Scheumann
Deutsches Institut fuer Entwicklungspolitik, Bonn, Germany
e-mail: waltina.scheumann@die-gdi.de

A. Kibaroglu et al. (eds.), *Turkey's Water Policy*,
DOI 10.1007/978-3-642-19636-2_14, © Springer-Verlag Berlin Heidelberg 2011

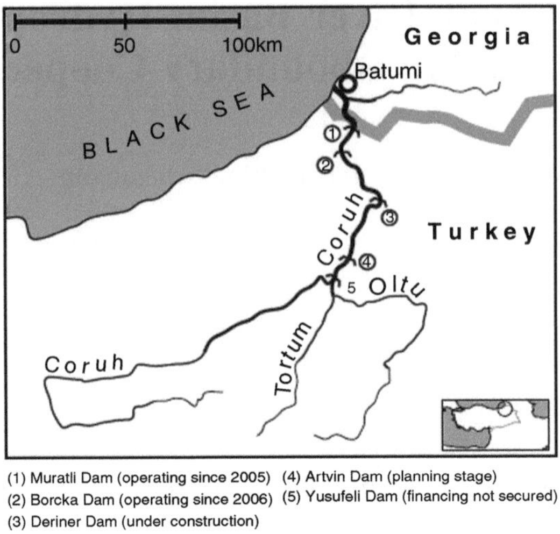

Fig. 1 Dams on the Coruh and its main tributaries on Turkish territory

from the inner mountain range. The Coruh valley located in the eastern part of Ispir is one of the deepest valleys in Turkey. Having passed the city of Yusufeli and the confluence with the Oltu River, the Coruh flows north and shapes a mountainous landscape with deep canyons. Passing through the cities of Artvin and Borcka, it leaves Turkish territory near the city of Muratli. Near Batumi, the capital of the Georgian semi-autonomous province Ajaria, the river empties into the Black Sea through a delta which is largely composed of alluvium that it has accumulated along its course.

Because of climatic conditions, the river carries plenty of water in all seasons albeit with remarkable seasonal variations (Yildiz 1999a). The river experiences a transitional climate between the Black Sea's mild and wet weather and the East Anatolia's cold climate. Turkey's average annual rainfall is 642mm, whereas the Coruh basin receives 475mm rain in an average year. According to long-term observations measured at the flow monitoring station in Muratli, the average flow rate is 202 cubic meters per second. The highest run-off measured at this station was 2,431 cubic meters per second and the lowest 37.6 cubic meters per second. Rainfall and especially snowmelt from the high mountains suggest that there is high water availability. In spring, total water flow reaches 221.38 million cubic meters (MCM) per year constituting 40.9 percent of the mean annual flow. The flow rate of the river in May alone equals approximately twice the amount it carries during the whole winter season.

According to Turkish long-term observations, annual flow rates range from 3.3 billion cubic meters (BCM) per year (1955) to 11.2 BCM/year (1968). The medium annual flow rate observed through long year measurements is 6.3 BCM/year, which corresponds to the 3.4 percent of the total water potential in Turkey (Kurucim 2002).[2] Apart from the comparatively high and variable flow rates, the river carries high levels of sediment and deposits (estimated at 5 MCM/year) which

[2]The World Bank published a somewhat higher estimation of the river's flow, i.e. 8.7 BCM/year (World Bank 2003).

Table 1 Cooperational context in the Coruh basin (Own compilation; Encon 2006; Jaoshvili 2003)

Coruh basin area: 21,100 km^2; mean annual discharge 6.3 BCM		
Riparian position	Basin area (percent of total)	Main water uses
Turkey upstream	19,200 km^2 (90 percent)	irrigation, hydropower, recreation / tourism
Georgia downstream	900 km^2 (10 percent)	Small-scale agriculture, recreation, fishery
Main agreements and covered issues		
Turkey - Georgia	1927 – border issues, river bank protection, water allocation, compensation requirements, joint commission	
	1996 – Meetings between DSI (Turkey) and GD Georgian Coastal Zone Protection	
	1998 – Series of meetings of Turkish and Georgian Experts on issues concerning the Coruh river, the construction of the Deriner dam, to initiate studies	
	1999 – Installation of two monitoring stations for river flow regime	
	2000 – technical cooperation, river bed changes	
	2002 – MoU on Cooperation for Obtaining Aerial Digital Maps of Coruh Basin in Georgian Territory for Determining Possible Downstream and Environmental Impacts of Hydraulic Structures being built on Coruh and its Tributaries	
	2002 – MoU between Representatives of Georgian and Turkish Governments for Cooperation	
	2003 – series of meetings	
	Agreed Minutes of Meeting held between delegation of Turkey and Georgia regarding the solution of problems connected to the construction of dams on the Coruh River	
	2006 – Meeting in Ankara for exchanging views on issues concerning the Coruh river; a MoU was signed	
Unsettled issues		
Quality	Exchange of data insufficient	
Sediment	Main disputed issue. Negative impacts of Turkish dams on the sediment regime and on coastal zones in the environs of Batumi (coastal erosion)	
Biodiversity	No comprehensive approach to the protection of freshwater ecosystems	
Water quantity	At present not a relevant transboundary issue. Current allocation rule (50:50) is not questioned	

stem from erosion in the Turkish mountain regions. Despite the rather small drainage basin area, the river has high hydropower production potential due to the topographic conditions, in particular, the sharp fall of the river from high mountains to sea level.

In contrast to reliable monitoring of the river's flow regime, accurate water quality data is lacking and/or is not easily accessible. Pollution of the river is relatively small, at least compared with other transboundary rivers in the region (e.g. Kura-Aras). At present, urban wastewater is a source of pollution because of insufficient wastewater treatment and solid waste disposal facilities. In addition – and limited to specific sites – industrial pollution plays a certain role, for example, discharge from the mining industry. In contrast, pollution from agriculture is of minor importance because farming is of small scale. Far from being pristine, the Coruh River, however, shows high value in terms of biological diversity and hosts plenty of important species and habitats. The river's valley is known for its rich flora with high endemism; some 2,500 species of vascular plants can be found, including 160 endemic ones. Large mammals are well represented in the region, as are important amphibians and reptiles.

2 Current and future water uses

Currently, only a minor share of the river's water is used for economic and social purposes. The most relevant uses are water withdrawals[3] and in-stream activities such as kayaking and boating. The latter has enjoyed an increase in recent years and provides significant added value to the regional tourist sector. Because of the topography of the basin, agriculture is only of minor importance and the expansion of irrigated areas (although mentioned in relevant planning documents) is estimated to play a rather minor role in development planning.

Today, the most important pressures on freshwater ecosystems come from the development of hydropower generation installations. Because of favourable topographic conditions, the Coruh River has – according to DSI estimates – the potential to provide some 13 percent of the usable hydroelectric power in the country, which to date remains largely untapped. The first hydropower plant in the basin that was completed and begun commercial operation was Tortum I (26 MW installed capacity) on the homonymous tributary. The other hydropower stations already in operation are Murgul, Muratli and Borcka hydroelectric power plants.

Initial studies concerning the hydropower production potential of the river had already been carried out by Turkish authorities in the late 1960s. The Coruh Master Plan was eventually finished in 1982 and was followed by the Coruh Basin Development Plan. Construction of the most downstream Muratli dam, located close to the Georgian border, started in March 1999, followed by the Borcka dam (1999) and preceded by the Deriner dam (November 1997). According to the Coruh Basin Development Plan, eleven dams will be built on the main stem of the river with a total installed capacity of 2,894 MW (see Table 2). The Deriner and the Yusufeli dams rank among the most important dams in Northern Turkey in terms of size and hydropower potential. The Turkish authorities consider the Coruh Basin Development Plan and associated dam constructions as vehicles to support economic development in north-eastern Turkey and to secure the country's growing energy demand. It is one of those rivers whose hydropower potential will be developed by private sector investments following the Electricity Market Law (No. 4628) of 2001 which foresees that the state refrains from financing hydropower dams, or hydropower components of multipurpose dams.

Although activities predominantly focus on hydropower generation and the supply of electricity, an irrigation component runs parallel to it. With the construction of the

[3]Groundwater is, however, an easy accessible substitute in several regions.

Table 2 Dams at Coruh river and its tributaries (as of November 2009) (T.C. Cevre ve Orman Bakanligi, Devlet Su Isleri GD, DSI Coruh Projeleri 26. Bolge Mudurlugu, Artvin, no year.)

Dams at main stem	Purpose / status	Project developer
Tortum Dam	26 MW operating since 1972	DSI
Murgul Dam	5 MW (run-off the river) operating since 1951	DSI
Muratli Dam	115 MW start of operation in 2005	DSI (Verbundplan, temelsu, VA TECH, Voest Hydro, Yuksel, Strabag)
Borcka Dam	300 MW start of operation in 2007	DSI (Verbundplan, temelsu, VA TECH, Voest Hydro, Yuksel, Strabag)
Deriner Dam	670 MW construction ongoing	DSI (ERG Inş.A.Ş., Technostroyexport, ABB Power Generation Ltd., Sulzer Hydro Ltd., Hydro Vevey Ltd., Stucky Consulting Eng. Ltd. and DECON)
Artvin Dam	332 MW planning stage	Auctioned acc. Law No. 4628
Yusufeli Dam	540 MW planning stage	DSI financing not secured
Arkun Dam	222 MW planning stage	Auctioned acc. Law No. 4628
Aksu Dam	120 MW planning stage	Auctioned acc. Law No. 4628
Gullubag Dam	84 MW planning stage	Auctioned acc. Law No. 4628
Ispir Dam	54 MW planning stage	Auctioned acc. Law No. 4628
Laleli Dam	99 MW planning stage	Auctioned acc. Law No. 4628
Dams at Berta tributary	**Purpose / status**	**Developer**
Bayram and Baglik dam	148 MW planning stage	Auctioned acc. Law No. 4628
Dams at Oltu tributary	**Purpose / status**	**Developer**
Oltu and Ayvali dam	190 MW planning stage	Auctioned acc. Law No. 4628
Dams at Barhal tributary	**Purpose / status**	**Developer**
Altinparmak dam	50 MW planning stage	Auctioned acc. Law No. 4628
Total	2,894 MW	

dam cascade, 30,000 hectares of land are planned for irrigation, mainly along the upper and middle sections of the river. This rather modest objective for agricultural development is due to the basin's topographical limits. A full realisation of the planned development of irrigated agriculture could, however, significantly increase agricultural water use and change settlement patterns in the area.

Georgia does not use water from the Coruh River for its domestic water supply or its industrial demands, and, as agricultural irrigation does not play any significant role either, fishing has assumed more importance.

3 Potential impact on downstream Georgia and the Black Sea

Since the planned dams will predominantly be used for hydropower generation, the impact of the infrastructure on annual water flows from Turkey to Georgia is rather limited and, unsurprisingly, transboundary water quantity questions are therefore not at the centre of political debates. Furthermore, Georgia is not dependent on the Coruh River for energy and water; this clearly reduces the potential for conflicts. The most serious transboundary impact of the dams will be the expected radical change of the sediment regime of the river.

So far, erosion problems along the Georgian Black Sea coast are the controversial issue between Turkey and Georgia. Georgian authorities and ecologists claim that the dams planned in Turkey, in particular the almost completed Deriner Dam, will stem the drift of the solid element of the river flow, in particular sand and other alluvial materials that constitute the characteristics of the river and shape the coastal region (RFE/RL 1998; Jaoshvili 2003). The coastline around Batumi is formed by these alluvial materials whereby the river flow deposits sand, stone and debris at the river mouth and the nearby coastal stretches, which then counteract the erosive action of the sea. Consequently, the most serious of the anticipated effects of the dams upstream in Turkey could possibly be increased coastal erosion that might not only threaten ecosystems and beaches in the vicinity of the river's delta but also fisheries, and urban areas in the agglomeration of Batumi.[4] Clearly, it is not only the River Coruh that is posing problems for the coastal protection in the region of Batumi: dams built on the Georgian rivers Enguri and Rioni have had the same effect in allowing erosion by the Black Sea to outpace the natural replenishment provided by the rivers.

The fact that the dams will change the sediment flow in one way or another is largely acknowledged by both Turkey and Georgia; however the expected and precise impact on the Georgian coast line and the possible acceleration of erosion in the Batumi region are not. In general, it is not easy to foresee the direct effects of the planned Turkish infrastructures because coastal erosion is a multi-faceted problem with a variety of causes and effects generated by a variety of human interventions. Thus, even within Georgia there is enormous variation in expense estimates required for additional coastal protection. While representatives of the Georgian National Ministry for Environmental Affairs are expecting costs for mitigation and prevention measures (e.g., artificial enhancement of sediment flow from other rivers, coastal protection measures) to be around US$ 150 million, others estimate a much lower financial burden.[5] For instance, rough calculations for more modest coastal protection measures amount to US$ 5 million.[6]

It is worth mentioning that environmental impact assessment studies for individual dam projects on the Coruh River have so far only been undertaken on request

---

[4]Besides reinforced coastal erosion, the dams will certainly have an impact on freshwater ecosystems and, inter alia, fish species in the river basin. However, the current debate on the transboundary consequences largely focuses on the erosion problem.

[5]Personal communication with the head of the water department of the Georgian Ministry for the Environment, November 2004.

[6]The leader of the Georgian Green Party, Giorgi Gachechiladze, has been cited with a suchlike assessment in the public media (The Messenger, 20 April 2005. http://www.messenger.com.ge Accessed 30 September 2005.

of external financiers (for instance, for the Yusufeli dam).[7] The development of numerous dams on the main stem of the river and its tributaries would well justify the issuing of a Strategic Environment Assessment as required by the European Union (EU) Strategic Environmental Assessment (SEA) Directive of 2001 in order to identify the accumulated impacts from all projects (e.g. dams, highway, and transmission lines) which may well extend beyond Turkish territory. However, the situation is complicated by the fact that private investors are invited for hydropower generating dams to be implemented under the Electricity Market Law 4628 of 2001. With the DSI granting Water Use Right Agreements to private applicants, licenses from Energy Market Regulatory Authority (EMRA) should be bound on the condition to secure the ecological transmissivity of the river and its tributaries including minimum in-stream flow requirements which too could benefit downstream uses.

4 Status of cooperation

There is no comprehensive bilateral agreement on the Coruh River between Turkey and Georgia but nevertheless – and somehow contrary to international perceptions of Turkish-Georgian water relations – rather close working relations with respect to the upstream Coruh River development (see Table 1). Both countries entered friendly and positive political relations after the demise of the Soviet Union and the declaration of Georgian independence in April 1991. Turkey and Georgia signed "The Friendship, Cooperation and Good Neighbourliness Agreement" in 1992. With this agreement, previous agreements and treaties between the two countries with respect to Turkey and the Soviet Union were also recognised. Today, Turkey is not only a principal political and a strategically important partner for Georgia in the region, but also a very important trading partner. Economic exchange between both countries has experienced an impressive increase in the last decade with the consequence that Turkish-Georgian trade constitute 17 percent of the total international trade volume of the Georgian economy.

The first, most relevant and most enduring water-related international agreement between the Soviet Union and Turkey goes back to the 1920s and covers, in general, water allocation and securing the borders of the boundary rivers. In 1927, Turkey and the Soviet Union signed the *Protocol on the Beneficial Uses of Boundary Waters* (the so-called Kars Protocol) which addressed, inter alia, the use of the Coruh River because it forms the boundary between the two states for 3 km (Kurucim 2002). The basic provisions of the arrangement are a fifty-fifty allocation of water (Article 1) and several regulations on infrastructure and dam building. A Joint Boundary Water Commission was established later. Since this agreement only applies directly to those stretches of the river forming a border, it does not cover all transboundary effects that might occur because of changes of the river

[7]Under specified conditions, dam projects are exempted from issuing an EIA (Turkish EIA Directive, Provisional Article 3), see Scheumann et al. in this volume.

flow elsewhere in the basin.[8] For instance, article 5 of the protocol demands compensation if a party suffers from dam building on a river that constitutes a border. However, it is rather unlikely that this (and other provisions of the protocol) applies to the dam projects under consideration, because this is not an agreement concerning transboundary resources flowing across the boundaries. The Soviet Union and Turkey signed another protocol which regulates necessary technical cooperation to avoid changes of the river bed in several border rivers, inter alia the Coruh River (see Annex 10). In the 2000s up until today, a series of meetings were held and memoranda of understanding signed, focusing on potential impacts in Georgia of dams constructed in Turkey.

In addition, the cooperation of Georgia and Turkey in the Black Sea Convention and other international activities to improve the environmental conditions of the Black Sea could serve as a context to facilitate cooperation in the Coruh River basin.

However, no treaty or protocol is in place to govern the use and protection of the transboundary Coruh River. Specific consultation and cooperation "records" have, however, been made for the Coruh dam development.

5 Outstanding issues and options for win-win solutions

As early as the 1980s, the Soviet government expressed concerns, via diplomatic channels, about the possible environmental impact of the planned dams and requested a joint investigation. The Soviet Union repeated this request in 1990 but due to the demise of the Soviet Union the diplomatic channels were no longer active. When Georgia expressed concerns about the Coruh River Development Programme in 1994, both countries entered a phase of bilateral technical cooperation in the form of a series of technical meetings in 1994 and 1995 (Yildiz 1999b). Even at this stage of consultation and negotiation, divergent problems emerged with different priorities. Turkey proposed to plan future dams in a bilateral manner and invited Georgia to enter into a broader Turkish-Georgian cooperation relating to joint energy projects. These joint developments were designed in such a manner that Georgia could receive compensation for potential damages from the already planned Coruh River development. In fact, the Turkish government was apparently not willing to consider a renunciation of the disputed dams on the Coruh River but proposed to broaden the negotiations. Georgia's government, in contrast, put the main emphasis on the negative environmental impacts of the already planned Turkish dams, and was neither prepared nor willing to negotiate bilateral cooperation on future joint dams, inter alia because of different priorities in energy policy.

The Coruh issue then entered a higher bilateral political agenda and was discussed during several political consultations on a ministerial level between

[8]For instance, Article 5 regulates the right of the parties to build a dam on the waters constituting the border; it further states that those parties that experience damage because of a dam on the boundary waters, should be compensated.

1997 and 1998. During an official visit by a Georgian delegation to Ankara in 1998, Turkey officially recognised Georgia's concerns. On that occasion, the Turkish delegation also stated that conditions were not suitable to sign an agreement concerning the environmental impact of the dams, because of incomplete and insufficient information. Moreover, the Turkish side renewed the idea of broadening water cooperation and embracing projects that would potentially have mutual benefits.

During the negotiations in 1997 and 1998, Georgia referred to the aforementioned coastal erosion at Batumi and region, and as a solution proposed a cost assessment of measures needed to alleviate the problem which would then be met by Turkey. However, Turkey's position on the impact of the dam maintained that dependable data was still lacking and future action to alleviate the possible effects should be determined in the light of reliable scientific evidence that could only be collated once the dam was installed. Turkey took over the financing of two stations in Georgia for monitoring the river flow and currents.[9] According to Georgian media reports, Georgia failed to meet its commitment to ensure the maintenance of the monitoring stations that had become damaged and that now lie derelict. This example illustrates that bilateral water cooperation on the Coruh is not only characterised by diverging priorities and objectives but also by very weak Georgian administration. However, the Turkish authorities pursued the monitoring of the river with the help of working groups that were dispatched to the site twice a year. Finally, the ceremony for the start of the construction of the Deriner Dam took place in 1998 and Georgia's then president Eduard Shevardnadze was one of the international participants.

Following the ceremony, Georgia and Turkey agreed upon the installation of a bilateral group of experts to monitor the effects of the dam on the coastal zone. However to date, these efforts have failed to provide any consensual scientific assessment. While representatives from the Georgian Environmental Ministry, national environmental non-governmental organizations (NGOs) and the Georgian Green Party stress the significance of the environmental impact, the Turkish ambassador in Georgia cited a significantly less convinced statement to the media: "To date, expert analysis has not revealed any indication of the dam's environmental impact as claimed by certain circles." (cited in Kupatadze 2005).

However, in the aftermath of the Georgian "revolution" and the election of President Sakashvili in 2004, the Coruh issue has re-entered the bilateral political arena. Furthermore, the imminent completion and filling of the dam brings the ecological question to the fore and re-addresses the counter-measures needed (Yerman 2004).[10] Recently, the Tbilisi authorities stated that they are taking the potential impact of the dams very seriously and that they are still seeking an adequate and satisfying agreement with Turkey. According to Georgian

[9]In this context it is worth remembering that Georgia does not dispose of a working river monitoring infrastructure.

[10]Lately a daily newspaper in Georgia, namely Akhali Veria, published articles on that issue by criticizing Turkish planned projects (see also Nazi 2001).

representatives, a possible solution might involve a neutral third party who would facilitate and mediate the joint environmental impact studies. Prevention and/or mitigation measures could then be accordingly designed and the costs allocated. Another aspect recently addressed by the Georgian authorities was the still controversial procedural question, where the most relevant issue relates to whether Turkey is obliged to carry out a comprehensive assessment of the transboundary environmental impact of specific dam projects planned, and if so, which procedural rules should be applied.[11]

The renewed political attention to the Coruh River must also be seen in the context of the recently changed political status in the semi-autonomous republic of Adjaria. For a couple of years Adjaria was out of Georgia's sphere of influence; however, in 2004 Georgia's new president brought Adjaria back under Georgian sovereign political influence.

Despite the unsatisfactory situation from a Georgian perspective, it is not envisaged that Georgia would risk damaging political relations with her strategically important neighbour Turkey. In addition, an escalation of ill will is unlikely because Turkey is also interested in a stable political and economic climate. In spite of the absence of an effective bilateral agreement, Turkey has already taken on the obligation of financing the monitoring of, and compensating for, the effects of dams on the other side of its border, a fact that points to particular caution being exercised in her relations with Georgia.

References

Encon (2006) Yusufeli dam and HEPP EIA Report, Appendix L, http://www.dsi.gov.tr/english/yusufeli/report/Yusufeli%20EIA%20-%20Appendices%20-%20RevF%20-%20July%202006/appendixL/Yusufeli%20EIA%20-%20Appendix%20L%20-%20RevF-%20July%202006.pdf. Accessed 28 May 2010

Jaoshvili S (2003) What may happen to the Adjarian coast after construction of dams on the Chorokhi river. Caucasus Environment, vol 3. Caucasus Environmental NGO Network. http://www.cenn.org/Magazine_1/Magazine_3/What_May_Happen.html. Accessed 12 June 2005

Kupatadze G (2005) Black Sea dams anger Georgia -Environmentalists claim flooding caused by Turkish dams is threatening Adjaria. The Messenger, May 3, 2005. http://www.falkor.org/news/dams.htm. Accessed 12 June 2005

Kurucim AY (2002) Coruh Havzasi Hidropolitigi, yayinlanmamis yuksek lisans tezi. Ankara: Hacettepe Universitesi

Nazi C (2001) Recommendations and proposals for ecologically safe use of the Chorokhi river. International Law #1-2001. Tbilisi: Institute of European Law and International Law of Human Rights. http://www.osgf.ge/interlaw/b0023%209-chinchaladze_nazi. Accessed 12 June 2005

RFE/RL (1998) Caucasus Report, vol 1, no 7, 14 April 1998. Radio Free Europe/ Radio Liberty. http://www.rferl.org/reports/caucasus-report/1998/04/7-140498.asp. Accessed 12 June 2005

[11]For instance, international NGOs and national Georgian environmental associations stated that, in spite of an already existing agreement between the two countries in the context of environmental protection, Turkey did not adequately consulted Georgia on the environmental impacts of the Yusufeli dam (see Kupatadze 2005).

T.C. Cevre ve Orman Bakanligi., Devlet Su Isleri GD, DSI Coruh Projeleri 26. Bolge Mudurlugu, Artvin, no year

The Messenger, 20 April 2005, http://www.messenger.com.ge. Accessed 30 September 2005

World Bank (2003) Water resources in Europe and Central Asia. Volume 1 – Issues and strategic directions. Volume 2 – Country water notes and selected transboundary basins. The World Bank, Washington, D.C.

Yerman E (2004) Coruh Havzasi ve Türkiye Gürcistan Iliskilerine. Unpublished Term Paper. Hacettepe University, Department of International Relations, Ankara

Yildiz D (1999a) Sinir Olusturan ve Sinirasan Su Kaynaklarimiz ve Kiyidas Ulkeler Arasinda Teknik Isbirligi Gereksinimi. Cevre ve Muhendis TMMOB Cevre Muhendisleri Odasi Yayini, no 18, 28-35

Yildiz D (1999b) Coruh Havzasinin Hidropolitigi Üzerine. Cevre ve Muhendis TMMOB Cevre Muhendisleri Odasi Yayini, no 18, 57-63

Kura-Aras River Basin: Burgeoning Transboundary Water Issues

Axel Klaphake and Annika Kramer

1 Introduction

The development of water cooperation between the basin countries of the Kura-Aras[1] river basin is attracting increasing political attention. Because of the complicated and partly unstable political relations between the riparian states, and the enormous water quantity and quality problems within the basin, the Kura-Aras basin was referred to as a "basin at risk" (Wolf et al. 2003) in some scientific studies. According to these studies, massive conflicts over water resources could occur in the years to come. However, the role of Turkey as an upstream country still appears understudied and barely considered.

2 Geographical and hydrological setting

The Kura-Aras basin is located in the South Caucasus and is, by far, the most important watershed in the region in terms of surface area, water flow, socioeconomic importance of water resources and preservation of freshwater ecosystems. The basin is an internationally significant river system, which is seriously degraded and continues to be threatened, whereby transboundary issues play a major role.

[1]Other names for the river Kura are Mtkvari river, Kuracay river, and Cyrus river. Other names for the Aras frequently used internationally are, inter alia, Araks and Araz.

A. Klaphake (✉)
Deutsche Gesellschaft fuer Internationale Zusammenarbeit, GIZ, Kampala, Uganda
e-mail: axel.klaphake@giz.de

A. Kramer
adelphi, Berlin, Germany
e-mail: kramer@adelphi.de

A. Kibaroglu et al. (eds.), *Turkey's Water Policy*,
DOI 10.1007/978-3-642-19636-2_15,

The watershed extends over 64 percent of the territory of the South Caucasus states and includes five countries, namely Turkey, Iran, Armenia, Georgia and Azerbaijan. The Kura-Aras basin covers the entire territory of Armenia, about 80 percent of the territory of Azerbaijan and about 52 percent of the territory of Georgia. Because of their comparatively small share of the watershed area, the river system is less crucial for Iran and Turkey at national levels, nevertheless there are regionally important water uses (CEO 2002).

The watershed consists of two main branches; the Kura and Aras rivers contribute 55 percent and 45 percent respectively to the total discharge (Revenga et al. 1998). The Kura River rises in Turkey – the source is a group of springs on Kyzyl-Gudiak Mountain at a height of 2,700m in the Anatolian highland of Turkey – and enters Georgia after some 210km. After 390km through mountainous terrain the river flows into the Azerbaijan steppes and finally discharges into the Caspian Sea. The river is fed by snow and ice melt water, underground sources and rain. The total watershed area of the Kura River (excl. the Aras watershed) is about 88,000 square kilometers with a total length of about 1,364km (UNECE 2007). The other main branch of the river system is the Aras River with a length of about 1,072km (UNECE 2007) and a total watershed area of 102,000 square kilometers. The Aras's spring also originates in Turkey, and after 300km the river forms several borders: between Armenia and Turkey, for a very short distance between Azerbaijan and Turkey, between Azerbaijan and Iran, between Armenia and Iran, and finally again between Azerbaijan and Iran. Eighty kilometers after crossing the border with Azerbaijan the Aras River joins the Kura River. Therefore, the Kura River and its main tributaries cover four full border crossings and four sections as the border river between different countries. Besides, there are over twenty minor tributaries which cross the border or discharge into the Aras River where it forms the border.

From the Turkish perspective, the Arpacay River, which is shared between Turkey and Armenia, is of particular importance for the two countries. It is formed through the merging of the Kars River originating in Turkey, and the River Ahuryan originating in Armenia. There are smaller transboundary waters crossing the Turkish-Iranian border within the Aras River catchment area also; the most important being Sarisu River which has a drainage area of 1,500 square kilometers in Turkey (Polat 2004). In total, about 40 river segments or tributaries have a transboundary character while crossing a border, being a border or ending at a border in the Kura-Aras basin (TACIS 2004). Figure 1 shows a map of the Kura and Aras rivers, their main tributaries and selected dams. Table 1 provides an overview of the context for cooperation on the rivers.

A variety of climates, precipitation conditions and landscapes can be observed (TACIS 2004).[2] As a general rule, annual rainfall within the basin declines from west to east but the Turkish part can be judged to be rather dry also. In Central Georgia, where the Kura enters from Turkey, average annual precipitation is

---

[2]Accordingly, the Kura basin is usually divided into six parts with different climates: South-west (from source to Georgian Borjomi), North (Borjomi-Tbilisi), Middle (Tbilisi-Khrami), Lower (below Mingachevir) and the Aras region.

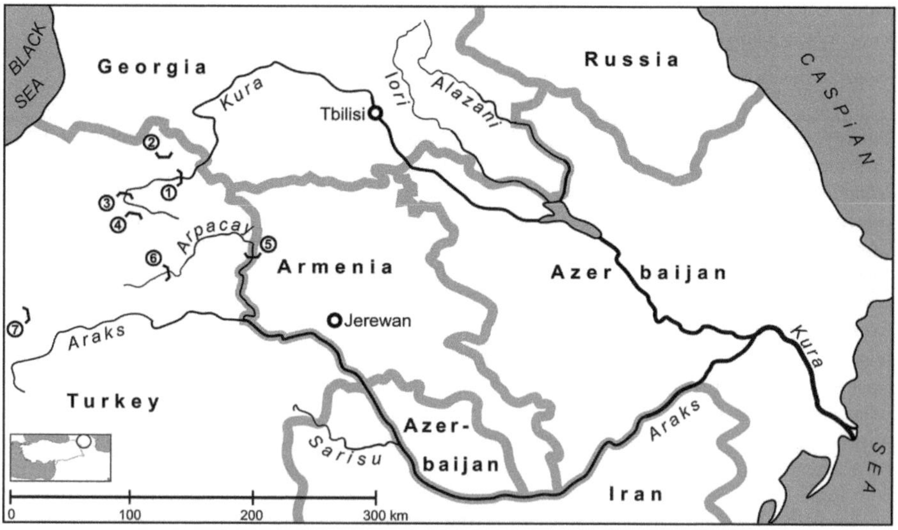

(1) Koroglu Dam (planned)
(2) Burmadere Dam (planned)
(3) Besikkaya Dam (planned)
(4) Durancam Dam (planned)
(5) Arpacay Dam
(6) Bayburt Dam
(7) Demirdoven Dam

Fig. 1 Map of Kura, Aras, main tributaries and selected Turkish dams

500mm but only 200mm in Azerbaijan, where the river flows into the Caspian Sea. Similarly, evaporation rates soar from west to east. The average discharge flow rate of the Kura River in Turkey, however, is 28.75 cubic meters per second; total annual water supply of the river in Turkey (up to the Georgian border) is about 1 Billion cubic meter (BCM). The total water flow of the River Aras in Turkey up to the Armenian border is 2.5 BCM (UNDP/GEF 2006a).

Because of the mountainous characteristics of the upper part of the basin and the specific climatic conditions, however, a huge seasonal variety in river flow is reported. The basin frequently experiences floods and mudflows with drastic economic consequences (UNDP/GEF 2006b). In flood periods, water discharges make up about 60 percent of annual discharges (TACIS 2004). On the other hand, drought periods in the Kura basin are also common and different regions are affected. Whilst even water-rich Georgia experiences seasonal droughts in several parts of the country and in exceptional years, Azerbaijan suffers from droughts and water shortages the most because more than 70 percent of the country's drinking water is dependent on the Kura and Aras rivers (Guluzada 2004, TACIS 2004). In addition, good quality groundwater resources are limited making Azerbaijan very vulnerable to scarcity. In contrast, Armenia has generally adapted well to permanent shortages, but water scarcity is still a factor in irrigation development and hydropower production (World Bank 2003). Armenia's advantage is its large stock of high quality groundwater. Georgia's drinking water is mainly drawn from the higher tributaries of the Kura River, with groundwater bodies with a satisfactory quality also used in parallel (UNDP/GEF 2006a). In the past, Turkey experienced

Table 1 Cooperational context in the Kura-Aras basin in brief

Kura-Aras Basin area: 190,000 km^2; mean annual discharge 32 BCM

Riparian position	Basin area (percent of total)	Main water uses
Turkey Upstream / border rivers with Armenia and Iran	27,208 km^2 (14 percent)	Irrigation, hydro power and domestic use
Georgia Upstream / downstream	33,763 km^2 (18 percent)	Irrigation, industrial and domestic use
Armenia upstream / downstream / border river with Turkey	34,257 km^2 (18 percent)	Irrigation, industrial and domestic use
Iran Upstream / downstream / border river with Azerbaijan	39,045 km^2 (21 percent)	Irrigation, hydropower, industry and domestic use
Azerbaijan Downstream / border river with Iran	55,632 km^2 (29 percent)	Irrigation, industry, hydropower and domestic use
Main agreements with Turkish involvement and covered issues		
Turkey - Georgia	1927 – border issues, river bank protection, water allocation, compensation requirements, joint commission 2000 – technical cooperation, river bed changes etc.	
Turkey - Armenia	1927 – border issues, river bank protection, water allocation, compensation requirements, joint commission 1964 – joint construction of the Arpacay-Dam, allocation of water and construction costs, joint commission to operate the infrastructure (see Annex 8) 1975 – instructions on the joint operation of the dam and reservoir on Arpacay/Ahuryan river (see Annex 9) 1990 – technical cooperation, river bed changes, joint hydropower facilities (see Annex 10)	
Turkey – Iran	1955 – basic principles of the water use in the border region, minimum water flow (1,8 m^3/s), water allocation (fifty-fifty allocation)	
Unsettled issues		
Quality	No agreement on water quality standards, exchange of data insufficient; rudimentary bilateral agreements (Georgia – Armenia, Azerbaijan – Georgia)	
Quantity	Main critical issue. No comprehensive agreement providing for a fair and economically sound allocation of water	
Biodiversity	Comprehensive approach to the protection of freshwater ecosystems is lacking	

temporary domestic droughts along the Aras River which led to plans to further develop water infrastructure.

As a general rule, one can state that the further downstream the rivers in the entire basin are, the greater the deterioration in water quality is, and with it the challenges of water management (World Bank 2003). The most important water quality issues are: i) discharge of untreated municipal wastewaters, ii) inflow of agricultural fertilisers, and iii) organic and heavy metal pollutants from industries (UNDP/GEF 2006a). Equally erosion and sedimentation are important water management issues which have an impact on downstream water uses. Natural erosion and sediment flow is aggravated by deforestation and flood irrigation in several regions (UNDP/GEF 2006a).

The basin is rich in terms of biological diversity. There are, for instance, unique alluvial forests along the Kura River, and many important wetlands in different parts of the basin. Currently, the basin has four Ramsar sites and 21 wetland-dependent Important Bird Areas (Jenderedjian 2006). Finally, the Kura-Aras basin is vital for the protection of the Caspian Sea's ecosystems because it is the second largest river which drains into the Caspian Sea (CEO 2002).

3 Current and future water uses

Various activities and water uses, including industrial, domestic, agricultural and hydropower, have always had adverse effects on the quality and quantity of the waters in the Kura-Aras basin. In total, about 60 to 70 percent of the water of the basin is allocated to agriculture, 20 to 25 percent to industrial use, and the remainder to drinking water (Polat 2004). Water withdrawal for household use is most relevant in Azerbaijan where water consumption already suffers from the poor quality of the Kura River with regard to organic pollution, hazardous waste, and sedimentation (CEO 2002).

Industrial water use in the basin is generally influenced by old and outdated technologies which cause unnecessarily high water demands and enormous levels of pollution. Mining is a major industry in the upper catchment and heavy metal pollution from tailings is a serious concern in some of the upper tributaries (UNDP/GEF 2006c). Major industrial pollution is further caused by the Georgian Rustavi industrialised region just downstream from Tbilisi and by a large group of copper-molybdenum enterprises in the city of Alaverdi (Armenia) (TACIS 2004), while there are only very few industrial sites (e.g. sugar plants) along the Aras River on Turkish territory (Polat 2004). As many countries in the basin have experienced significant economic decline since 1990, industrial pollution has decreased, at least temporarily. However, abandoned contaminated industrial sites can potentially release pollutants over large periods of time.

Because of climatic conditions, agriculture in all riparian states depends largely on irrigation. For this purpose, a total of 130 reservoirs and dams were built within the basin, mainly located on the tributaries. The main branches of the rivers (Kura and Aras) are regulated by two major reservoirs only, namely the Mengechevir reservoir on the Kura River in Azerbaijan and the Nakhichevan reservoir on the Aras River shared by Azerbaijan and Iran. Both are mainly used for irrigation but also for hydropower generation (TACIS 2004). Today, Armenia, Georgia and Azerbaijan, in particular, have large but inefficient irrigation systems (based on dam storage) which suffer huge water loss and salinisation. Negative effects cumulate downstream where poor water quality goes hand-in-hand with quantity problems as the rivers enter the Caspian Sea. In recent years, however, because of some restructuring of the agricultural sector in former Union of Soviet Socialist Republics (USSR), several irrigation systems broke down, which lead to a temporary decrease in water use; albeit the decline in industrial water use has been more drastic since the early 1990s in the Caucasus states (CEO 2002).

Water development in Turkey

It is necessary to distinguish between the two main branches of the Kura -Aras river system when looking at Turkish plans to build new dams in the region (Polat 2004). At present, the Kura River has no important water infrastructure (e.g. dams, reservoirs or large scale irrigation systems). Some 2,984 hectares of land (6 percent of the irrigable land in the basin) is irrigated by individual farmers with spring waters and by the General Directorate of Rural Services (GDRS) through the waters of the 27 small lakes in the Turkish Ardahan province. However, the Durancam, Besikkaya, Burmadere Dams are due to be built. Thus, 50,670 hectares of land are planned to be irrigated. In addition, there are plans to develop the Koroglu Dam for hydropower generation.

The Aras basin plays a much more important role for agricultural irrigation in Turkey. The Serdarabat Regulator, located where the Aras River enters into the Igdir Plain, diverts water both to Turkey and Armenia and with a regulated flow, irrigated agriculture can be practiced in the Igdir Plain (Yildiz 1999). Generally, Aras River waters are vital to the livelihood in the Igdir Plain and to the middle Asian lowlands which often experience dry climates. The Arpacay and Demirdoven dams were built in Turkey on the headwaters of the Aras River and its tributaries. The Arpacay Dam, located where the river forms the border between Armenia and Turkey, has an active storage capacity of 510 Million cubic meters (MCM) per year. Half of its water is used by Armenia, the other half by Turkey for irrigation in the Igdir Plain. The Demirdoven Dam was built between 1988 and 1995 and serves an irrigation area of about 8,293 hectares. At present, the total irrigated area in Igdir, Kars-Akyaka, Kars-Alabalik, Karayazi-Koycegiz is equivalent to 48,094 hectares. Additionally, construction of the Bayburt Dam and of irrigation systems (such as the East Igdir, Arpacay Plain, the Cildir Lake, the Pasinler Demirdoven Dam) fed by the Kars River continues and will irrigate 33,221 hectares. Furthermore, there are thirteen other dam projects in the pipeline (Polat 2004, Yildiz 1999).[3] With a total storage capacity of these dams at around 4,110 MCM the total irrigated area of all the projects will be 17,903 hectares. Other planned irrigation projects are the Susuz Project, the Tuzluca Project, the Digor Project and the Cildir Project. These projects will irrigate 185,458 hectares of land.

[3]Namely: Soylemez, Karakurt, Kuloglu, Denizgolu, Denizhan, Katranli, Asbuga, Alabalik, Kars, Gecit, Dolayli, Varli, and the Cildir lake.

Most of the projects mentioned above are in rather early planning stages and their realisation is not yet assured because of the Turkish government's limited financial resources and the restructuring of the Turkish administration and agricultural sectors. The Bayburt Dam is the only large infrastructure project under construction[4] because of the importance of its water supply has for domestic and industrial purposes in the dry Kars region, whereby the associated irrigation target is rather modest (about 5,200 hectares). Due to rather unfavourable hydrological conditions (at least compared to other Turkish river basins), hydropower generation projects are of minor importance and the majority of dams planned would be predominantly constructed for irrigation. In contrast, Iran's policy of developing hydropower is potentially much more important because the Iranian Government has plans to develop a number of dams on the Aras River; several of them are planned as shared projects with their neighbours Azerbaijan and Armenia. Furthermore, 15 large irrigation projects are under consideration by Iran to irrigate over 370,000 hectares in the Aras basin (UNDP/ GEF 2006d).

4 Potential downstream impact of Turkish water resources development

The downstream riparian countries, Armenia, Georgia and Azerbaijan have identified the major transboundary issues in the Kura-Aras basin to be: variation and reduction of hydrological flow, deterioration of water quality, ecosystem degradation in the river basin and increased flooding and bank erosion (GEF 2008). Even though the transboundary effects of current and planned Turkish activity and development within the river basin has yet to be seriously assessed, one can expect negative effects on water quality and quantity (Polat 2004, CEO 2002). In principle, the impact of Turkish activities on both water quality and quantity is more relevant in the Aras river basin while the hitherto planned activities on the Kura River are less important.

Construction of dams and the development of irrigation in Turkey on the scale described, even if not all the projects would see the light of day, would certainly affect flow patterns, water availability and freshwater ecosystems such as wetlands and alluvial plains in the downstream states, mainly in Armenia, Iran and Azerbaijan. Reduced hydrological flows and alteration of the hydrological regime will further impact on the anadromous fish species in the Kura-Aras rivers and the Caspian Sea which already suffered from the inundation of the Mingechaur dam reservoir on the Kura River; it has resulted in a loss of spawning grounds for mature sturgeon and blocked important migration paths (GEF 2008, Barannik et al. 2004).

---

[4]Bayburt Dam is envisaged to be finalised in 2011. Personal correspondence with Salim Fakioglu, Devlet Su Isleri (DSI) Investigation and Planning Department, 26 July 2010.

Today, water availability is already a matter of concern between Turkey and Armenia but also between Turkey and Iran. In this context, the Sarisu River is the one that has attracted a certain amount of political attention, because Turkey could not always provide the promised water flow.

With regards to transboundary water quality issues, riparian states in the Kura-Aras basin have claimed transboundary impacts caused by upstream neighbours in various cases. However, for an unbiased and scientific validation of such claims, improved water quality data and monitoring would be necessary. Based on an initial assessment of available data for the three downstream riparians, several claims have been identified as being unjustified (UNDP/GEF 2006a). On the Kura River, most important transboundary pollution sources stem from Georgian industrial sites and urban agglomerations.

Water pollution from the Iranian territory is barely documented, and relevant data from Turkey are also not easily accessible (UNDP/GEF 2006a). It can be expected, that Turkey causes only a minor share of the river's pollution. The Turkish provinces of Erzurum, Ardahan, Kars and Igdir lying upstream in the basin show comparatively good environmental conditions, even though watershed degradation, erosion and agricultural pollution (chemicals, pesticides) are issues of concern (Polat 2004). Planned developments in Turkey might change this picture, as intensive agriculture on irrigated land usually has an impact on water quality because of salinisation and the use of fertilisers and pesticides.

There are also several engineering aspects of river development along the borders that would demand improved bilateral cooperation. For instance, Turkey intends to facilitate and improve use of the Aras River's water from the joint Serdarabat Regulator which was constructed for irrigation purposes. Since the envisaged technical measures on the Turkish side would imply removal of the riverbed along a short stretch of the river, approval by the Armenian government and technical cooperation with the Armenian authorities would be necessary (Yildiz 1999).[5]

Generally speaking, it is obvious that long-term sustainable development within river basins and the preservation of freshwater ecosystems demands international and multilateral cooperation of the basin states. Long term efforts and initiatives to manage the Kura-Aras basin and to coordinate the riparian states would certainly also require the involvement of Turkey. The same is true for the protection of the

[5]Turkey and Armenia utilise the regulated waters of the Aras River from the joint regulator of Serdarabat for irrigation purposes. Just prior to the Serdarabat regulator on the left bank of the river in Armenian territory, Armenia withdraws water, whereas from the right bank of the river Turkey withdraws water. However, to be able to use that water, Turkey first needs to collect the water in the sediment ponds as the Aras brings large amounts of deposits to the right bank where the inner bend of the river is situated. DSI has tried to make the necessary arrangements to provide good quality and an equal amount of water to Turkey; yet, DSI underlines that, in order to find a complete solution to water withdrawals on the Turkish side, the river bends will have to be removed. This would in fact result in the rearrangement of the riverbed. Yet, as the Aras (Arpacay) river forms the boundary between Turkey and Armenia, such adjustments in the riverbed should be agreed upon by both sides. Existing regulations within the bilateral agreements do not address this matter. Hence, technical collaboration is needed to respond to the challenge (Yildiz 1999).

Caspian Sea, which significantly suffers from river pollution. In addition, flood and drought management are issues with high transboundary relevance.

5 Status of cooperation

In face of the challenges described, coordination and cooperation between the five basin countries is rudimentary. There is neither an agreement nor a joint body covering the entire Kura-Aras river basin, and only few bilateral agreements exist.

Bilateral commissions between Armenia and the Islamic Republic of Iran, as well as between Azerbaijan and the Islamic Republic of Iran act on the basis of the Agreement between the USSR and Iran of 1957 (UNECE 2009). With regard to coordination between Georgia, Azerbaijan and Armenia, the previous intra-Soviet-Union mechanism of cooperation, information sharing etc. ceased to function after the break-up of the Soviet Union. Rather rudimentary bilateral agreements on cooperation in the environmental field were concluded in 1998 between the ministries of Georgia and Armenia as well as Georgia and Azerbaijan. More recently, transboundary water cooperation between Georgia and Azerbaijan has been further detailed in 2007 through a Memorandum of Understanding (MoU) between the Georgian and Azerbaijan Environmental Ministries, that included the establishment of a working group on exchanging information and joint monitoring of transboundary waters (Adeishvili 2007). Broader political concerns mean that a formal basin-wide steering group, international commission, or any other high-level and politically endorsed entity is not possible now.

There are also only rudimentary agreements between Turkey and its neighbours (Polat 2004). The history of Turkish cooperation within the basin goes back to the 1920s and the early days of the Soviet Union. The most important boundary rivers shared between Turkey and the Soviet Union were the Posof (which now forms the border with Georgia), the Arpacay and the Aras (now shared with Armenia). All these rivers used to form boundaries between Turkey and the USSR. In 1927, Turkey and the Soviet Union signed the 'Protocol on the Beneficial Uses of Boundary Waters' (so-called Kars-Protocol) which concerns, inter alia, the use of the Arpacay and the Aras rivers. The basic provisions of the arrangement are a fifty-fifty allocation of water (Article 1) and several regulations on infrastructure and dam building. A Joint Boundary Water Commission was established as the responsible authority to settle any dispute (Akbulut 2003). The protocol is still in force and today, bilateral commissions on boundary waters (between Armenia and Turkey as well as between Georgia and Turkey) still act on the basis of the 1927 protocol (UNECE 2009).

An important agreement between Turkey and the USSR is the Protocol on the Joint Construction of the Arpacay Dam dating from 1964 (see Annex 8). This protocol provides a set of rules concerning joint dam construction the waters of which would be shared on a fifty-fifty basis. Both states are free to use their water for irrigation purposes and may build a hydropower plant on their respective territories. In addition, article 18 of the protocol regulates quantitative water use downstream of the dam up to the Iranian border. Equally, issues such as the allocation of construction costs and the compensation for land losses are addressed in the protocol; the same is true for the founding of a joint dam commission to operate the infrastructure. The 1964 protocol was later followed by the "Cooperation Agreement on the Construction of a Dam on the Bordering Arpacay (Ahuryan) River and the Construction of a Dam Lake". The cooperation agreement, which was signed on 26 October 1973 and officially ratified in 1975 (see Annex 9), assures the basic principles that were already outlined in the 1964 protocol. The provisions of the 1975 agreement play a key role in the regulation of boundary waters; they explicitly provide for regulations concerning the tributaries and make several clarifications omitted in the 1927 agreement.

As a last agreement signed between Turkey and the USSR, a protocol was signed on 7 March 1990 (see Annex 10). According to this, "...any shift in the riverbeds of the Arpacay, Coruh, Posof and Caksu rivers would be prevented jointly or the necessary facility to adjust the watercourses would be built in collaboration" (Esenyel 2001). In addition, the protocol addresses technical issues relating to the construction of joint hydro-technical facilities however, the respective installations have not yet materialised.

Under "The Protocol on the Joint Utilization of the Waters of the Sarisu and the Karasu River" which was signed on 18 November 1955 by Iran and Turkey, water use rules were defined and basic principles such as a "fair use of the waters in the border region" were laid down. Article 10 of the protocol states that, "the parties may develop irrigation facilities on their portion of the river after agreeing on the need of using water for irrigation". Moreover, Turkey promised to release 1,8 cubic meters per second water from the Sarisu River to Iran under all circumstances. In addition, both states agreed on a fifty-fifty allocation of the water of the Karasu River by reaching an agreement on the needs and requirements of their irrigation projects and by adhering to the existing border regime (Polat 2004).

Against the background of existing agreements that predominantly address border issues, water infrastructure development and respective water withdrawals at border rivers, comprehensive agreements on transboundary water issues are clearly lacking. For instance, water quality questions and freshwater ecosystem preservation are not addressed. In addition, the existing agreements lack implementation and monitoring structures, the same is true for procedural rules, such as transboundary (environmental) impacts assessments, information sharing, notification etc.

Over the past years, several projects related to technical aspects of the management of the Kura-Aras river basin have been initiated and funded by different international organizations or foreign governments.[6] It has been observed, that

[6]Some of the recent completed or ongoing projects were funded, inter alia, by i) United States Agency for International Development (USAID): Water Management in the South Caucasus Project (2000-2004) and South Caucasus Water Program (2004-2008) http://www.scaucasuswater.org/; ii) United Nations Development Programme (UNDP) Reducing Transboundary Degradation in the Kura-Aras Basin co funded by the Swedish International Development Agency (SIDA) (2003-2005) and by the Global Environmental Facility (GEF) (2005-2007) see http://kura-aras.iwlearn.org/;

although these projects are very much related to each other, share common goals and sometimes have overlapping actions, there is little or no cooperation between the different organizations (Vener and Campa 2008).

These recent initiatives are conspicuous by the absence of Turkey and Iran. Almost all of them focus on fostering cooperation between Armenia, Georgia and Azerbaijan. Only the United Nations Development Programme/ Global Environmental Facility (UNDP/GEF)-funded project[7] included Iran, and in its current follow-up phase, supported by the Environment and Security Initiative, one explicitly aims to integrate Turkey[8]. However, Iran and Turkey are both involved in some joint dam projects with their respective neighbours. Since 2004, a Turkish – Armenian Interstate Commission exists on the use of the Arpacay Dam mentioned above (UNECE 2009). Iran, in addition to two dams that were jointly built at the time with the USSR, and are now shared with Azerbaijan, has plans to construct two more dams with Azerbaijan and further projects with Armenia (UNDP/GEF 2006d).

6 Conclusion: Outstanding issues and options for win-win solutions

Contrary to the high importance that the management of the transboundary water of the Kura-Aras basin has for its downstream South-Caucasian riparian, the topic does not receive much political attention in Turkey, and the interest in water conflict matters is rather low. Georgia, as the main polluter of the Kura River, appears little concerned about Turkey's influence on the river. Georgia lacks the basic monitoring capabilities required to either assess the water quality of the Kura River or to quantify the pollutants stemming from the Turkish stretch of the river.[9] As far as the Aras River is concerned, the planned development of dams might have a serious impact on Armenia's water usage and to a lesser extent on Iran and Azerbaijan as well, but these issues have not yet been comprehensively addressed in the political arena. However, it is assumed by some observers that the impact of Turkish dams on the Aras River was possibly discussed during negotiations for the

iii) NATO South Caucasus River Monitoring Project (2002-2008), see http://www.kura-Aras-natosfp.org/; iv) The European Union: EU Technical Assistance to the Commonwealth of Independent States (TACIS) Joint Rivers Management Programme (2002-2004) and its follow-up Transboundary River Management Phase II for the Kura River basin (2008-2010); see http://www.kuraarasbasin.net; v) The German Federal Ministry for the Environment, Nature Conservation and Nuclear Safety: "Transboundary cooperation for hazard prevention in the Kura-river basin" (2003-2006), see http://www.kura.iabg.de/ergebnisse_engl.htm All accessed 18 January 2010.

[7]For further information refer to http://kura-aras.iwlearn.org/.

[8]For further information refer to http://europeandcis.undp.org/environment/georgia/show/3D22B3E0-F203-1EE9-BF6A485C55606B6C. Accessed 18 January 2010.

[9]Personal communication with the head of the water department of the Georgian Ministry for the Environment, November 2004.

Baku-Tiflis-Ceyhan Oil Pipeline Project where, inter alia Turkey, Azerbaijan and Georgia are party to the agreement (Yildiz 1999).

The water flow of the Sarisu River is a matter of concern for Iran and Turkey at present. Iranian authorities expressed their concern on this matter when Turkey could not provide the water quantity they had assured for dry periods; they could only in fact provide 1 cubic meter per second or 1.5 cubic meters per second (Polat 2004).

In this context, it is worth mentioning the very complicated political situation in the basin, which turns cooperation into a political and diplomatic challenge. For instance, Armenian–Turkish relations have been strained by a number of historical and political issues. Only recently has there been some easing of the tension, and an accord between Armenia and Turkey was signed by the foreign ministers of the two countries on 10 October 2009, designed to allow the opening of borders and to set up a formal diplomatic relationship. Turkish–Iranian relations are characterised by a low degree of affinity as the two countries - far from being traditional allies - have serious differences over terrorism issues. In contrast, historically, Turkey has always had close relations with Azerbaijan and good and cooperative relations with Georgia.

However, different forms of cooperation on natural resource management do exist in the basin. While Turkey is barely involved in these initiatives notably an Interstate Commission already exists between Armenia and Turkey - and it is hoped that cooperation on such technical issues as water resources will become increasingly possible.

References

Adeishvili M (2007) Fostering a dialogue between riparian states for development and establishment of initial legal and institutional frameworks for increased cooperation and joint management in the Kura-Aras river basin. Presentation held at the UNECE, UNDP and OSCE workshop on River basin Organizations, Water Governance and Regional Networking, 23-26 October 2007, Almaty, Kazakhstan

Akbulut I (2003) Guvenlik Boyutunda Sinir Asan Sularin Onemi Nedir? Turkiye'nin Sinir Asan Sular Konusundaki Taraf Oldugu Hukuksal Duzenlemeler Kapsaminda Bolge Ulkeleriyle Arasindaki Anlasmazliklara Kars Ne Gibi Tedbirler Almalidir, Tez, Istanbul, Genelkurmay Baskanligi, Harp Akademileri Komutanligi.

Barannik V, Borysova O, Stolberg F (2004) The Caspian Sea region: Environmental change. Ambio, vol 33, no 1–2, Feb. 2004

CEO (Caucasus Environment Outlook) (2002) Caucasus Environment Outlook Report 2002. Completed through financial assistance provided by UNDP and the Swiss Agency for Environment, Forests, and Landscape. http://www.grid.unep.ch/product/publication/download/Caucasus_Environment_Outlook_2002.pdf Accessed 28 May 2010

Esenyel O (2001) Turkiye'nin Su Potansiyeli ve Potansiyelin Kullanilmasi. Tez, İstanbul: T.C. Harp Akademileri Komutanligi Kara Harp Akademisi Komutanligi

GEF (2008) Reducing transboundary degradation in the Kura-Aras basin. Project Identification Form. GEF Agency Project ID: PIMS no 2272 http://www.gefweb.org/uploadedFiles/Projects/Work_Programs/April_2008_Work_Program/PIF percent20Document percent201375.pdf. Accessed 10 February 2010

Guluzada LM (2004) Kura river - Transboundary watercourse of Caucasus. Paper presented at the Conference on Integrated Water Management of Transboundary Catchments: A Contribution from TRANSCAT. Venice, Italy, 24-26 March 2004. http://www.feem-web.it/transcat_conf/conf_papers/Guluzada.pdf. Accessed 28 May 2010

Jenderedjian K (2006) Transboundary management of Kura basin wetlands as an important step towards waterbird conservation in the South Caucasus region. In: Boere GC, Galbraith GA, Stroud DA (eds) Waterbirds around the world. The Stationery Office, Edinburgh, UK, pp 731-732. http://www.jncc.gov.uk/worldwaterbirds Accessed 18 January 2010

Polat MH (2004) Aras-Kura Havzasinin Hidropolitik ve Stratejik Degerlendirilmesi, Ankara: Hacettepe Universitesi Hidropolitik ve Stratejik Arastirma Merkezi.

Revenga C, Murray S, Abramovitz J, Hammond A (1998) Watersheds of the world: Ecological value and vulnerability. World Resources Institute, Washington, DC

TACIS (The European Union's Technical Assistance to the Commonwealth of Independent States) (2004) Joint River Management Programme on Monitoring and Assessment of Water Quality on Transboundary Rivers. Final Report.

UNDP/GEF (2006a) Water quality in the Kura-Aras river basin. TDA Thematic Report for the project RER/03/G41/A/1G/31: Reducing Trans-boundary Degradation of the Kura-Aras River Basin. http://kura-aras.iwlearn.org/project-reports/undp-gef-component/ Accessed 18 January 2010

UNDP/GEF (2006b) Diagnostic analysis of floods in the Kura-Aras river basin. TDA Thematic Report for the project RER/03/G41/A/1G/31: Reducing Trans-boundary Degradation of the Kura-Aras River Basin. http://kura-aras.iwlearn.org/project-reports/undp-gef-component/ Accessed 18 January 2010

UNDP/GEF (2006c) Assessment of pollution from land-based sources in the Kura-Aras river basin. TDA Thematic Report for the project RER/03/G41/A/1G/31: Reducing Trans-boundary Degradation of the Kura-Aras River Basin, http://kura-aras.iwlearn.org/project-reports/undp-gef-component/ Accessed 18 January 2010

UNDP/GEF (2006d) Transboundary diagnostic analysis Aras river basin - National report of Islamic Republic of Iran. Report for the project RER/03/G41/A/1G/31: Reducing Trans-boundary Degradation of the Kura-Aras River Basin. http://kura-aras.iwlearn.org/project-reports/undp-gef-component/ Accessed 18 January 2010

UNECE (2007) Our waters: Joining hands across borders. First Assessment of Transboundary Rivers, Lakes and Groundwaters. Economic Commission for Europe Convention on the Protection and Use of Transboundary Watercourses and International Lakes. New York and Geneva: United Nations. http://www.unece.org/env/water/publications/pub76.htm Accessed 18 January 2010

UNECE (2009) River basin commissions and other Institutions for transboundary water cooperation. Economic Commission for Europe Convention on the Protection and Use of Transboundary Watercourses and International Lakes. ECE/MP.WAT/32. New York and Geneva: United Nations. http://www.unece.org/env/water/documents/CWC percent20publication percent20joint percent20bodies.pdf Accessed 18 January 2010

Vener BB, Campana ME (2008) Conflict, cooperation, and the new 'Great Game' in the Kura-Aras basin of the South Caucasus. Proceedings, Universities Council on Water Resources/National Institutes for Water Resources Annual Conference, 22-24 July 2008 Durham, NC. http://water.oregonstate.edu/projects/2008/Vener_Campana.pdf Accessed 18 January 2010

Wolf AT, Yoffe SB, Giordano M (2003) International waters: Identifying basins at risk. Water Policy, vol 5, no 1:29-60

World Bank (2003) Water resources in Europe and Central Asia. Volume 1 – Issues and strategic directions. Volume 2 – Country water notes and selected transboundary basins. The World Bank, Washington, DC

Yildiz D (1999) Sinir Olusturan ve Sinirasan Su Kaynaklarimiz ve Kiyidas Ulkeler Arasinda Teknik Isbirligi Gereksinimi. Cevre ve Muhendis TMMOB Cevre Muhendisleri Odasi Yayini, no 18:28-35

Euphrates-Tigris Rivers System: Political Rapprochement and Transboundary Water Cooperation

Aysegul Kibaroglu and Waltina Scheumann

1 Introduction

Water-related development projects on the Euphrates and Tigris rivers have been highly contested over the last four decades and have caused relations between the riparian states, i.e. Turkey, Syria and Iraq, to become highly strained and serious crises occurred. All co-riparian states are unilaterally strengthening their efforts to develop water resources to increase their hydropower potential, and to extend their irrigated agricultural areas. These activities pose the main threat to their mutual relations, and to date, the riparians have failed to achieve a common agreement. Since major non-water issues are now solved, or are at least approached, in a more pragmatic manner, the prospects for joint initiatives have improved. Figure 1 shows a map of the two rivers, their main tributaries and selected dams. Table 1 and Table 2 provide an overview of the context for cooperation on both rivers.

2 The Euphrates and Tigris rivers system[1]: geographical and hydrological setting[2]

The Euphrates and its tributaries drain an enormous basin of 444,000 square kilometers of which 33 percent lies in Turkey, 19 percent in Syria, and 46 percent in Iraq. On the other hand, the Tigris and its tributaries drain an area of 387,600

[1]Internationally, 'Euphrates' and 'Tigris' are the names used. In Turkish, the Euphrates is called Firat, and Al-Furat in Arabic. Tigris is named Dicle in Turkish, and Dijla in Arabic.

[2]This section draws from Kibaroglu 2002a.

A. Kibaroglu (✉)
International Relations Department, Okan University, Istanbul, Turkey
e-mail: Aysegul.Kibaroglu@okan.edu.tr

W. Scheumann
Deutsches Institut fuer Entwicklungspolitik, Bonn, Germany
e-mail: waltina.scheumann@die-gdi.de

A. Kibaroglu et al. (eds.), *Turkey's Water Policy*,
DOI 10.1007/978-3-642-19636-2_16,

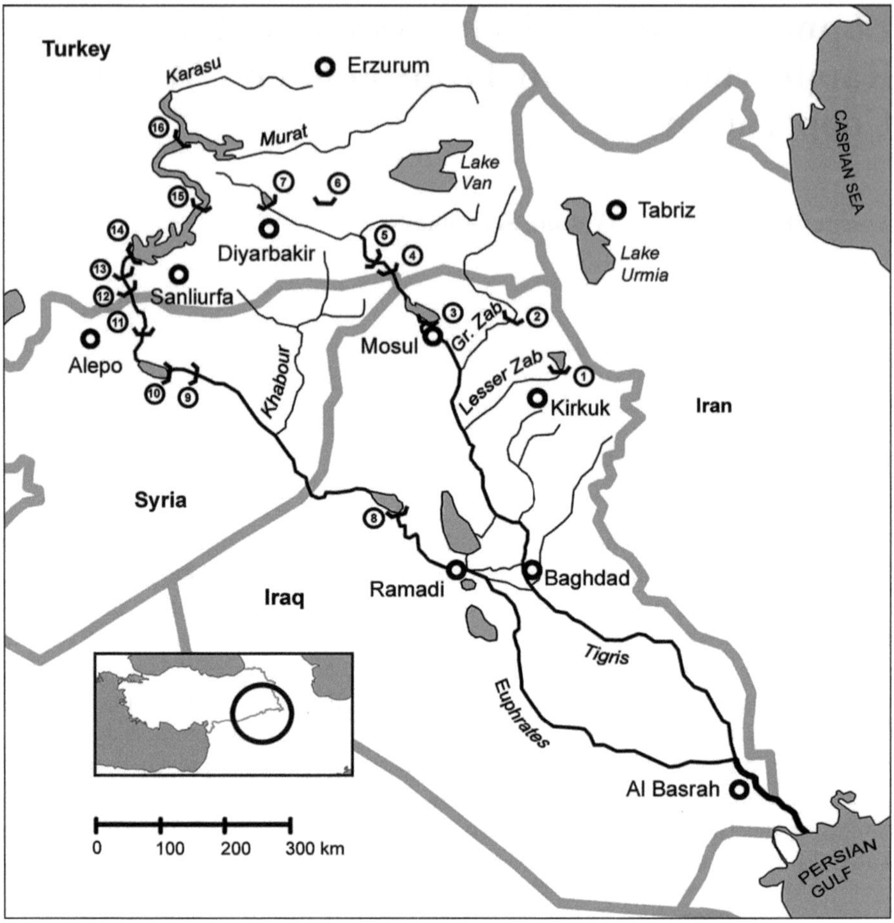

(1) Dukan Dam
(2) Bekhme Dam (unfinished)
(3) Mosul Dam
(4) Cizre Dam (planned)
(5) Ilisu Dam (planned)
(6) Batman Dam
(7) Kralkizi Dam
(8) Haditha Dam
(9) Al Baath Dam
(10) Tabqa Dam
(11) Tishreen Dam
(12) Karkamis Dam
(13) Birecik Dam
(14) Ataturk Dam
(15) Karakaya Dam
(16) Keban Dam

Fig. 1 Map of Euphrates, Tigris, main tributaries and selected dams

square kilometers of which 15 percent lies in Turkey, 0.3 percent in Syria, 75 percent in Iraq, and 9.5 percent in Iran. Both rivers originate in Turkey, scarcely 30km from each other, flow through Syria and Iraq, and form the Shatt-al-Arab watercourse north of Basra in Iraq before discharging into the Persian Gulf. These twin rivers have extremely high seasonal and multi-annual variances in their flow, and severe drought and destructive flooding have been common phenomena for millennia.

Table 1 Cooperational context in the Euphrates River

Euphrates Basin area: 444,000 km^2; mean annual discharge 32 BCM		
Riparian position	Basin area (percent of total) Contribution to annual discharge	Main water uses
Turkey upstream	146,520 km^2 (33 percent) 28.922 BCM (90 percent)	irrigation, hydropower, flood control
Syria downstream	84,360 km^2 (19 percent) 3.213 BCM (10.0 percent)	irrigation, hydropower
Iraq downstream	204,240 km^2 (46 percent) 0.0 BCM (-)	Irrigation, inhabitants of the Mesopotamian Marshes
Main agreements and covered issues		
Turkey - Syria	1987 - interim protocol, water quantity to be released at Turkish-Syrian border (see Annex 12) 2001 - Joint Communiqué 2003 - Implementation Protocol: exchange of expertise, training, joint projects; Joint Technical Committee 2009 - Memorandum of Understanding in the Field of Efficient Utilization of Water Resources and Combating of Drought 2009 - Memorandum of Understanding in the Field of Remediation of Water Quality	
Syria – Iraq	1990 - water sharing between Syria and Iraq	
Unsettled issues		
Quantity	No final tripartite agreement. No common approach	
Groundwater	Overuse on both sides	
Quality	At present, subordinate	
Protection/ restoration	Mesopotamian marshlands	

The *Euphrates* has two main sources, the Murat and Karasu rivers, which drain the high plateau to the north-west of Lake Van. The Keban Dam was built in the early 1970s where the streams meet in Kharput. Downstream of the Keban Dam, the main river stem joins numerous springs of various discharges; it then joins the Tohma tributary upstream of the Karakaya Dam and receives incremental contributions from the Kahta River (upstream of the Ataturk Dam) and from the Nizip tributary (downstream from the Birecik Dam). The Euphrates then first flows south-eastwards, then south-west and breaks through the mountains in a gorge near Hilvan, and crosses the Syrian border at Karkamis.

In Syria, the Euphrates has two tributaries, the Balikh and the Khabour Rivers. The Khabour sub-basin together with its transboundary tributaries and its springs is the most complicated element of the system; various branches of the Khabour originate either from Turkey or from Syria and are estimated to have a significant potential of 200 million cubic meters (MCM) per year. From there, over the remaining 1,000km of its course in Iraq, it gains no further increments of water. In Iraq, at a point 360km from the border, the Euphrates reaches its 100,000 square kilometers giant plain at Ramadi. Further downstream, near Nasiriye, the river

Table 2 Cooperational context in the Tigris River

Tigris Basin area: 387,000 km^2; mean annual discharge 52 BCM		
Riparian position	Basin area (percent of total) Contribution to annual discharge	Main water uses
Turkey upstream	57,600 km^2 (14.9 percent) 20.840 BCM (40 percent)	irrigation, hydropower
Syria - border with Turkey / Iraq	1,000 km^2 (0.3 percent) –	
Iraq downstream	292,000 km^2 (75.3 percent) 26.571 BCM (51 percent)	irrigation (diverts water through Thartar Canal to Euphrates), hydropower
Iran - upstream on one tributary	– 4.689 BCM (9 percent)	
Main agreements and covered issues		
Turkey-Iraq	1946- flow regulation of the Tigris and Euphrates rivers and of their tributaries (see Annex 11)	
Turkey-Syria	2009 - Memorandum of Understanding on the Establishment of a Pumping Station in the Territories of Syrian Arab Republic for Water Withdrawal From the Tigris River	
Disputed issues		
	No consensus on procedure No consensus on whether Euphrates-Tigris forms one single watercourse system Dispute over Ilisu Dam construction	

becomes a tangle of channels some of which drain into the shallow Lake of Hammar and the remainder joins the Tigris at Qurna.

The *Tigris* originates from a small mountain lake, south of the city of Elazig in eastern Turkey, and flows through the basaltic district of Diyarbakir. It forms the border between Turkey and Syria, and Iraq and Syria. Its two major tributaries are the Great Zab and Lesser Zab, which join the river downstream of Mosul. The contribution of the Tigris tributaries to the river's potential is very significant and amounts to roughly 50 percent of the Tigris flow at Baghdad.

Downstream from Baghdad the river's slope is flat and it becomes exceedingly tortuous with the Tigris joining the Euphrates to form the Shatt al-Arab watercourse north of Basra. Most of the water in the lower part of both the Euphrates and Tigris is lost in a wide area of salinated swamps and marshlands. The combined area of lakes and swamps at the head of the Persian Gulf varies from 8,288 square kilometers at the end of the dry season to 28,490 square kilometers during spring floods.

In summer, the mean annual flow of the Euphrates is 32 billion cubic meters (BCM) per year of which about 90 percent is drained from Turkey, whereas the remaining 10 percent originates in Syria. As for the Tigris, the average total discharge is determined as 52 BCM/year, of which approximately 40 percent comes from Turkey, whereas Iraq and Iran contribute 51 percent and 9 percent, respectively.

Estimates for the total flow of the Tigris-Euphrates and their tributaries vary between 68 BCM and 84.5 BCM.

The catchment areas of both rivers experience a sub-tropical Mediterranean climate with wet winters and dry summers. As the snow melts in spring, the rivers are in spate, augmented by seasonal rainfall, which reaches its maximum between March and May. This climate prevails in south-eastern Turkey, as well as in northern Syria and Iraq. Winter precipitation ranges between 400 and 600mm and allows rain-fed cultivation of winter grain though supplementary irrigation raises yields and allows multiple cropping. In the Mesopotamian Plain annual rainfall is rarely above 200mm. The summer season is hot and dry, with midday temperatures approaching 50°C resulting in high evaporation and daytime relative humidity as low as 15 percent. Evaporation also reinforces water salinisation and water loss in major reservoirs in Turkey and Syria, as well as in Lake Habbaniya and the Thartar Canal in Iraq.

3 Water resources development

At present, irrigated agriculture - the greatest user of water - is unequally developed in the three riparian states. Iraq has used the Euphrates to irrigate 1 to 1.3 million hectares for a long time now. Syria started in the 1960s, and intensified irrigation in the Upper Euphrates after the completion of the Tabqa Dam in the mid 1970s. Prior to the completion of the Ataturk Dam (1990), irrigation in south-east Turkey was limited to groundwater and extended to about 114,000 hectares. A major threat to water resources, and to the riparians relations, is the envisaged enlargement of areas to be irrigated with water withdrawn from the Euphrates and the Tigris in all three countries: about 1.7 million hectares in Turkey as part of the Southeastern Anatolia Project (GAP), 640,000 hectares in Syria and 500,000 hectares in Iraq.

Variation in the flow of both rivers ranged from conditions of severe drought to destructive flooding before upstream reservoirs were built in Turkey that are capable of smoothing out such variances and providing a dependable year-round flow downstream. However, since the 1960s Turkey, Syria and Iraq have invested in large-scale water development projects, the largest of which is Turkey's GAP. A series of dams were built, first in Iraq, then in Syria and Turkey to provide irrigation water and to generate hydropower. The major dams on the Euphrates are, Keban, Karakaya, Ataturk, Birecik and Karkamis in Turkey; Tabqa, Al-Baath and Tishreen in Syria, three more dams can be found on the Khabour River (Khabour Dam, Eastern Khabour Dam, Western Khabour Dam) in Syria. Because a large portion of Iraqi territory rarely exceeds 300m in altitude, the topography limits the possibility of impounding the Euphrates behind large dams. However, since 1988 the Thartar Canal has linked the Tigris with the Euphrates in Iraq, in this way using the Tigris' water for irrigation. As is the case with Syria, most of Iraq's land is low-lying and afflicted by deposits of gypsum and salt; both of these are not compatible with irrigation (Kibaroglu and Unver 2000).

4 Incidents of crises over the Euphrates

As a result of supply-led developments, the water demands of the riparians exceed the actual amount of water that can be supplied by the Euphrates and Tigris rivers. If all irrigation projects envisaged are realised by 2040 (the completion date for all projects), total demand would far exceed supply.[3] Although consumption targets are very subjective, they nevertheless form the basis of the riparians' claims to the rivers' water. Rapidly increasing populations in these countries and the importance of food production have given further impetus for the utilisation of the rivers.

During the 1970s, 1980s and 1990s a number of crises occurred in the region, following the unilateral development of several water resource projects. Turkey started impounding the Keban reservoir in February 1974 at the same time as Syria had almost finalised construction of the Tabqa Dam. The impounding of both reservoirs in the following two years escalated into a crisis in 1975[4] with Iraq accusing Syria of reducing the river's flow to intolerably low levels, while Syria blamed Turkey. The Iraqi government was not satisfied with the Syrian response, and mounting frustration resulted in mutual threats. This was averted when Saudi Arabia mediated and ensured that extra water was released from Syria to Iraq.

Another major crisis occurred in the early 1990s when the Ataturk Dam in Turkey was filled. On 13 January 1990, Turkey temporarily reduced the flow of the Euphrates River in order to fill the Ataturk reservoir (January was chosen because it was the month with no demand for irrigation water). Turkey notified its downstream neighbours before November 1989 of its intension and in a communication it explained the technical reasons behind the action and also provided a detailed programme for the replenishment of the losses. Turkey also released twice the usual amount of water for two months prior to terminating the flow and sent delegations to the region to explain the need for the action, and the measures taken. The work was finished in three weeks, as opposed to the one month initially planned. However, the Syrian and the Iraqi governments still registered official complaints, and consequently called for an agreement to share the waters of the Euphrates, as well as for a reduction of the impounding period.

Finally in 1996, after Turkey started construction of the Birecik Dam, an after-bay dam on the Euphrates, both Syria and Iraq sent official notes to the Turkish government in 1995 and 1996 indicating their objections to construction on the grounds that it would affect the quantity and quality of waters flowing into Syria and Iraq. The issue became an international affair when Syria and Iraq requested that Arab League countries cease financial aid to Turkish projects and boycott European companies that had financed the dam (Scheumann 2003, 750). The Birecik Dam was not designed for consumptive purposes, but to regulate

[3]Iran's demand is not included; its supply amounts to 9 percent, i.e. 4.7 BCM/year. For details, see Belül 1996.

[4]The situation was exacerbated because impounding took place during a period of continuously dry weather.

the water levels of the Euphrates when power generation at the Ataturk Dam would be at peak.

5 The negotiation process and status of cooperation

Negotiations between Turkey and Iraq on the development of the Euphrates' water originally started in the 1940s[5]. However, since the early 1960s a new series of technical negotiations has attempted to foster new dialogue and information sharing for the region. The following sections highlight these negotiations. It appears that from this time until negotiations came to a close in the early 1990s, the riparians hardly changed their positions (Kibaroglu 2002b).

In the 1960s, the three riparians entered a new phase of their relationship over water, upon Turkey's decision to construct the Keban Dam on the Euphrates. The downstream riparians, particularly Iraq, insisted on guaranteed flows (350 cubic meters per second at minimum) to be released by Turkey during the impounding period. Hence, a first meeting was held in June 1964 with Turkish and Iraqi experts attending. At the end of negotiations, Turkey guaranteed to undertake all necessary measures to maintain a discharge of 350 cubic meters per second immediately downstream from the dam, provided that the natural flow of the river was adequate to supply this discharge. This was communicated to Syria and Iraq the same year. Moreover, during this meeting, Turkey proposed the establishment of a Joint Technical Committee (JTC), which would inspect each river to determine its average yearly discharge. The JTC would determine the irrigation needs of the three countries through joint field studies and would be authorised - by calculating the needs of the riparians for present and future projects - to prepare a statement of the main principles and procedures in order to facilitate an agreement on water rights.

Following this first technical meeting between Turkey and Iraq, a few more ad hoc meetings were held. Among these the most notable one - the first tri-partite negotiation - was held in Baghdad in 1965 where the three delegations exchanged technical data on the Haditha (Iraq), Tabqa (Syria) and Keban (Turkey) dams. In line with a Turkish proposal, Syria suggested that it would be beneficial to commission a JTC study of the water requirements of the irrigable lands, and subsequently to examine the possibility of covering possible shortages of water supplied by the Euphrates by diverting a part of the Tigris River's water to the Euphrates. Iraq strongly opposed this proposal and insisted on negotiating only on the waters of the Euphrates.

---

[5]One of the most important legal texts between Iraq and Turkey on the water resources of the Euphrates and Tigris rivers and tributaries is the Protocol annexed to the 1946 Treaty of Friendship and Good Neighbourly Relations (see Annex 11). The protocol provides a framework for the two parties to deal with their respective interests along the river system. It emphasised mainly the urgency of building up flood control works on the Euphrates and Tigris rivers and underlined the positive impact of storage facilities to be sited in the Turkish territory.

During the 1970s, delegations from the three countries gathered on several occasions to exchange information about technical issues relating to the reservoirs. No agreement was reached, and Turkey and Syria unilaterally determined the impounding programmes for their reservoirs.

In the early 1980s, the Turkish development plans created a new demand for cooperation. This time Iraq proposed the formation of a permanent JTC. At the end of the first meeting of the Joint Economic Commission between Turkey and Iraq in 1980, a JTC was established which Syria joined in 1983, whereupon Turkey, Syria, and Iraq held sixteen meetings up to 1993 (Kibaroglu 2003).

The mandate given to the JTC was defined as determining the methods and procedures, which would lead to a definition of a reasonable and appropriate amount of water that each country would need from both rivers. The main items on the JTC's agenda were the exchange of hydrological and meteorological data, the sharing of information on progress achieved in the construction of dams and irrigation schemes in the three countries, and the discussion of initial plans for the filling of the Karakaya and Ataturk reservoirs.

However, after sixteen meetings, the JTC could not fulfil its mandate, and the talks became deadlocked. The major issues that led to the deadlock related to both the subject and the object of negotiations: whether the Euphrates and the Tigris could be considered a single water system, or whether the discussions should be limited to the Euphrates.[6] The wording of the JTC's final objective, i.e. reaching common terminology, was also problematic: whether to formulate a proposal for the *'sharing'* of *'international rivers'*, or to achieve a trilateral regime to determine the *'utilisation of transboundary watercourses'*. Iraq and Syria consider the Euphrates an *international* river and insist on an immediate sharing agreement under which its waters would be shared on the basis of each country's stated water needs. On the other hand, Turkey regards the Euphrates and Tigris as forming a *single transboundary river basin* where the waters should be *allocated* according to the identified needs.

During negotiations it emerged that the water potential was unable to meet the declared demands of the three riparians. And, more importantly, there were also uncertainties and inadequacies relating to the data on water and land resources. In response to Syrian and Iraqi demands to formulate urgent 'sharing arrangements' dependent on criteria put forward by them, Turkey proposed the *Three Stage Plan for Optimum, Equitable and Reasonable Utilization of the Transboundary Watercourses of the Tigris-Euphrates Basin* in 1984. The Three Stage Plan encompasses joint inventory studies of land and water resources of the region and

[6]The Turkish side regards the Euphrates and Tigris as one river system because both rivers form the Shatt al-Arab watercourse. This opinion is reinforced by the existence of the Thartar Canal, which was built by Iraq: it connects the Tigris with the Euphrates and diverts water from the Tigris to the Euphrates. This view is, so far, not shared by Iraq and Syria. With respect to these contradicting views, Article 2a of the UN Water Convention reads as follows: "'Watercourse' means a system of surface and groundwaters constituting by virtue of the physical relationship a unitary whole and normally flowing into a common terminus."

the estimation of water needs for the competing sectors, agriculture in particular. It was expected that this would provide the basis for optimum allocation of the available water resources related to the determined needs (Kibaroglu and Unver 2000). With the Three Stage Plan, Turkey also called for the establishment of a joint body to collect, handle and exchange data regarding water and land resources so that annual and seasonal variations could be incorporated in the estimates made, in order to determine allocations. Along with reaching a set of agreed upon criteria in data-sharing, it was hoped that negotiations could move on to coordinating development projects and create joint projects. However, the Turkish Three Stage Plan was coolly received by Iraq and Syria, and they continued to demand fixed water quotas (Kibaroglu 2007).

Syria and Iraq use mathematic formulae to define their water quotas. Syria proposed that the co-riparians should declare their demands for each river separately, i.e. the Tigris and the Euphrates rivers. If the claims exceed a river's discharge, the deficit will be proportionally deduced from each share. The Iraqi mathematic formula is somehow different and admits that each riparian should declare its claims for the realised projects, for those under construction and, finally, for any that are planned. Each country's water quota would be defined subsequently, i.e. first for the projects in operation, then for those under construction, etc., with the realised projects having priority over planned projects.

Although an agreement was not reached over procedures or over water quotas, in 1987 and 1990, two bilateral accords were concluded (see Box 1) which were largely products of the then prevailing political atmosphere.[7] They were, however, not the results of JTC negotiations, but were initiated at the highest political levels. Both are acknowledged as interim agreements by all riparians.

Box 1 Bilateral accords concerning the Euphrates River
The Turkish-Syrian Protocol of 1987 (see Annex 12)

The Turkish-Syrian Joint Economic Commission meeting on 17 July 1987 had an important effect on water issue negotiations. *The Protocol on Matters Pertaining to Economic Cooperation,*[8] signed by Turkey and Syria at the conclusion of the meeting, incorporated provisions for water, the temporary nature of which was recognised. Article 6 of the Protocol reads as follows:

During the filling up period of the Ataturk Dam reservoir and until the final allocation of the waters of the Euphrates among the three riparian countries the Turkish side undertakes to release a yearly average of more than 500 m^3/s at the Turkish-Syrian border and in cases where monthly flow falls below the level of 500 m^3/s, the Turkish side agrees to make up the difference during the following month.

(continued)

[7]See Scheumann (2003) for the relevance of non-water issues as disturbing factors.
[8]United Nations Treaty Series 87/12171, 17/7/1987.

As a basis for comparison, the long-term average flow of the Euphrates is about 1,000 m^3/s at the Turkish-Syrian border.

The Syrian-Iraqi water accord of 1990

Syria and Iraq perceived the interruption to the flow of the Euphrates (from impounding actions at the Ataturk Dam) as the first of many similar disruptions resulting from The Southeastern Anatolia Project (GAP) activities, and consequently signed a bilateral accord in 1990. The Joint Minutes (1) read as follows:

The Iraq water share on the border region between Iraq and Syria is 58 percent as a fixed annual total percentage (water year) of the water Euphrates River allowed to pass in Syria through the border with Turkey, and the Syrian share of water is the remainder quantity 42 percent of the water of Euphrates River allowed to pass through the border between Turkey and Syria.

However, the role of the JTC should not be underestimated; even if its meetings were infrequent and if it appeared that little substantive progress was made on the question of water allocation, it served as a useful channel of communication. Even though the JTC originated from the Joint Economic Commission, it focused on water allocation only. Its ultimate aim of ensuring cooperation and coordinated management of water resources could not be fulfilled because the riparians were persistently claiming their water rights.

6 International concerns on GAP

After the completion of the major dams on the Euphrates in Turkey such as the Keban, Karakaya, Ataturk, Birecik and Karkamis, the Ilisu Dam (which is a component of the GAP scheme) became a controversial issue, not just among the riparians, but between Turkey and export credit agencies (ECAs) and international non-governmental organisations. The dam, sited on the Tigris River, is expected to create a reservoir with a volume of 10.4 BCM and a surface area of 313 square kilometers. The Turkish authorities consider it to be a key project as it is their largest remaining power installation. The Ilisu and Cizre dams combined will produce about 5 billion kWh per year, and generate more than 400 million US$ for the Turkish economy. Hydropower generation is planned with an installed capacity of 1,200 MW with expected yearly electricity production of 3,800 GWh (Altinbilek 2000).

An international consortium of ECAs from Switzerland, the United Kingdom, Germany, Italy, Austria, Japan, Portugal, Sweden and the US, coordinated by the

Swiss Export Risk Guarantee, considered funding the project. The project itself and the policies of the ECAs were strongly criticised by environmental and human rights groups[9] on social, environmental and cultural grounds. In response, in December 1999 the ECAs announced that four conditions would have to be met by the Turkish Government before the project would receive export credit support. These conditions were as follows (quoted from European Rivers Network 2000):

1. Draw up a resettlement programme which reflects internationally accepted practice and includes independent monitoring;
2. Make provision for upstream water treatment plants capable of ensuring that water quality is maintained;
3. Give an assurance that adequate downstream flows will be maintained at all times;
4. Produce a detailed plan to preserve as much of the archaeological heritage of Hasankeyf as possible.

In October 2000, less than one year later, an international Fact-Finding Commission visited the area to assess the progress made. The Commission concluded that "the conditions have yet to be met, and that the prospect that they will be met in the near future is remote". Shortly before the report was released, a Swedish company which had a 24 percent stake in the consortium withdraw from the project, followed by Balfour Beatty and all the other foreign companies in the consortium in late 2001.

From a Turkish perspective, the Commission was criticized for not having paid enough attention to on-going archaeological rescue activities,[10] and to the Ilisu Dam Lake Area Sub-regional Development Plan project which were initiated by GAP Regional Development Administration (GAP RDA) back in the early 1990s. Both projects had to be deferred due to the state of emergency in the region for almost a decade. Thereafter, the salvage project for the documentation and protection of the archaeological heritage of the area started in 1998 with funds provided by GAP RDA. Educational institutes from within Turkey collaborated with international teams from the US, Germany, Italy and France to devise a comprehensive schedule for the work. Since then archaeological sites in the area have been extensively surveyed and recorded, and excavations and relief works have commenced (GAP 2005). However, the Ilisu Dam Lake Area Subregional Development Plan could only start in 2002, which caused a delay by changing resettlement projects through the development of preferable spatial alternatives (GAP RDA 2001).

The Turkish government reacted critically to the campaign which, it claimed, was led by UK-based activist groups. The Ministry of Foreign Affairs stated that the

[9]Friends of the Earth, the International Rivers Network, the Center for International Environmental Law, and the Washington Kurdish Institute (http://www.ilisu.org.uk); see also WCD Thematic Review, Regulation, Compliance and Implementation (2000) http://www.adb.org/Water/topics/dams/pdf/tr54main.pdf Accessed 28 May 2010.

[10]I.e. the project in Hasankeyf which is the major ancient town on the Ilisu Dam site.

dam would neither reduce the flow of the river nor cause pollution (Ministry of Foreign Affairs 2004). The *Turkey Country Report* to the Third World Water Forum also claimed that the actual facts were somewhat different than those asserted by the Fact-Finding Commission. With reference to the transboundary (downstream) issues involved, the report reads:

> The Ilisu Dam is not designed for irrigation, only for power generation: The water passing through the turbines has to flow back into the river bed. River water flowing into Iraq and Syria will not be polluted because the use of water for hydropower is non-polluting. As a result of Ilisu, new sewage treatment facilities will be built in the towns upstream, thus improving water quality. Ilisu will act as regulator holding back water during the winter floods and releasing it during the summer droughts (Republic of Turkey 2003, 76).

After the halting of the project in 2001, the Turkish Government has undertaken a renewed effort in July 2005 to seek funding based on an updated Environmental Impact Assessment (EIA) Report and an updated Resettlement Action Plan. In 2005, the project was awarded by DSI at a total price of EUR 1.1 billion to a Turkish-German-Swiss-Austrian construction consortium. German, Austrian and French private banks signed financing contracts in August 2007, and the Turkish Undersecretary of Treasury accepted the credit agreements in January 2008. The governments of Germany, Austria and Switzerland approved export credit guarantees for the project of approximately EUR 500 million.

It states that Turkey has informed and consulted the downstream neighbours and shared all the project-related documents with them in October 2006. It also states that scientific studies would be conducted to determine minimum flows to be released as well as keeping the water quality standards to be released to downstream neighbours during the filling and operation of the dams (DSI 2009). The Iraqi minister of water resources, Latif Rashid, complained to the Europeans about Turkey's Ilisu Dam project, asking them to prevent Ankara from going ahead with it. This came at a time when Turkey's environment minister, Veysel Eroglu, vowed to release significantly more water to Iraq (Kibaroglu 2009).

The Turkish Prime Minister Recep Tayyip Erdogan conducted a ceremonial groundbreaking for the dam in August 2006, and it was announced that the project would be completed by 2013. The construction of the project started in May 2008. But German, Swiss and Austrian ECAs called for a halt to the construction in December 2008, amid concerns that the project was failing to meet 150 international standards and World Bank safeguard policies for resettlement and environmental and cultural preservation. After a 180-day review period, they withdrew their pledged US$610 million in export credit in July 2009.

DSI posted detailed reports to explain that the Project Implementation Unit (PIU) had fulfilled at least 47 tasks out of 89 tasks indicated in the protocol signed jointly by the ECAs and DSI.[11] They also referred to the June 2009 report of the Committees of

[11]See for details the DSI website http://www.dsi.gov.tr/ilisu/ilisu_hasankeyf.swf http://www.dsi.gov.tr/ilisu/coe_reports/EMG_June_2009_Report.pdf http://www.dsi.gov.tr/pdf_dosyalar/ilisu_baraji_bilgilendirme_notu.swf Accessed 20 December 2009.

Experts[12] who acted as monitoring units and advisors to the PIU, which included a consensus decision on those tasks related to resettlement plans and environmental management to be completed by the PIU, whereas no consensus was reached on the implementation progress of the protection of cultural heritage. Shortly after the announcement of the funding loss, both the Minister of Environment and Forestry and the Prime Minister asserted that the dam would be built, and that Turkish public and private banks would step in to provide the financial services.

A further issue dating back to the early 2000s concerning the Ilisu Dam and other GAP dams were their anticipated negative impacts on the Mesopotamian Marshlands in Iraq. The Euphrates-Tigris rivers' system used to divide into many channels at Basra, forming an extensive marshland area. The marshes were, however, largely drained by Saddam Hussein's government in the 1990s as a means of driving out the rebellious Marsh Arabs. The study "The Mesopotamian Marshlands: Demise of an Ecosystem" of the United Nations Environment Programme claims:

> The Mesopotamian marshlands, which until recently extended over an original area of 15,000 to 20,000 km^2, have been devastated by the combined impact of massive drainage works implemented in southern Iraq in the late 1980s/ early 1990s and upstream damming (UNEP 2001, ix).[13]

Since the 2003 invasion of Iraq, drainage policy has been reversed and the Ministry of Water Resources in Iraq has embarked on a large programme of engineering to reorganise the whole drainage area, by removing many engineering installations and irrigation schemes and restructuring agricultural practices in the region in order to replenish the marshes. These efforts were supported by the Japanese Ministry of Foreign Affairs. It has provided funds through United Nations Environment Programme's (UNEP) Post Conflict Assessment Unit and is engaged through funding for GRID-Europe to develop an Iraq Marshlands Observation System. This is a decision-making support tool, to develop and implement a monitoring system to systematically acquire, analyse and exchange information about the Marshlands ecosystem; to develop information products and services based on the data gathered to support management of the restoration process; and to evaluate the success of wetland restoration and its impacts on the regional environment, including that of the northern Persian Gulf.[14] The Iraqi programme has so far managed to reclaim about 30 percent of the lost marshland.[15]

Despite these developments, the European Parliament, in line with general international opinion, has requested that Turkey "be sensitive to the water

[12]External Monitoring Group (2009) Ilisu Dam and HEPP Project. External Monitoring Group Report, June 2009. http://www.dsi.gov.tr/ilisu/coe_reports/EMG_June_2009_Report.pdf Accessed 28 May 2010.

[13]See also UNEP 2003.

[14]http://www.grid.unep.ch/activities/sustainable/tirgis/mmps.php Accessed 29 September 2005.

[15]Personal correspondence with Prof. Mukdad Ali, Baghdad University, College of Science, March 2005; see also www.grid.unep.ch/activities/sustainable/tigris/index.php accessed 28 May 2010 and UN Chronicle 2002 Issue 2 http://www.un.org/Pubs/chronicle/2002/issue2/0202p44_mesopot... Accessed 29 September 2005.

requirements of these countries, with particular reference to the lower Mesopotamian Marshes in Iraq and Iran, where water flows have been significantly reduced by the construction of the Ataturk Dam" (European Parliament 2004).

While the rate of marshland diminution, and its causes, has yet to be assessed and reviewed more accurately,[16] the case once more establishes the need to harmonise and coordinate basin-wide development efforts, not just by considering in-stream flows and sectoral water demands but also by looking at all uses and users. The Ilisu Dam should be taken by Turkey as an opportunity to re-consider the fact that environmental and social issues are more adequately dealt with at the planning and implementation phases. There is a need for improved participation as early as possible in the planning stage, and possibly in designing resettlement programmes. It is noted that European firms and ECAs apply the non-objection rule which makes approval of projects by riparian states conditional.

7 Recent developments and prospects for cooperation

7.1 Intergovernmental level

Relations between Turkey and Syria improved considerably following the Adana Protocol signed on October 20, 1998,[17] and new and promising initiatives have been undertaken since then. In 2001, Turkey's GAP RDA made contact with Syria by sending a delegation on the invitation of the Syrian General Organization for Land Development (GOLD), part of the Syrian Ministry of Irrigation. As a result, a joint communiqué was signed between GOLD and GAP RDA on 23 August 2001. Once again the water issue was relegated to the technical level and was handled by intergovernmental networks composed of technocrats. GAP-GOLD cooperation is based on the common understanding of the sustainable utilization of the region's land and water resources through conducting joint rural development and environmental protection projects, joint training programmes, expert and technology exchanges, and study missions. Syrian and Turkish delegations paid visits to each others' project sites. During these periods they had the opportunity to exchange experiences pertaining to the positive and negative impacts of the decades old water and land resources development projects. Unlike the technical negotiations in the 1960s, the GAP-GOLD dialogue included multiple issues such as urban and rural

[16]The UNEP/DEWA/GRID website considers that "positive signs of environmental recovery have been emerging [. . .] visible in new satellite images taken in May 2003." (UNEP/DEWA/GRID, 2004).

[17]Bilateral relations between Turkey and Syria had long been uneasy. Two principal sources of friction were Syria's extensive logistical support to the separatist terrorist organization, the Kurdistan Workers' Party (PKK) and Syrian irredentist claims to the province of Hatay in Turkey. Despite official denials by Damascus, Syria's support of subversive actions against Turkey since the early 1980s have been widely known and documented.

water quality management, rural development, participatory irrigation management and agricultural research.[18] Even though the dialogue between these two leading institutions could not materialize in concrete project implementation or regular exchange programs, it has served as a semi-formal consultation mechanism; it paved the way for forthcoming cooperative initiatives taken by other related government offices and the ministries in the years 2008 and 2009.

Furthermore, the improved political and economic relations among the riparians since the late 1990s have produced fruitful impacts on water-based development efforts in the region. Significant progress in the economic relations of Syria and Turkey can be observed in major sectors such as e.g. agriculture, energy and health. A series of government, private sector and civil society delegations have paid numerous mutual visits, reaching productive understandings and agreements on trade and economic matters. These culminated in the signing of the Free Trade Agreement in 2004, a real breakthrough in the advancement of bilateral economic relations. The years 2003 and 2004 witnessed the signing of two framework cooperation agreements on health and agriculture, respectively. Both agreements underlined the importance of enhanced cooperation and development in the two neighbouring countries. They included, among other things, discussion of water-related issues, such as soil and water conservation in agricultural practices and combating water-borne diseases (Kibaroglu 2006).

7.1.1 High-level cooperation between the riparians

Recently, the Turkish government has embarked upon cooperative foreign policy initiatives involving her southern neighbours, Syria and Iraq, in particular. The political reasons behind these initiatives can be analyzed at contextual, regional, bilateral and domestic levels. This is beyond the focus of this chapter. However, below the reader may find how the political will at the highest level in Turkey also refers to initiatives which relate to transboundary water development and management in the Euphrates, Tigris and the Orontes rivers.

Turkey and Iraq signed the Joint Political Declaration on the Establishment of the High-Level Strategic Cooperation Council on 10 July 2008. Accordingly, the first ministerial meeting between Turkey and Iraq under the title of the High-Level Strategic Cooperation Council, a mechanism of joint meetings of the Iraqi and Turkish cabinets, was jointly led by Turkish Foreign Minister and his Iraqi counterpart occurred on 17-18 September 2009 in Istanbul. The Turkish Foreign Minister was accompanied by seven executive ministers of the cabinet, including the ministers of trade, energy, transportation, agriculture and environment, while the Iraqi Minister was accompanied by nine executive ministers of the Iraqi Cabinet,

[18] A joint Communiqué between the Republic of Turkey, Prime Ministry, Southeastern Anatolia Project Regional Development Administration (GAP) and the Arab Republic of Syria, Ministry of Irrigation, General Organization for Land Development, 23 August 2001, Ankara, Turkey, on file with the authors.

who are the counterparts of the seven Turkish ministers, as well as three deputy ministers. The meeting between Turkish and Iraqi ministers came a day after Turkey and Syria signed a deal to create a similar mechanism of strategic cooperation during a visit by the Syrian President Bashar al-Assad to Istanbul on 16 September 2009.

A shadow cast over bilateral relations between Iraq and Turkey due to the presence of the outlawed Kurdistan Worker's Party (PKK) organization in northern Iraq gradually dispersed in 2008, while Turkey also improved its relations with the regional Kurdish administration in northern Iraq following a long hiatus after the U.S. invasion of Iraq. Seeds of multidimensional bilateral cooperation between Ankara and Baghdad were actually sown during Erdogan's July 2008 visit. Then, Erdogan and Maliki signed a strategic partnership agreement committing Turkey and Iraq to cooperation in the fields of politics, economy, energy, water, culture and security. The formation of the High-Level Strategic Cooperation Council was outlined during that visit.

According to the strategic partnership agreement signed between Ankara and Baghdad, the High-Level Strategic Cooperation Council will meet at least once a year, with the prime ministers of the two countries presiding over the meeting. The ministerial level mechanism, meanwhile, will meet at least three times a year, while technical delegations will meet four times a year. Decisions made at the High-Level Strategic Cooperation Council will be implemented through an action plan. Barham Salih, Iraq's former deputy prime minister, had called the agreement "the starting point of the Middle East common market" and likened improving relations between Iraq and Turkey to the relationship of France and Germany in the 1950s.[19]

On the other hand, the first High-Level Strategic Cooperation Council meeting between Turkey and Syria took place in Damascus on 22-23 December 2009. A short while after the Turkish-Syrian Strategic Cooperation Council Agreement was signed on 13 October, and following the ministerial meeting which took place in Aleppo and Gaziantep with the contribution of various ministers, the Council's meeting at the prime ministerial level gives clues about the progressive approach adopted in Ankara-Damascus relations, and possibly paves the way beyond good wishes towards more institutional and concrete steps and processes.

The protocols, projects and the revised memoranda, which were taken into consideration on 13 December, have been under study by delegations since then. On 22 December, at the Council in Damascus, the high level diplomats from both countries came together and made the final revisions for the protocols and projects which would be signed at the High-Level Strategic Cooperation Council. It was hoped that the signing of the protocols would not only strengthen Turkish-Syrian relations, but would also represent the first concrete results for the Strategic Cooperation Council. Some commentators argue that the Turkish-Syrian Strategic Cooperation Council Meeting is an indication of the fact that relations between Ankara and Damascus are developing towards a goal of integration. In the same

[19]"Iraq, Turkey want to integrate economies, transform Mideast," Today's Zaman, E-Gazette, September 18, 2009.

manner, the Prime Minister and the Minister for Foreign Affairs in Turkey pointed out that the Strategic Cooperation Council not only promotes relations between two countries, but also aims at finding joint solutions for regional and international problems.

These cooperative initiatives taken at the highest political level have made possible to deal with prolonged disputes in Turkish-Syrian relations. Thus, by the leadership of two ministers: the Minister of Irrigation, Syria and the Minister of Environment and Forestry, Turkey, a commission composed of technocrats and diplomats of the two countries met on 8 December 2009 in Ankara to prepare the framework and contents of an agreement concerning building a joint dam on the Orontes/Asi River. During the meeting, other issues related to water management and use in the Euphrates-Tigris rivers system were discussed, including, the construction of a water pumping station to be located on the Syrian side of the Tigris River that would enable Tigris waters to be used for irrigation in northeastern Syria as well as additional agreements on improving water quality in transboundary rivers and joining efforts to fight against drought (Ayhan 2009).

7.1.2 New water protocols - new water use rules

Among the forty eight Memorandum of Understandings (MoU) which were signed between Turkey and Iraq on 15 October 2009, one was on water. With that protocol the two sides agreed to exchange hydrological and meteorological information as well as exchanging expertise in these fields. Both sides also emphasized the utilization and management of regional water resources in an efficient manner.

On December 23 and 24, 2009 Turkey and Syria signed fifty MoUs at the first meeting of the High-Level Strategic Cooperation Council in Damascus including four MoUs related to regional waters:

- The MoU Between the Government of the Republic of Turkey and the Government of the Syrian Arab Republic for the Construction of a Joint Dam on the Orontes River Under the Name "Friendship Dam." According to this, both countries will share costs of the dam, which will be built at the border. It will produce energy for both sides and irrigate 20,000 hectares of agricultural area in Turkey and 10,000 hectares in Syria. Although the details of the dam will be ironed out in the feasibility study, it is expected to be approximately 15m high and have a capacity of 110 MCM of water storage. Of that total, 40 MCM will be used to prevent flooding and the rest for energy production and irrigation. The foundation of the dam is expected to be laid out in 2010.
- The MoU Between the Government of the Republic of Turkey and the Government of the Syrian Arab Republic on Establishment of a Pumping Station in the Territories of Syrian Arab Republic for Water Withdrawal From the Tigris River. With this protocol, the quantity of water drawn annually from the Tigris River by Syria, when the flow of water is within the average, will be 1.25 BCM. The water

withdrawals are arranged according to monthly flows, and it is indicated that pumping will be done when time and place allows.[20]

- The MoU between the Government of the Republic of Turkey and the Government of the Syrian Arab Republic in the Field of Efficient Utilization of Water Resources and Combating of Drought.
- The MoU between the Government of the Republic of Turkey and the Government of the Syrian Arab Republic in the Field of Remediation of Water Quality.[21]

7.1.3 JTC meetings-revitalized

On 22 March 2007, on an occasion to inaugurate an international conference in Antalya, Turkey, the Turkish Energy and Natural Resources Minister invited the Syrian Minister of Irrigation and Iraqi Water Resources Minister to discuss how to set up a cooperative framework to deal with regional water issues. The Ministers decided that periodic meetings of the JTC, held between 1982 and 1992 before being suspended, would be reconvened. Hence, a series of JTC meetings were conducted since then: the first one being convened in Syria on May 7-11, 2007, followed by a tripartite ministers meeting on January 10-11, 2008 in Syria. At another JTC meeting on February 24-25, 2009 in Istanbul, officials decided that they would share data (current and historical) regarding meteorological patterns and water quality in the Tigris and Euphrates rivers. Another JTC meeting took place in Syria in 2009.

Moreover, in March 2008, the three ministers agreed to establish a water institute in Turkey. It was decided that each riparian appoints fifteen water engineers with an aim to conduct studies on water use efficiency and better water management in the region. The institute would map water resources in the region and draw a report on measures that the respective countries must take for effective management of these resources. The engineers from the three countries have been meeting to exchange information and know-how. In this context, the first training activity was on modern irrigation practices in the region; the second one was about construction and safety of the dams. An interest has arisen within the group to study climate change and its impacts on regional waters. The training institute in Istanbul hosted the third training programme for the experts on this issue.[22]

[20]In 2002, a bilateral agreement between Syria and Iraq was signed concerning the installation of a Syrian pump station on the Tigris River for irrigation purposes. The quantity of water drawn annually from the Tigris River, when the flow of water is within the average, will be 1.25 km^3 with a drainage capacity proportional to the projected surface of 150,000 hectares. Personal communication with the Turkish officials at Ministry of Foreign Affairs and DSI, January 2010.

[21]"Joint Statement of the First Meeting of the High-Level Strategic Cooperation Council between Syria and Turkey," December 24, 2009, Syrian Arab News Agency.

[22]Personal communication with officials at DSI, Ministry of Environment and Forestry, March 2010.

Additionally, at the Fifth World Water Forum convened in Istanbul in March 2009, Turkey and Syria called for Iraq to conduct negotiations regarding reasonable utilization and allocation of the transboundary rivers. Two months later, in May 2009, the Iraqi Parliament adopted a resolution which stipulated that it would block any agreements with Syria or Turkey that did not include recognition Iraqi water rights. The decision taken by the Iraqi Parliament can be interpreted as "despite the recent instances of cooperation, the underlying problems remain." Yet, the following month, in June 2009, both Turkey and Syria agreed to help alleviate the drought in Iraq by increasing water flow.

On 3 September 2009, Turkey, Syria and Iraq held a meeting at which they decided to cooperate to initiate water training programmes and to monitor and exchange information related to climate change and drought conditions in the three countries. In addition, they agreed to erect new ones and modernize water flow gauging stations. After talks between the Iraqi Foreign Minister and the Turkish Environment Minister, Turkey also agreed to provide 550 cubic meters per second of water from the Euphrates River to Iraq until 20 October 2009. The three states scheduled their next meeting for January 2010 in Baghdad, which is currently postponed due to parliamentary elections in Iraq.

7.2 Non-governmental level: ETIC - a network of professionals[23]

One significant development in the region is the Euphrates-Tigris Initiative for Cooperation (ETIC) established in May 2005 by a group of scholars and professionals from the three major riparian countries.[24] The overall goal of the initiative is to promote cooperation among the three riparians to achieve technical, social and economic development in the Euphrates-Tigris region. The composition and the role of ETIC remarkably fit the epistemic community theory and its role in institutional bargaining. Epistemic communities are a "network of professionals with recognized expertise and competence in a particular domain and an authoritative claim to policy-relevant knowledge within that domain or issue-area" (Haas 1992, 349).

The origin of ETIC may be traced to early meetings among the concerned scientists from Iraq, Syria, Turkey and the United States in 2004.[25] This group of

[23]This section is mainly drawn from Kibaroglu A (2008) The Role of Epistemic Communities in Offering New Cooperation Frameworks in the Euphrates-Tigris Rivers System. Journal of International Affairs. vol 61, no 2, 191-195.

[24]The co-author of this chapter, namely Aysegul Kibaroglu is the co-founder of the Euphrates-Tigris Initiative for Cooperation (ETIC).

[25]As a spin-off from a project conducted by the International Center for Peace at the University of Oklahoma, some Iraqi, Syrian and Turkish participants in the said project have decided to launch a cooperation initiative, in collaboration with the University of Oklahoma and Kent State University. See http://www.ou.edu/ipc/etic/.

dedicated scholars has been meeting with flexible agendas. During the first stage of these gatherings, the participants shared information concerning national water policies and raised the significance of water issues in the countries' socio-economic development targets. In a short period of time, the members of the group have been able to develop a common understanding of the existing conditions, pressing problems and needs in the region. In doing so, these concerned scientists have decided to turn their expertise and experience into the joint initiative of the ETIC.

ETIC is a track-two effort, meaning that it is voluntary, non-official, non-binding, non-profit seeking and non-governmental. It is not affiliated with any government, but it aims to contribute positively to efforts, official and unofficial, that will enhance the dialogue, understanding and collaboration among the riparians of the Euphrates-Tigris system. As a multi-riparian initiative, ETIC has been unique in that it looks beyond water rights, per se, to themes related to environmental protection, development and gender equity, water management, governance and grass-roots participation in a holistic, multi-stakeholder framework.[26]

The ETIC members contend that awareness of socio-economic development is compulsory to understand the real dynamics of the region. Hence, the vision of the ETIC is defined by the founders as "ensuring the quality of life for people in all communities, including rural and urban areas, is improved, and harmony among countries and with nature in the Euphrates-Tigris region is achieved". The aim is to promote cooperation for technical, social and economic development in the Euphrates-Tigris region.[27] In line with its vision and overall goal, ETIC prepares and implements joint training and capacity building programmes[28] as well as research and projects[29] with an aim to respond to the common needs and concerns of the people in the region.[30] Towards this end ETIC has built partnerships with international organizations (United Nations Educational, Scientific and Cultural Organisation - UNESCO, UN Food and Agriculture Organization - FAO and United Nations Development Programme - UNDP); with non-governmental organizations (Stockholm International Water

[26]Summary statement presented at the conclusion of the XII World Water Congress, November, 26 November 2005, New Delhi, India; ETIC Newsletter 1, no. 3, ETIC workshop synthesis document (World Water Week, Swedish International Water Institute, Stockholm, 21 August 2006). On file with the authors.

[27]See http://www.etic-org.net/. Accessed 30 May 2010.

[28]ETIC organized a training program in 2006 on dam safety in collaboration with the United Nations Educational, Scientific and Cultural Organization (UNESCO) for professionals from Iran, Iraq, Syria and Turkey. ETIC organized a workshop on Knowledge Technology in March 2009 in Gaziantep, Turkey for participants from Iraq, Syria and Turkey. The last training workshop was organized by ETIC in Aleppo in January 2010 on Geographical Information Systems and their implementation in natural resources management. See http://www.etic-org.net/. Accessed 30 May 2010.

[29]ETIC has been pursuing a research activity entitled "Collaborative Planning and Knowledge Development in the Tigris-Euphrates Region". The stakeholders in this activity are Iraqi, Syrian and Turkish universities faculty members. On file with the author.

[30]ETIC Newsletter (2006) vol 1, no 4. On file with author.

Institute (SIWI) and Advancing the Blue Revolution Initiative)[31] and with the universities (Bahcesehir University, University of New Mexico and the American University in Beirut).

8 Challenges for long-term and sustained cooperation

Delli Priscoli and Wolf analyze the transformation of water conflicts in four stages (2009, 97). In line with their analyses, it is possible to claim that water dispute in the Euphrates-Tigris river basin has already passed *Stage I*, where negotiations were often *adversarial* with an emphasis on water rights. Trust building has been accomplished between Turkey and Syria. However, there are still some concerns between Turkey and Iraq. Even though a new water protocol was signed between Iraq and Turkey, and the negotiation mechanisms (at the prime ministerial and ministerial levels) have been either newly established or revived such as the case of the JTC, Iraq preferred to act unilaterally regarding the deliberations on the Ilisu Dam and also took the decision on assurance of the water rights to the parliament. Yet, the riparians have managed to establish various levels of contacts starting from the highest political level to the concerned bureaucracy as well as engaging business and private entities. Above all, they were innovative in convening a "joint cabinet of ministers" to seal that the riparian governments act very closely with each other in taking decisions related to regional economy and social-cultural affairs. It is also possible to assert that the negotiations have moved forward to *Stage II*, where negotiations have progressed to a *reflexive* stage with an emphasis on needs. Turkey has released 550 cubic meters per second (more than the average amount promised in the 1987 Protocol) of the Euphrates flows to Syria, to enable the Syrian and Iraqi governments to cope with the recent droughts. Moreover, Turkey has been responsive to the needs of Syria from the Tigris River and signed the MoU on the pumping of the Tigris boundary waters by Syria for irrigating about 150.000 hectares of land in northeastern Syria.

It has been observed recently that the riparians through the leadership of highest level politicians preferred functional cooperation adopting a benefit-sharing approach. Thus, one productive approach to the development of transboundary waters is to examine benefits in the basin from a regional perspective. The Euphrates-Tigris case supports the observation that when negotiations focused solely on water sharing, upstream and downstream differences were reinforced, which made water gains and losses more prominent. The opportunities to broaden the scope of the negotiation agenda to involve other sectors beyond water (energy, trade), should be seized in the current political atmosphere. This regularly has required the riparians to get past looking at water as a commodity to be divided – a zero-sum, rights-based approach – and rather to develop an approach that equitably allocates not the water but the benefits derived there from a positive-sum,

[31] See http://www.dai.com/work/project_detail.php?pid=190. Accessed 21 July 2010.

integrative approach. Adding sectors can widen the zone of possible agreement and can make implementation more manageable. Multi-resource linkages may offer more opportunities for creative solutions to be generated, allowing for greater economic efficiency through a "basket of benefits."

However, the biggest challenge now is to operationalise the numerous agreements (MoUs) related to various aspects of cooperation ranging from trade, investment, water, environment, agriculture, education, energy to security issues by establishing a regime framework, which could mobilize the relevant actors and actions to implement benefit-sharing projects and distribute the benefits equitably. Hence, the success or the failure of the recent high level cooperation between the riparians will be tested through the systematic analyses of the changes in the socio-economic status of regional people: whether their social and economic status are better-off in terms of increasing income levels and distribution of the benefits fairly.

Even though the political will is expressed and initial actions have been taken at the highest level, it all depends on the institutional capacity of the riparians to implement these cooperative agreements. Hence, a regime framework with its institutions (principles, norms, rules and decision-making procedures) should be built to make the recent cooperative initiatives sustainable and permanent. In this context, track two (unofficial, professional networks) initiatives such as the ETIC should be supported as it provides necessary scientific cooperation to operationalise the cooperative agreements related to water and socio-economic sectors. ETIC is flexible to bring diverse stakeholders together. It adopts a holistic, development focused, multi-sectoral approach as opposed to one aiming at sharing the river flow. The latter has proven to be divisive and unproductive.

References

Altinbilek D (2000) The Ilisu Dam Project. In: Water and development in Southeastern Anatolia: Essays on the Ilisu Dam and GAP, Proceedings of a seminar held at the Turkish Embassy in London, February 2000, pp 29-38

Ayhan V (2009) Turkish-Syrian Strategic Cooperation Council's First Prime Ministers Meeting. Center for Middle Eastern Strategic Studies (ORSAM). http://www.orsam.org.tr/en/showArticle.aspx?ID=125 Accessed 30 December 2009

Belül ML (1996) Hydropolitics of the Euphrates-Tigris basin. M.Sc. Thesis, Graduate School of Natural and Applied Sciences, Middle East Technical University, Ankara

Delli Priscoli J, Wolf AT (2009) Managing and transforming water conflicts. University Press, Cambridge

DSI (2009) Ilisu Baraji Bilgilendirme Notu (Ilisu Dam Information Note), June 2009. DSI, Ankara

European Parliament (2004) Turkey's progress towards accession. European Parliament resolution on the 2004 regular report and the recommendation of the European Commission on Turkey's progress towards accession (COM(2004)0656 - C6-0148/2004 - 2004/2182(INI)) 15 December 2004

European Rivers Network (2000) Ilisu Dam Fact Finding Mission: preliminary findings. RiverNet NEWS on Turkish Rivers, 16 October 2000. http://www.rivernet.org/turquie/prs00_01.htm Accessed 30 May 2005

External Monitoring Group (2009) Ilisu Dam and HEPP Project. External Monitoring Group Report, June 2009. http://www.dsi.gov.tr/ilisu/coe_reports/EMG_June_2009_Report.pdf Accessed 28 May 2010

GAP (2005) Devam Etmekte Olan Projeler. http://www.gap.gov.tr/Turkish/Kultur/kuldp.html Accessed 27 May 2005

Haas PM (1992) Introduction: Epistemic communities and international policy coordination. International Organization, vol 46, no 1:1-35

Kibaroglu A (2002a) Building a regime for the waters of the Euphrates-Tigris river basin. Kluwer Law International, London, The Hague, New York

Kibaroglu A (2002b) Settling the dispute over the waters of the Euphrates-Tigris river basin. In: Bogardi J, Castelein S (eds) Selected papers of the International Conference From Conflict to Co-operation in International Water Resources Management, Delft, Holland, 20-22 November 2002. UNESCO-IHP, pp 329-343

Kibaroglu A (2003) Settling the dispute over the waters of the Euphrates-Tigris river basin. Proceedings of the 1st International Conference on Hydrology and Water Resources in Asia-Pacific Region. Kyoto, Japan, 13-15 March 2003, pp 219-225

Kibaroglu A (2006) Cooperation for development: Emerging frameworks for sharing benefits in the Euphrates–Tigris river basin. Boğaziçi Journal: Review of Social, Economic and Administrative Studies, Special Issue on Political Theory, vol 20, no 1-2:135-152

Kibaroglu A (2007) Socioeconomic development and benefit sharing in the Euphrates-Tigris river basin. In: Shuval H, Dweik H (eds). Springer, Berlin, pp 185-193

Kibaroglu A (2008) The role of epistemic communities in offering new cooperation frameworks in the Euphrates-Tigris rivers system. Journal of International Affairs, Spring/Summer 2008, vol 61, no 2:191-195

Kibaroglu A, Unver O (2000) An institutional framework for facilitating cooperation in the Euphrates-Tigris river basin. International Negotiation, vol 5, no 2:311–330

Kibaroglu A, Klaphake A, Kramer A, Scheumann W, Carius A (2005) Cooperation on Turkey's transboundary waters. Research report, German Federal Ministry for Environment, Nature Conservation and Nuclear Safety, Berlin

Kibaroglu M (2009) What does the future hold for Turkey and Iraq. Bitterlemons, 9 July 2009, edn 26, vol 7

Ministry of Foreign Affairs (2004) Ilisu Dam. Ministry of Foreign Affairs, Republic of Turkey. http://www.mfa.gov.tr/MFA/ForeignPolicy/MainIssues/WaterIssues/ILISUDAMsub3.htm Accessed 27 May 2005

Republic of Turkey (2003) Turkey country report. Prepared for the 3rd World Water Forum March 2003. Republic of Turkey, Ankara

Scheumann W (2003) The Euphrates issue in Turkish-Syrian relations. In: Brauch HG, Liotta PH, Marquina S, Rogers PF, El-Sayed Selim M (eds) Security and Environment in the Mediterranean. Conceptualising Security and Environmental Conflicts. Springer, Berlin, pp 745-760

Southeastern Anatolia Project Regional Development Administration (2001) Southeastern Anatolia Project Status Report 2001. English version on CD-Rom

UNEP (2001) The Mesopotamian Marshlands: Demise of an ecosystem. Early Warning and Assessment Technical Report UNEP/DEWA/TR.01-3 Rev. 1, prepared by Hassan Partow. Division of Early Warning and Assessment, UNEP, Nairobi, Kenya

UNEP (2003) Desk study on the environment in Iraq. UNEP, Nairobi, Kenya

UNEP/DEWA/GRID (2004) Water returns to the desiccated Mesopotamian Marshlands. Geneva: UNEP/DEWA/GRID. http://www.grid.unep.ch/activities/sustainable/tigris/2003_may.php Accessed 27 May 2005

Orontes River Basin: Downstream Challenges and Prospects for Cooperation

Waltina Scheumann, Ilhan Sagsen, and Ece Tereci

1 Introduction

The Middle East is one of the most water scarce regions of the world. This is why water issues have - among other factors - influenced the countries' relationship. Turkey, which is considered a relatively water-abundant country if compared with her southern neighbours, had not considered water as a major foreign policy issue until the 1960s. Following Turkey's initiation of water development projects on the Euphrates and Tigris Rivers in mid-1970s and during the 1980s, transboundary water issues emerged as a Turkish foreign policy concern, particularly in her relations with Syria.

The Orontes River, which is called "Asi" and "Nahr al-Asi" in Turkish and Arabic respectively, is among the controversial waters that Ankara and Damascus had been arguing about. It passes through three countries' territories, i.e. Lebanon (upstream), Syria (midstream) and Turkey (downstream) where it discharges into the Mediterranean Sea. While Lebanon and Syria have settled their issues on water allocation, Turkey and Syria have yet to find a consensus.

Both countries have yet to agree on the status of the rivers concerned. The official Turkish argument is that rivers which cut across Turkey are "transboundary" waters over which Turkey claims sovereign rights while emphasising the principle of "equitable utilization". Syria, on the other hand, puts forward the argument that these are "international rivers" the waters of which should be "shared".

W. Scheumann (✉)
Deutsches Institut fuer Entwicklungspolitik, Bonn, Germany
e-mail: waltina.scheumann@die-gdi.de

I. Sagsen
Department of International Relations, Middle East Technical University, Ankara, Turkey
e-mail: isagsen@metu.edu.tr

E. Tereci
Turkish Ministry of Energy and Natural Resources, Ankara, Turkey
e-mail: ecetereci@gmail.com

A. Kibaroglu et al. (eds.), *Turkey's Water Policy*,
DOI 10.1007/978-3-642-19636-2_17,

This rhetoric coupled with other political issues such as Syria's territorial claims over Hatay (a southern province of Turkey) and Syria's support of the terrorist organization PKK (Partia Karkeren Kurdistan- Kurdistan Workers' Party) against which Turkey has been fighting for a long time, subsequently led to a long-term hostile atmosphere (see Turan in this volume).

Because of Syria's policy of harbouring the Kurdistan Worker's Party (PKK), the two countries' relations have been very tense. The 1990s witnessed a serious crisis between Turkey and Syria: it reached a breaking point in 1996 when Ankara decided to suspend all official relations with Damascus, following Syria's refusal to expel the PKK leader Abdullah Ocalan, after Turkey's official request to do so. However, Syria's support of the PKK which had almost brought the two states to the edge of war, started to change in 1998 with the signing of the Adana Security Protocol. Bashar Assad's ascension to presidency in 2000 can be regarded as the beginning of a relatively peaceful period between these two neighbours. Since then, although the two countries have not launched fully fledged cooperation regarding water issues, they have achieved significant cooperation in agriculture, transportation, energy, health and environment. Thus, the positive atmosphere which began with the presidency of young Assad also enabled some progress regarding the solution of the water problem. A joint dam project which is planned to be built on the Orontes River by the two countries indicates the countries' rapprochement concerning water.

2 Hydro-geographical features of the Orontes River

Rising in Lebanon in the springs of Labweh near the city of Baalbek in the northern part of the Bekaa Valley, the Orontes flows in a northerly direction, parallel to the coast and the Lebanon and anti-Lebanon Mountains, enters Syrian territory near the town of Hermel and drains into the Qattaneh reservoir. It then passes through the cities of Homs and Hamah, and crosses the fertile Al-Ghab region. As it forms the Turkish-Syrian border for 31km, it flows into Turkey and discharges into the Mediterranean Sea in the Turkish province of Hatay. Out of its total length of 448km, 35km are in Lebanon, 325km in Syria, and 88km in Turkey (Arisoy and Turkoglu 1998).[1] The catchment area covers 37,900 square kilometers of which 49.94 percent lie in Turkey, 44.32 percent in Syria and 5.74 percent in Lebanon (TFDD 2002). The Afrin River which is the major tributary of the Orontes, and the Karasu River rise in the northern part of the basin, namely in the Akcadag, Karadag and Sof mountains in Turkey. While the Karasu flows through Turkish territory and forms the border between Turkey and Syria for a short distance, the Afrin crosses Syrian territory before it re-enters Turkey and flows into Lake Amik (see map).

[1] According to Bazza and Najib, its total length is 485km, of which 336km are within Syria (2003, 7).

The three rivers – Orontes, Afrin, Karasu – have an estimated combined mean annual discharge potential of 2.4 billion cubic meters (BCM). When the Orontes River enters Turkey it has a yearly potential of 1.4 BCM, the potential of Karasu is 0.4 BCM/year, and the annual discharge rate of the Afrin River is 0.6 BCM/year (Fakioglu 2010).

3 Water resources development in the riparian countries

The water of the Orontes River and its tributaries are intensively used in all riparian countries for irrigation purposes, domestic water supply and to service industry. However, the main strain on water resources comes from Syria's and Turkey's unilateral development plans to expand irrigated areas, and from Syria discharging untreated industrial and domestic wastewater into the river making downstream use problematic. Figure 1 shows a map of the Orontes River, its main tributaries and selected dams, and Table 1 provides an overview of the context for cooperation on the river.

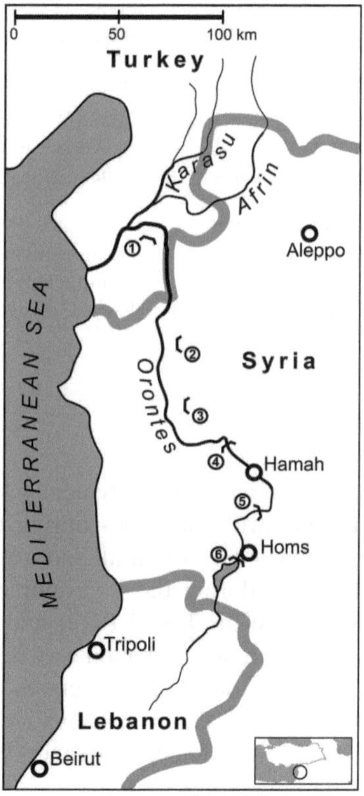

(1) Yarseli Dam (5) Al Rastan Dam
(2) Zeizoun Dam (6) Qattaneh Dam
(3) Afamia Dam
(4) Mahardeh Dam

Fig. 1 Map of the Orontes River with its main tributaries, and selected dams

Table 1 Cooperational context in the Orontes River

Orontes basin area: 37,900 km²; mean annual discharge 2.8 BCM		
Riparian Position	Basin area (percent of total) Contribution to annual discharge	Main water uses
Lebanon Upstream	2,175 km^2 (6 percent) 0.3 BCM (11 percent)	Domestic water supply, irrigation, hydropower
Syria mid-stream	16,797 km^2 (44 percent) 1.2 BCM (43 percent)	Domestic water supply, irrigation, hydropower
Turkey Downstream	18,972 km^2 (50 percent) 1.3 BCM (46 percent)	Domestic water supply, irrigation, hydropower, flood control
Agreements and covered issues		
Turkey - Syria	1939 – demarcation of the Thalweg of the rivers as the border	
	2004 – Free Trade Agreement	
	2004 – start of planning joint dam project	
	2007 – Turkish Energy Minister, Syrian Water Works Minister and Iraqi Water Resources Minister met in January and agreed to conduct periodic meetings of the Joint Technical Committee	
	2007 – JTC met in May	
	2008 – JTC met in January	
	2009 – 23-24 December, Turkey and Syria signed at the first meeting of the High-Level Strategic Cooperation Council in Damascus the Memorandum of Understanding related to construction of the joint (Friendship) dam	
	2008/09 – installation of two modernized flow measuring stations in Syria by Turkey to facture early warning	
Syria - Lebanon	1994 - allocation of water (80 MCM/year for Lebanon)	
Unsettled issues		
Quantity	No agreement between Syria and Turkey on water allocation, and no joint contingency plan (Syrian Zeizoun dam break in 2002)	
Quality	Not addressed	

Lebanon

Lebanon, which has a typically Mediterranean climate with heavy winter rains and dry and arid conditions during the rest of the year, is in a relatively favourable position as far as rainfall and water resources are concerned. However, water is not available during the dry summer months. Lebanon's main sources of irrigation waters are the Litani River and the Litani-Awali water resource system, not the Orontes. In fact, Lebanon uses relatively little water for irrigation from the Orontes. Country-wide, 87,500 hectares are irrigated with surface water (as of 1993), of which about 11,500 hectares are located in Northern Lebanon.[2] According to the bilateral agreement signed in 1994 between Lebanon and Syria concerning the

[2]Detailed data on water uses from the Orontes river are not available.

sharing of Orontes' waters, Lebanon's annual share amounts to 80 million cubic meters (MCM) out of 420 (or 510) MCM.[3]

Lebanon plans to build a multi-purpose dam on the Orontes River, the project's objectives being as follows: (i) to supply water to the cities of Hermel and Baalbek; (ii) to provide irrigation water for 6,100 hectares of land; (iii) to construct the Asi Dam to irrigate land in the Bekaa Valley, and (iv) to install a hydropower plant. Syria objected to the project at the beginning, but later conceded. However, it has still not been implemented due to a lack of finance (Canatan 2003, 11-12).

Syria

Syria has heavily developed surface (and ground) water resources in the Orontes basin where annual rainfall ranges between 300 and 800mm, and annual evaporation between 1,200 and 2,000mm. Unlike the rest of the country, harnessing the water of the Orontes started during the French Mandate (1930s) and was given further impetus with Decree No.3 of 1972, which initiated the construction of multipurpose dams. In the 1950s, Syria started to systemically drain the Al-Ghab marshes in order to open up land for irrigation. The Orontes River bed was enlarged and deepened, and dams were built to regulate the flow of the river and to provide water for irrigation. The Al-Ghab Project was carried out between 1958 and 1967 and covered 46,000 hectares. The project was considered very important for Syria's economy, as it allowed the further settlement of farmers in the region.

According to a World Bank study (World Bank 2001), the Orontes provides 20 percent of Syria's total estimated water use volume (and 8 percent of Syria's overall water supply) and ranks second to the Euphrates River (Daoudy 2006, 44). The same study notes that water use per sector from the Orontes is as follows: agriculture 82 percent, domestic water supply 8 percent and industry 10 percent.

Targeting at food security, the opening up of land for irrigation has been a means to achieve it, and policies have prioritized water allocation for agriculture. Figures given by Daoudy show that in the early 2000s, agriculture (not only in the Orontes basin) received more than 90 percent, dropped to 80 percent during a severe drought lasting from 1999 to 2000, and recovered to 89 percent from 2000 onward until 2006 (Daoudy 2006, 42).

Syria has built a number of dams on the Orontes River with a total storage capacity of 736 MCM, such as the Al-Rastan Dam, the Qattaneh reservoir, the Mahardeh Dam, the Afamia dams and the Zeizoun Dam. The large Homs-Hama canal which starts from the Qattaneh reservoir provides water for 23,000 hectares of land; the Mahardeh reservoir supplies water to the Asharneh plain, and the Rastan reservoir irrigates areas in the Asharneh and Al-Ghab plains. After completion, the

[3]The FAO Aquastat Database estimates surface water flow to Syria at 510 MCM/year through the Orontes River and the bordering El Kebir river (FAO Aquastat 2008).

Afamia, Zeizoun and Kastoun dams will provide water for another 72,000 hectares. However, it is said that the cumulated reservoir capacities still do not satisfy demand, and groundwater is pumped for irrigating another 20,000 hectares.[4]

Turkey

In the early 1940s, the Amik Lake had a few natural drainage canals and consequently fields and nearby villages were frequently flooded. Hence, in the early 1970s - through the drainage and land reclamation works of the General Directorate of State Hydraulic Works (DSI) – by draining of the Amik Lake, the Amik Plain developed into a significant agriculture asset in the region. However, shortcomings in the drainage works undertaken in the 1970s resulted in continuous flooding.

There are twelve development projects in the Turkish part of the Orontes basin including its tributaries, four of which are in operation, two under construction and six in the investigation phase. The projects are designed to regulate the flow of the river and its tributaries in order to provide water for irrigation, domestic needs, to generate hydropower and to protect land and settlements from floods (Kibaroglu et al. 2005).

The projects in operation irrigate 14,067 hectares of land and produce 17.0 GWh/year of energy with a capacity of 3.30 MW. The projects currently under construction will irrigate 8,019 hectares and provide 0.95 MCM/year of drinking water. The planned projects will irrigate 77,489 hectares, protect 20,000 hectares of land from floods, provide 36.43 MCM drinking water per year, and, with an installed capacity of 1.60 MW, produce about 62.77 GWh/year. If the projects under construction and in planning are implemented, the total irrigated area will cover 99,575 hectares, 180 GWh/year of energy will be produced, 37 MCM/year drinking water will be provided, and 20,000 hectares will be protected from floods.

However, Turkish authorities argue that these plans cannot be realized with the amount of water they currently receive from Syria: "Syria has been making use of 90 percent of the total flow which reaches an annual average of 1.2 BCM at the border. Out of this total capacity, only a meagre amount of 120 MCM enters Turkey, after it has been heavily used by Syria. This amount will further decrease to the range of 25 MCM if the planned reservoirs of Ziezoun and Kastoun in Syria are built and start operation" (Ministry of Foreign Affairs 1996). The Syrian irrigation schemes referred to are the Hama-Homs (20,000 hectares) and the Al-Ghab (70,000 hectares). Furthermore, Syria might be able to satisfy its drinking water and hydropower demands with the Jisr el Shugur Embankment, the Al-Rastan, Mahardan, Zeizoun and Kastoun dams on the Orontes.

[4] All figures either related to irrigated area, storage capacity, overall water availability and use, have to be taken with care.

4 Status of cooperation

So far, conflicts over the use of the Orontes' waters occurred between Turkey (downstream) and Syria (midstream). From the Turkish point of view, the present challenge to cooperation stem from agricultural water demand and planned irrigation projects in both countries, and from the quality of water when it enters Turkey. The chronological record of bilateral cooperation between the riparian countries reads as follows:

On 19 May 1939, Turkey and Syria signed the *Final Protocol to Determine Syria-Hatay Border Limitation* which stipulated that the waters of the rivers (i.e. Orontes, Karasu and Afrin), where they constitute the boundary between Syria and Turkey, will be utilised in an equal manner. The Thalweg lines[5] of the rivers should demarcate the Turkish-Syrian border (Inan 1994) but no specifications were made on how to use the rivers' waters.

In 1950, Syria approached the World Bank to obtain funding for the Al-Ghab Project. An agreement was signed between the two parties in the same year. The World Bank assumed that water usage in the Orontes river basin would not be jeopardised as a result of the project, and that the control of winter floods would be beneficial for all riparians, while the summer flow would provide enough water to irrigate all areas in the region. However, the World Bank also considered the concerns of the Turkish side and organised a meeting between Turkish and Syrian experts in Syria. On this occasion, the Turkish representatives claimed that Turkey would face frequent floods during construction, and that the project would leave no water for Turkey during irrigation seasons.

In 1962, Syria assigned the development of the Orontes River project to the Dutch company Netherlands Engineering Consultants NEDECO. According to Caponera (1993), the plan was drafted without taking Turkish needs into consideration. During a meeting between the engineers of the two riparian states, the Turkish delegation offered a draft protocol which stated that a river basin development plan for the whole basin should be developed to include measures to mitigate flood hazards, to study the feasibility of a dam on the border to irrigate the Amik Plateau and to install flood warning systems in the basin. However, the meeting ended without agreement.

In the same year, a Syrian-Lebanese joint committee was established to deal with the Orontes. The committee in question allocated 100 MCM/year for Hermel and Ka'a regions in Lebanon. However, in a meeting which was held later in 1968, the committee decided to re-consider the annual water needs of both countries. As a result, Syria offered to build a complex of dams and embankments within her borders in order to facilitate Lebanon's water management. Although Syria and Lebanon decided in 1972 to annually allocate 80 MCM to Lebanon, this agreement never became operative (Salha 1995). However, on 20 September 1994 *Bilateral Agreement Concerning the Usage and Sharing of the Waters of the Al-Asi River*

[5]Thalweg is the line following the lowest part of a valley, whether under water or not.

(Orontes) between the Syrian Arab Republic and the Lebanese Republic, 80 MCM/year were earmarked for Lebanon (Ministry of Foreign Affairs 1996). Its major points are:

- Both countries agreed to consider the Orontes River as "common waters"
- The headspring of the river and the boundaries of the headspring area were defined together with the average annual discharge (403 and 420 MCM)
- Lebanon's annual share of the river is 80 and Syria's 340 MCM/year
- It was agreed that in case the annual discharge of the river falls below 400 MCM, Lebanon's share will be proportionally reduced
- The control and management of the river's flow within Lebanese territory was assigned to a technical committee to be formed by both countries
- The management of the headspring and the tributaries of the river within Lebanese territory would be under the responsibility of both countries, although it was to be financed by Syria only. The supervision of the parties' commitment to the agreement was to be left to a joint arbitrary committee. In cases in which this committee fails to solve any conflict, the problem would be transferred to a higher committee to be established then
- All the artesian wells and facilities existent on the Lebanese part of the river basin as of 24 September 1994 would be counted. Syria would be notified about any further installations and related usage, and the amount of water used from these wells would be reduced from Lebanon's share of 80 MCM (Canatan 2003, 82).

Turkish bureaucrats and the Turkish public alike reacted critically to this agreement because downstream Turkey was excluded, and were neither notified nor consulted (Salha 1995).

Negotiations between Turkey and Syria were more complicated than the relations between Syria and Lebanon for the following reasons:

Since the start of negotiations between Turkey, Syria and Iraq under the mandate of the Joint Technical Committee in the early 1980s, Turkey and Syria adopted conflicting strategies with regard to the subject of negotiation (see Kibaroglu and Scheumann in this volume). While Turkey insisted that negotiations would encompass all regional transboundary waters including the Orontes, the Euphrates and the Tigris, Syria refused to formally discuss the Orontes River with Turkey. Since Syria claimed the Turkish province of Hatay – through which the Orontes River flows and discharges into the Mediterranean – as Syrian territory,[6] Syria regarded the Orontes River as a 'national river' which flows on Syrian territory and drains into the Mediterranean Sea without crossing Turkey. Any negotiation would have been tantamount to acknowledging Turkey's sovereignty over the Hatay region.

However, since the signing of the Adana Security Protocol between Turkey and Syria in October 1998, there have been a number of promising mutual official visits

[6]Until recently, Syria claimed the former Ottoman sub-province of Alexandretta (presently the Turkish province of Hatay) as Syrian territory. Hatay appeared as part of Syria on all official Syrian maps, and consequently the Orontes River was regarded as an internal affair of Syria.

which indicate an improved bilateral dialogue and a new trust in the region (Scheumann 2003; Altunisik and Tur 2006). As a result of this rather recent rapprochement, the two riparians have further improved their economic relations and have signed the first Free Trade Agreement on 22 December 2004 which actually recognizes state boundaries (Sagsen 2006). During the official visit of the Syrian President Bashar Assad to Turkey in 2004 (the first visit ever made by a Syrian president), the two countries recognized the borders of each other with the *Agreement on Avoidance of Double Taxation* and *Agreement on Reciprocal Promotion and Protection of Investment* (FERB 2004). These trade agreements with corresponding assurances to open Syrian trade missions in Hatay were considered to imply de facto recognition of the current border. Turkish authorities interpret the signing of the Free Trade Agreement as Syria's acknowledgment and recognition of Turkey's borders including the province of Hatay. A Turkish diplomatic source said that Damascus lifted its reservations to signing the trade deal after an "accord" was reached on affirming Turkey's sovereignty in the southern province of Hatay, formerly Alexandretta, to which Syria had claims.[7]

Syrian Information Minister Mahdi Dakhl-Allah expects to double the size of bilateral trade which currently stands at around US$1 billion annually. Better trade ties with Syria would open up the Arab markets for Turkey (FTA 2004). Furthermore, there has been considerable progress about the possibility of opening a Syrian consulate in Gaziantep, a Turkish province in southeast Anatolia.

During Erdogan's visit to Syria in 2004, the press recorded that the Turkish Prime Minister indicated his cooperation, and promised technical assistance to the Syrian Prime Minister Otari. He further proposed a joint dam project to be built on the Orontes River, on which both sides reached a concord.[8] The purpose of this dam would be to provide water to irrigate 20,000 hectares in Turkey and 10,000 hectares in Syria, to produce hydropower for Turkish and Syrian needs and to improve flood control (Sen and Celik 2004; Radikal 2004). It was agreed that a joint technical delegation would be formed to study the technical issues pertaining to the construction of the joint dam. A Turkish-Syrian delegation visited the Orontes basin in Syria to examine the topographical and geological characteristics of the region as well as the places likely to be affected by the dam's construction (DSI 2005). As a result of these studies, certain sites have been identified for the construction of the dam (DSI 2006).

On December 23 and 24, 2009 Turkey and Syria signed at the first meeting of the High-Level Strategic Cooperation Council in Damascus, 50 agreements including four Memoranda of Understanding related to water. One was the Memorandum of Understanding between the Government of the Republic of Turkey and the Government of the Syrian Arab Republic for the Construction of a Joint Dam on the Orontes River Under the Name "Friendship Dam". According to this, both

---

[7]Hürriyet Daily News 14 April 2005. http://www.hurriyetdailynews.com/h.php?news=turkish-syrian-summit-affirms-historic-changes-in-the-region-2005-04-14. Accessed 15 December 2009.

[8]Hürriyet Daily News 2 January 2005. http://www.hurriyetdailynews.com/h.php?news=prime-minister-erdogan8217s-return-visit-to syria-2005-01-02. Accessed 15 December 2009.

countries will meet the costs of the dam. Although the details of the dam will be ironed out in the feasibility study, its height may reach 15m and its storage capacity about 110 MCM. Of that total, 40 MCM will be used to prevent flooding and the rest for energy production and irrigation. The foundation of the dam will be laid in 2010. As part of the technical cooperation, both sides agreed to install and operate modernised flow measuring stations that serve as early warning mechanisms.

5 Outstanding issues and prospects for cooperation

The present challenge for cooperation between Syria and Turkey on the Orontes' waters stem from conflicting agricultural water demands, planned irrigation projects in both countries and the quality of the water which arrives in Turkey. At present, basic data relating to water resource potential and actual water use in both riparian countries is still contested. An agreement has yet to be reached on how to arrive at water sharing and harmonisation of national development plans. The issue of water quality requires heavy investment in water treatment facilities since, until recently, Syria was virtually without domestic wastewater treatment facilities and untreated wastewater was disposed of directly into rivers (World Bank 2001, 33).[9] Organic pollution concentrations and levels of ammonia were dangerously high, and the rivers' ecosystem is seriously deteriorated.

It seems that the proposed joint *Orontes Friendship Dam* will materialise. One may hope that the seeds of cooperation observed in the Orontes river basin may pave the way for further confidence-building measures between Turkey and Syria.

Concerns over water quality and nature protection are not addressed at the regional level although the waters reaching the Orontes Delta located in Turkey are heavily polluted. Waste from an oil refinery built in 1957 near the Qattineh reservoir, discharges from a fertilizer factory built in 1976 in the west of the Qattineh reservoir near the city of Homs, wastewater from other industrial complexes, return flows from agricultural lands and the discharge of untreated wastewater from the city of Homs are major contaminating sources. In places where the river is used for domestic needs and irrigation, epidemics such as typhoid, dysentery and cholera have been observed. Near Hamah, where the villagers are dependent on the river to satisfy their water needs, 90 percent of diseases observed are water-borne (Canatan 2003, 28), a fact Syrian authorities and experts are well aware of.[10] Since polluted waters cross the Turkish - Syrian border, Turkish authorities warned Syria via diplomatic channels and demanded the treatment of these waters before they enter Turkish territory (DSI 2006).

---

[9]Canatan (2003, Annex 6) lists industrial plants which discharge untreated waste water into the Orontes; the list includes e.g. sugar processing industry, plastic, textile, ceramics, concrete, and steel.

[10]Yahia Bakour, for instance, the General Director of the Arab Region Center for Agricultural and Developmental Studies and Consultancy (ARCADSAC) (see Bakour 1992)

The Mediterranean Association to Save the Sea Turtles (MEDASSET) has proposed to protect a coastal strip, which includes the Orontes Delta (Turkish territory), as a breeding grounds for the Green Turtle (*Chelonia mydas*). The MEDASSET report to the Standing Committee of the Convention on the Conservation of European Wildlife and Natural Habitats proposed the protection (in particular) of the beaches to the north and south of the mouth of the Orontes River where the highest nesting density can be found (MEDASSET 2003). If supported by Turkish authorities, the MEDASSET initiative may have repercussions for in-stream water flows and on water quality.

A further field of cooperation is the joint development of contingency plans for responding to emergencies and an early warning system in order to notify of either a natural or man-made disaster. One such example is the burst of the Syrian Zeizoun Dam in June 2002. The water held behind the dam (about 70 MCM) not only submerged five Syrian villages and destroyed the livelihood of around 10,000 Syrian people, but flooded some 1,200 hectares of cultivated land in Turkey, and submerged a large area of the town Altinozu (AFP 2002).

References

AFP 05 June 2002 (www.reliefweb.int/rwarchive/rwb.nsf/db900sid/ACOS-64D62N?OpenDocument&rc=3&emid=AC-2002-0338-SYR. Accessed 3 Dec 2009

Altunisik MB, Tur O (2006) From distant neighbors to partners? Changing Syrian-Turkish relations. Security Dialogue, vol 37, no 2:229-248

Arisoy AI, Turkoglu Y (1998) Turkiye'nin Sinirasan ve Sinir Olusturan Sularinin Siyasi Cografya Acisindan Degerlendirilmesi. Seminer, Gazi Universitesi, Ankara

Bakour Y (1992) Planning and management of water resources in Syria. The country experiences with water resources management. WB Technical Paper no 175, The World Bank, Washington D.C.

Bazza M, Najib R (2003) Towards improved water demand management in agriculture in the Syrian Arab republic. FAO First National Symposium on Water Management and Rationalization of water Resources Use in Agriculture, University of Damascus, 28-29 April

Canatan E (2003) Turkiye-Suriye Arasinda Asi Nehri Uyusmazligi. Masters thesis, Hacettepe University, Ankara

Caponera D (1993) Legal aspects of transboundary river basins in the Middle East: The Al Asi (Orontes), the Jordan and the Nile. Natural Resources Journal, XXLIII, no 3:629-663

Daoudy M (2006) The optimization of irrigation in Syria: Institutional, economic and environmental challenges. Symposium Proceedings on Irrigation Modernization: Constraints and Solutions. 28-31 March 2006, Damascus, Syria, IPTRID/FAO, pp 39-56

DSI (2005) Turk Heyeti Suriye'yi Ziyaret Etti. Basin Bultenleri, DSI Genel Mudurlugu, Basin Musavirligi, 18 May 2005. DSI. http://www.dsi.gov.tr/basin/suriyeturk.htm. Accessed 15 February 2006

DSI (2006) Personal communication with an official, 30 June 2006, Ankara

Fakioglu (2010) Personal communication with Deputy Director of Investigation and Planning Department, DSI, June 2010, Ankara

FAO Aquastat (2008) Lebanon. Aquastat - FAO's Information System on Water and Agriculture. FAO. http://www.fao.org/nr/water/aquastat/countries/lebanon/index.stm. Accessed 16 December 2009

FERB (2004) Dinner invitation organized by DEIK/Turkish-Syria Business Council on the occasion of the visit of the Syrian Arab Republic Prime Minister Mr. Muhammed Naci Otri. Foreign Economic Relations Board, 14 July 2004

FTA (2004) Turkey, Syria sign free trade agreement. Bilaterals. http://www.bilaterals.org/article.php3?id_article=1081, 23.12.2004. Accessed 15 December 2009

Hurriyet Daily News (2005a) Prime Minister Erdogan's return visit to Syria. Hurriyet Daily News 2 January 2005. http://www.hurriyetdailynews.com/h.php?news=prime-minister-erdogan8217s-return-visit-to-syria-2005-01-02. Accessed 15 December 2009

Hurriyet Daily News (2005b) Turkish Syrian summit affirms historic changes in the region. Hurriyet Daily News 14 April 2005. http://www.hurriyetdailynews.com/h.php?news=turkish-syrian-summit-affirms-historic-changes-in-the-region-2005-04-14. Accesed 15 December 2009

Inan Y (1994) Sinirasan Sularin Hukuksal Boyutlari. A.U. Siyasal Bilgiler Fakultesi Dergisi, XLIX, Ocak-Haziran: 243-253.

Kibaroglu A, Klaphake A, Kramer A, Scheumann W, Carius A (2005) Cooperation on Turkey's transboundary waters. Research report for the German Federal Ministry for Environment, Nature Conservation and Nuclear Safety. Adelphi Research, Berlin

MEDASSET (2003) Conservation status of Green Turtle (*Chelonia mydas*) in the Eastern Mediterranean: Kazanli, Akyatan, Samandag (Turkey). Report to the 23rd Meeting of the Standing Committee of the Convention on the Conservation of European Wildlife and Natural Habitats (Bern Convention). T-PVS (2003) 14

Ministry of Foreign Affairs (1996) Water issues between Turkey, Syria and Iraq. Ministry of Foreign Affairs, Republic of Turkey. http://www.sam.gov.tr/perceptions/Volume1/June-August1996/WATERISSUESBETWEENTURKEYSYRIAANDIRAQ.pdf. Accessed 16 December 2009

Radikal (2004) Barisa yeni adim. Radikal, 23 December 2004. http://www.radikal.com.tr/haber.php?haberno=138161. Accessed 16 December 2009

Sagsen I (2006) The issue of management of the waters of the Euphrates and Tigris Basin in international context. Masters Thesis, Department of International Relations, Middle East Technical University, Ankara

Salha S (1995) Turkiye, Suriye ve Lubnan Iliskilerinde Asi Nehri Sorunu, Dis Politika Enstitusu, Aralik, pp 25-28

Scheumann W (2003) The Euphrates issue in Turkish-Syrian relations. In: Brauch HG, Liotta PH, Marquina A, Rogers PF, El-Sayed Selim M (eds) Security and environment in the Mediterranean. Conceptualising security and environmental conflicts. Springer, Berlin pp 745-760

Sen E, Celik H (2004) Sam'da Asi, Dicle jestlesmesi. Zaman, 23 December 2004. http://www.zaman.com.tr/haber.do?haberno=124263. Accessed 16 December 2009

TFDD (Transboundary Freshwater Dispute Database) (2002) International river basin register. Oregon State University. http://www.transboundarywaters.orst.edu/. Accessed 15 December 2009

World Bank (2001) Syrian Arab Republic irrigation sector report. Rural Development, Water, and Environment Group, Middle East and North Africa Region, Report no 22602-SYR. http://www-wds.worldbank.org/servlet/WDSContentServer/WDSP/IB/2001/09/21/000094946_01090804014647/Rendered/PDF/multi0page.pdf. Accessed 15 December 2009

Cooperation on Turkey's Transboundary Waters: Analysis and Recommendations

Aysegul Kibaroglu, Axel Klaphake, Annika Kramer, and Waltina Scheumann

1 Epilogue

In Part II, we tried to analyze and assess the status of cooperation on Turkey's transboundary rivers. While the politics of transboundary water resources in the Euphrates and Tigris river basin have already been extensively discussed in international literature, other transboundary river basins, such as the Orontes, Kura-Aras, Coruh or Meric have, in general, received far less political and scientific attention. Hence, the chapters are intended to provide a thorough assessment of crucial water management challenges on major Turkish transboundary rivers, the current state of cooperation and unresolved disputes.

It is observed that, recently, Turkey's water policies towards and its role in the neighbourhood, namely Europe, the Middle East and Caucasus have been changing considerably. However, international knowledge on the current state of water cooperation and unresolved disputes on Turkish transboundary rivers is lacking. Among these evolving relations the one with the European Union (EU) within the framework of on-going accession negotiations deserves special attention. The goal of EU accession implies that Turkey is obliged to adopt and implement the entire

A. Kibaroglu
International Relations Department, Okan University, Istanbul, Turkey
e-mail: Aysegul.Kibaroglu@okan.edu.tr

A. Klaphake
Deutsche Gesellschaft fuer Internationale Zusammenarbeit, GIZ, Kampala, Uganda
e-mail: axel.klaphake@giz.de

A. Kramer
adelphi, Berlin, Germany
e-mail: kramer@adelphi.de

W. Scheumann
Deutsches Institut fuer Entwicklungspolitik, Bonn, Germany
e-mail: waltina.scheumann@die-gdi.de

A. Kibaroglu et al. (eds.), *Turkey's Water Policy*, 313
DOI 10.1007/978-3-642-19636-2_18, © Springer-Verlag Berlin Heidelberg 2011

body of European Environmental Law, covering many far reaching legal requirements (for example, the Water Framework Directive – WFD - with significant implications for the member states' international water cooperation), and a couple of international environmental agreements where the EU is the contracting party. Against this backdrop cooperation on Turkish transboundary rivers becomes an important issue in accession negotiations and demands careful analysis of achievements and shortcomings to date. In the EU-Turkey accession partnership, the transboundary water issue has already been identified as a priority issue, which demands short-term considerations and progress.

Methodologically, Part II is drafted by putting Turkey in the centre of analyses. Turkish needs, policies and projections regarding transboundary waters are scrutinized. While drafting Part II, the bulk of research (collecting and analysing literature, documents, databases and personal communications) has been conducted in Turkey. Yet, international literature as well as accessible data and information from riparian countries have also been utilized when deemed necessary. Clearly, consolidation of adequate data represents a challenge particularly when it comes to the analyses of state and inter-state behaviour, and, as a result, some of the assessments are rather intuitive and would need further empirical data. However, our results illustrate the variety of issues that are (or were) disputed on Turkish transboundary waters and allow for some general conclusions to be drawn concerning prospects for sustainable transboundary water cooperation.

2 Controversial issues, hitherto agreements and conflict intensity

Our analyses in Part II reveals significant and challenging disputes over all rivers under scrutiny that need new approaches and strategies to overcome cooperation obstacles. However, the rather alarming descriptions of Turkish transboundary water disputes by many scholars and politicians indicating a potential for serious water conflicts appears exaggerated and does not realistically mirror the current situation, even in the most marked water quantity dispute over the Euphrates-Tigris rivers system.

In terms of hydro-political constellations, we have the classic constellation of Turkey being the upstream riparian on several rivers (Euphrates-Tigris, Coruh, Kura), a mixed situation because of many transboundary tributaries (Aras) and Turkey as the downstream riparian on the Orontes and Meric rivers. In addition, there is variation concerning the objects of dispute. While, for instance, the Euphrates-Tigris, the Sarisu (Aras basin) and the Orontes rivers mainly concern guaranteed river flow, the issue of conflict on the Coruh River is sediment flow. Between the Meric riparians, flood protection is a matter of concern in addition to water quantity and quality issues. However, water quality generally plays a minor role, while quantity / water flow issues still dominate, mirroring respective weak national water quality provisions or their weak enforcement.

At present, most disagreements relate to the building of dams which influence river usability downstream. The Turkish water development policy, with its emphasis on hydropower production and irrigation projects by means of infrastructure (e.g. dams) was outlined in several chapters (Part I). In principle, we can assess disputes at transboundary waters as the external consequences of the internal economic development strategy, putting strong emphasis on the production of agricultural commodities and on achieving independency from energy imports. Apparently, the Turkish Southeastern Anatolia Project (GAP) on upstream Euphrates-Tigris is the most important single Turkish development programme on its transboundary waters which caused the most downstream objections and, in several phases, serious tension between the riparians. At present, the downstream riparians, particularly internal opposition in the Iraqi parliament, are demanding augmented, guaranteed river flow and have expressed concerns about the Ilisu Dam project. Massive river development programmes are under construction or are planned on other transboundary rivers (e.g., Coruh, Aras, Orontes) but the probability that all the planned works will materialise is not easy to assess.

Even though it is generally not easy to anticipate future water dispute constellations, it can be expected that water quantity issues will gain in importance because of population and economic growth in Turkey and her neighbour states. Importantly, the riparian countries to, for example, the Euphrates, Tigris and Orontes rivers are pursuing the same development path, where water resource development is perceived to be crucial for economic growth in particular in the agricultural and energy sectors. Consequently, an accentuation of water quantity rivalry cannot be excluded per se. This assessment relates not only to the Euphrates-Tigris rivers but also to other rivers such as Coruh or Kura-Aras. For instance, realisation of all currently planned dams along the Aras River could not only seriously affect downstream water users and freshwater ecosystems, particularly in Azerbaijan, but also cause disputes with neighbouring Iran. The situation on all the transboundary rivers studied, strongly suggests the need for joint efforts to assess and coordinate planning and management in order to harmonise basin-wide development, where, apart from water sector demands (energy, agriculture), in-stream flows and ecosystem protection should also be taken into account.

Transboundary water quality disputes are looming in several basins too, inter alia in the Orontes. Since water quality has seriously deteriorated in several transboundary waters, joint initiatives to improve the situation are clearly required. Another important ecological issue is the protection of freshwater ecosystems. However, joint activities by riparians are either lacking or are in an early stage and need additional support. For example, wetland protection issues actually play a role in the Meric basin, and the biodiversity value and the protection status of the river's delta could make this issue even more important in the future. Equally, freshwater ecosystems are of pivotal importance in the Kura-Aras basin too, as there are lake areas and wetlands of high biodiversity value in the downstream stretches of the rivers. Clearly, the already initiated effort to restore the Mesopotamian marshlands in Iraq could potentially ignite a new demand for coordinated transboundary water management in the Euphrates-Tigris rivers system.

With regard to regional cooperation, there are, at least, rudimentary forms of cooperation and agreements for all rivers. As outlined in the analysis, all these agreements are bilateral and predominantly concern water quantity or border issues. Turkey and its neighbours have not yet agreed on more comprehensive forms of cooperation that would tackle the different aspects of water use and needs (quality, quantity, flood protection, preservation of ecosystems and prevention of accidents) in an integrated manner and could potentially facilitate negotiations by linking different water management issues. Most of these agreements lack an effective organisational back-up at least in the form of joint monitoring. On the Euphrates-Tigris rivers, for instance, a number of crises have arisen due to the lack of regulated consultation mechanisms among the riparians. Between Turkey and the South-Caucasian riparians, there are merely some outdated agreements that do not provide for adequate regulations on the issues of the day. In all cases, questionable data – or simply lacking data – regarding stream flow, water removals, return flow, present water use etc. play an important role in the negotiation processes. In general, available data and information are incomplete and not regularly exchanged between the riparians. That, in fact constitutes a major limitation for proper assessment and management of water and land resources in the basins.

In all aspects, the tense political relations between Turkey and the riparian states have seen a shift towards a more favourable political environment recently. For instance, bilateral relations between Turkey and Syria improved considerably after 1998 and have already proved to be favourable for dealing with water-related disputes over the Orontes and the Euphrates-Tigris rivers. Since 2007 many political initiatives were taken to deal with transboundary water issues in a cooperative manner. In that context, existing institutions such as the Joint Technical Committee (JTC) have been revitalized, furthermore a series of Ministers' meetings were held thus opening a new channel of communication. Above all, the highest level political will was sealed with the signing of Memoranda of Understanding which address water shortage problems, efficient use of water, drought management, water quality and joint water development (dam) projects. With the regime change in Iraq, there is an evolving international context too which brings, on the one hand, new downstream claims for water, but, on the other hand, offers new prospects for cooperation. Even though, the political conflict between Armenia and Turkey has not yet been solved, the two countries have managed to sustain working boundary water relations (inherited from the Turkey-Soviet Union period) since the construction of the Arpacay (Ahuryan) Dam in 1986. Furthermore, political relations between Greece and Turkey are, in the meantime, much more pragmatic, allowing for a deepening cooperation on water issues as well. Similarly water cooperation between Turkey and Bulgaria is already more intensified, and this could benefit from the positive atmosphere caused by Bulgaria's joining of the EU (2007) and the prospect of EU membership for Turkey.

Although we assessed and highlighted the risk of disputes escalating in several rivers because of increasing demand for domestic, irrigation and energy production purposes, there are also a number of water development projects and changing water demands, which can potentially contribute to an easing of conflicts. For instance, the drastic decline in irrigated agriculture and industrial water use has eased the intensity of water conflict in the Kura-Aras basins. An ease in water rivalry can also be expected from the decline of irrigated agriculture in Bulgaria. A lack of financial resources has caused Syrian irrigation programmes to fall well behind schedule, and initial Iraqi plans are not likely to be implemented soon either. Finally, the future of the various ambitious Turkish development plans is not easy to predict because of increasing financial and institutional difficulties.

The position of Turkey is strongly characterised by her national economic background of increasing water development and utilisation. As mentioned previously, the current problems with transboundary waters must be interpreted in the context of the Turkish national development plan. Turkey is attempting, inter alia, to increase the production of agricultural commodities, to increase the economic well-being in structurally under-developed parts of the country and to satisfy growing energy demands by exploiting its hydropower. This national water development approach, apparently characterised by a classic engineering approach, is complemented by a clear articulation of interest in transboundary water development. Clearly, other riparians, e.g. Syria on the Orontes, Bulgaria on the Meric, follow a similar economic development path and follow comparable water use patterns which may then lead to conflicts over water consumption.

Generally speaking: The Turkish Government follows a combined strategy of cooperation (e.g., building joint dams with Bulgaria, Georgia and Syria, joint training measures with Syria) with a rather reserved position concerning the conclusion of regional water sharing agreements. During the Cold War, Turkey pursued a unilateral water resources development strategy, which must be interpreted in the context of the often very tense relations with the other riparians (e.g. with Syria and Iraq) who, for their part, have for a long time followed a veto-strategy by trying to prevent Turkey from achieving its water resource development plans. The main aspects for building cooperation at the transboundary level can be summarised as following:

In the past, *Turkey's position regarding international water law* was widely perceived as being reluctant, and the fact that Turkey voted against the United Nations (UN) Convention on the Law of the Non-navigational Uses of International Watercourses (1997) supports this view. However, Turkey acknowledges several basic principles of international customary water law. According to the Turkish position, the principle of equitable and reasonable utilisation should serve as a guiding rule for the allocation of transboundary waters and the settlement of conflicts. Consequently, Turkey pleads for the *limited territorial sovereignty* doctrine but objects to the doctrine of *co-sovereignty of the riparians*, which would strengthen downstream interests (according to the Turkish position) in an asymmetrical manner. However, it seems that Turkey's reservations mainly stem from a reluctance to agree on far-reaching procedural rules (e.g., compulsory mechanisms for dispute settlement; detailed procedures for prior notification). As our analyses of cooperation over the rivers show, this does not mean that Turkey rejects any transboundary cooperation. Interestingly, the historical bilateral agreements which concerned riparians include mechanisms for conflict resolution.

Against this background, the *EU accession partnership* is one of the rare documents entailing a Turkish commitment to pursue international water cooperation not only in line with the EU's WFD but also with international conventions to which the EU is a party. This commitment concerns, for instance, the United Nations Economic Commission for Europe (UNECE) Convention on the Protection and Use of Transboundary Watercourses and International Lakes (1992) which implies a broad range of procedural rules. Herewith, the accession perspective clearly allows for a changed situation regarding Turkey's willingness to accept the basic provisions of regional water law. In addition, an implementation of international and EU regulation on transboundary Environmental Impact Assessments (Espoo-Convention, EU EIA Directive) would certainly show significant progress.

At present, *Turkish cooperation is largely of a bilateral nature* and mainly based on bilateral protocols that predominantly tackle technical issues and the allocation of water rights. While such an approach might show an advantage for potential negotiations, because of an apparently limited number of riparians and topics, an enlargement of the states cooperating and the issues under scrutiny might offer the countries new mutually beneficial issues and could make the negotiation of linkage issues easier. For instance, an increase in topics to be negotiated on the Euphrates and Tigris rivers on the basis of water-related development sectors could probably make the necessary improvements easier to achieve.

Transboundary water allocation is a complicated and politicised issue making the development of a conflict resolution strategy a challenge. Turkey – like most of the other riparians – traditionally did not want third-party involvement in water negotiations. From today's perspective it seems rather unlikely that this attitude will drastically change in the near future even if EU accession will demand new approaches to transboundary water management of Turkey. Thus, transboundary water issues were already identified as problematic issues during the preparation of the EU-Turkey accession negotiations and always characterized as weak in the European Commission's (EC) Annual Progress Reports.

In most cases, an *adequate data base is still not yet in place*; this is strongly demanded by Turkey. However, this fact can be interpreted from different perspectives: On the one hand, a reliable data base is virtually a *conditio sine qua non* for successful water cooperation, where the Turkish insistence on an improved information base is more than understandable. For instance, cooperation on the Euphrates / Tigris rivers clearly suffers from the lack of complete data and limited data exchange and joint investigations. On the other hand, the demand for data could be used as a pretext to protract negotiations and new commitments. Irrespective of whether the latter assessment is correct or not, a deepening of water cooperation, demands a new culture of information exchange and joint studies of all basins. However, a weak administrative set-up in the riparian states might prevent progress.

Turkey's position on transboundary water issues is also characterised by initiatives and / or proposals to *jointly investigate water use and water needs* in respective countries, instead of merely negotiating water rights. This paradigm shift is probably best illustrated by the Turkish offer to build joint dams with Georgia, Bulgaria and Syria that could serve the energy needs of both countries, and the proposed Three-Stage-Plan for the Euphrates-Tigris rivers system. The latter would contribute to water allocations that take into account water needs for agriculture, population, industrial water use etc. and the basin-wide costs and benefits of the different management options. Taking this Turkish proposal seriously, the offer could contribute to a sustainable water management strategy. However basin-wide and needs-based coordination is highly challenging in political terms, including open questions of distribution and institutionalisation. But in the long-term, the shift from water-rights negotiation to a needs-based approach is highly relevant in the context of water scarcity in international basins.

Turkey has already come up with more concrete *joint dam development* project proposals in the river basins as initiatives for enhancing mutual benefits related to hydropower and irrigation. Joint water storage projects such as the Serdarabad regulator (already in operation) on the Aras River (Arpacay), the Suakacagi Dam (in planning and negotiation stage) on the Meric River (Tunca/Tunca), and the recently proposed Friendship Dam on the Orontes River (being dealt in current Turkish-Syrian technical talks) are all examples of Turkish initiatives for joint water development driven by a pragmatic and workable approach to transboundary water cooperation in these river basins.

Although there are only preliminary examples available, Turkey appears to be much more open to *international cooperation concerning environmental issues*, such as nature protection. This is illustrated by recent initiatives relating to cooperation on nature protection between Turkey and its Caucasian neighbours. Potentially, such initiatives could serve as starting points for broader cooperation in the basins and the development of a more integrated management perspective.

Finally, a trend has developed in Turkish water management strategies suggesting that health issues, water quality aspects and integration of stakeholders into the decision-making processes have gained in importance. In general, one can observe a somewhat divided national discourse on water management with one 'development centric' branch emphasising water infrastructure development for national economic and energy development reasons, and in contrast, the other rather environment-orientated green branch focusing on environmental and social issues, and calling for improved environmental regulations, concerning freshwater ecosystems etc. Consequently, the predominant emphasis on dam building and water infrastructure development as a major instrument of Turkish water policy is increasingly debated with stakeholders. In the long term, this might slow the pace of infrastructure development in the country because political positions which challenge schemes designed solely for infrastructure development, are gaining in importance. In this sense, future infrastructure developments to meet the increasing demand for water-produced services could be counterbalanced by equally important issues such as environmental protection and social harmonisation.

3 Cooperation and outstanding issues: similarities and differences

The state of cooperation and the outstanding disputed issues on the many Turkish transboundary rivers are characterised by both similarities between the river basins, in particular in terms of external and internal political forces, and context specific factors and developments.

All river basins under scrutiny can be found in regions that have always been traditionally characterised by some political tension between Turkey and its neighbours. These political circumstances aggravated past water conflicts which otherwise could have been solved much more easily if the political climate had been more favourable. "Water disputes" or "water conflicts" were overlaid, or at least influenced, by multifaceted interstate conflicts involving other core political issues, such as terrorism, recognition of borders or territorial disputes.

In addition, in situations where Turkey was the upstream country, classical upstream-downstream conflicts occurred, characterised by the divergent interests of the riparian states. The respective Meric and Orontes river cases are important exceptions, with Turkey as a downstream riparian. The knowledge that Turkey is upstream at several important transboundary rivers (Euphrates-Tigris, Coruh etc.) goes hand-in-hand with the widespread international perception of (strong) upstream states aggravating conflicts, or being reluctant to cooperation, and has also clearly contributed to Turkey's rather dubious 'international water cooperation reputation'. However, there are basic forms of cooperation in both constellations which illustrate that cooperation can principally be developed in upstream-downstream constellations too, and that location is not the only and not necessarily the decisive factor in explaining whether and when cooperation takes place.

Furthermore, water cooperation in many of the transboundary basins has to take place in the context of limited water availability in several (or even all) riparian countries, making the allocation of water quantity, i.e. agreements on guaranteed river flow, an important and potentially discordant issue. Because of natural conditions and basic political decisions on national development, Turkey, as well as her neighbours, heavily relies on water for irrigation and hydropower production with water being an important and, in some respects, strategic resource for the national economy. Consequently, regional water negotiations are frequently exacerbated and governed by states insisting on their sovereign right to water because of their economic needs. Equally, comprehensive transboundary agreements or treaties that could help regulate potentially inharmonious claims by riparian states are not in place. There are a limited number of bilateral (and sometimes outdated) protocols and other arrangements. Finally, river basin organisations or committees that might serve as fora for the accommodation of water conflicts are not available either, or they are unable to fulfil their mandate.

Despite these similarities, the picture of water disputes and cooperation at the individual transboundary river basins is far more complex and multi-faceted because of huge differences in terms of the hydro-geopolitical constellations, the causes of individual water conflicts, the dynamics of the respective bilateral political relations and, finally, the range of agreements available and the organisational approaches to the disputed issues.

In this context, analyses of transboundary waters provide the following detailed results.

The Meric-Ergene river basin

Bilateral relations in the Meric basin have improved over the last decade, providing a political context for negotiations and for settling major water-related conflicts. Due to the increasing number and magnitude of the floods, transboundary collaboration in flood protection still needs concerted attention, and conflicting claims by Turkey and Bulgaria to water for the purpose of irrigation can be an impediment to the implementation of irrigation projects in Turkey. It remains to be seen whether the planned joint dam project will be realised and whether it offers an appropriate solution for conflicting water needs. No agreements exist yet on water quality in the basin, and upstream water pollution is increasingly perceived as an issue for Turkey and Greece. EU-membership for all three riparian countries offers a good incentive to increase transboundary cooperation. The prospect of joint nature conservation activities and a legal framework for the protection of wetlands provide further incentives for collaboration in water resource management. Such collaboration could also contribute to good neighbourly relations between the riparian countries, and among the communities living along the border regions.

The Coruh river basin

Water cooperation on the Coruh River benefits from generally good political relations between the two riparians, i.e. Georgia and Turkey. However, apart from some outdated and obsolete agreements between the Soviet Union and Turkey, there is still no adequate legal or organisational (institutional) approach to water management in place. The expected negative impact of the many Turkish dams under construction and in the planning stage on the Georgian Black Sea coastline is at present a matter of concern. The conflict surrounding the Coruh River is not about water sharing, but about impacts on the sediment regime from damming the river and, as a result, an increasing erosion in the Batumi vicinity. This 'sediment' conflict was bilaterally addressed in the late 1990s by the establishment of several technical committees and some cooperative moves were made, however, at least from the new Georgian government's point of view, an agreement acceptable to both sides is yet to be reached.

The Kura-Aras river basin

The Kura-Aras river basin is influenced by a much more complicated political constellation than, for instance, the Coruh River, because the large number of riparians goes hand-in-hand with weak and, in some cases, much more tense political relations. At present, Turkey is barely involved in ongoing international efforts to bring the riparian states together and to improve coordination within the basin. While this appears somewhat justified because of minimal transboundary impact caused by current Turkish water use (e.g. on the Kura River) and expected political impediments caused by the Turkey-Armenia disagreements, increasingly diverging interests between Turkey and other basin countries cannot be completely ruled out in the long term. Several riparian states, in particular Turkey and Iran, have plans to develop water infrastructure for hydropower generation and/or irrigation. However, Turkey is interested in long-term cooperation in the basin, and political relations with the other riparian states (e.g. Armenia) are expected to improve through small-scale confidence building measures.

The Euphrates and Tigris Rivers

Water dispute in the region clearly stems from uncoordinated water development projects. So far, the agreements on the Euphrates are all bilateral (Turkey-Syria, Syria-Iraq), as is the Joint Communiqué (Turkey-Syria) and the Memorandum of Understandings (Turkey-Syria and Turkey-Iraq). Recently, the Turkish government has embarked upon cooperative foreign policy initiatives involving her southern neighbours, Syria and Iraq, in particular. The political will expressed and sealed at the highest level in Turkey for broader cooperation with southern neighbours are reflected in official understanding and cooperative initiatives related to transboundary water development and management in the Euphrates and Tigris rivers system. Trust building has been accomplished between Turkey and Syria. However, there are still some concerns between Turkey and Iraq. Even though a new water protocol was signed between Iraq and Turkey, and the negotiation mechanisms (at the prime ministerial and ministerial levels) have been either newly established or revived, as was the case of the JTC, Iraq preferred to act unilaterally regarding the deliberations on the Ilisu Dam and also took the decision on the assurance of the water rights to parliament. Yet, the riparians have managed to establish various levels of contacts starting from the highest political level to the concerned bureaucracy as well as engaging business and private entities. However, the biggest challenge now is to operationalise the numerous agreements (Memorandum of Understandings) related to various aspects ranging from trade, investment, water, environment, agriculture, education, energy to security issues by establishing a framework regime which could mobilize relevant actors to implement benefit-sharing projects and distribute benefits equitably.

The Orontes river basin

While Syria and Lebanon agreed on water allocations in 1994, Turkey and Syria only settled the demarcation of the border in 1939. However, in December 2009, Turkey and Syria signed the Memorandum of Understanding for the construction of a joint dam on the Orontes. The purpose of this dam would be to provide irrigation water and hydropower for Turkish and Syrian needs. Both countries focused on developing water resources for irrigation, hydropower, drinking water and infrastructure for flood control. However, as perceived by the Turkish side, the present challenge to agree on the Orontes' waters stems from agricultural water demand, the planned irrigation projects in both countries and the quality of the water which arrives in Turkey. There is a need for reliable data on the water resource potential and the actual use in both riparian countries. The proposal to protect a coastal strip including the Orontes Delta on Turkish territory as a breeding ground for the Green turtle will have repercussions on in-stream water flows and on water quality. To solve the latter, investment in water treatment facilities is essential.

4 Recommendations for enhancement of cooperation

The following aspects might serve as preliminary points for the development of regional and basin-specific supporting strategies:

Meric-Ergene river basin

Even though agreements exist between Bulgaria and Turkey and between Bulgaria and Greece, the exchange of information, for instance, early warnings, and the operation of dams during floods have not been satisfactory. International support should be sustained to the three riparian countries to establish a joint programme for flood warning and control.

The same refers to the prevention of hazards. A joint programme for transboundary cooperation on hazard prevention could, for example, be the scope of an EU Twinning project.

Several initiatives touch on the issue of transboundary cooperation in the basin. Building on these initiatives or providing technical or financial support to them could improve the status of ecosystems and biodiversity in the basin area and foster cooperation between the riparian countries in water management.

Water quality in the basin suffers from the discharge of insufficiently treated domestic wastewater, mainly in Bulgaria and Turkey. Solving this problem requires major investment. Financial support as provided by EC programmes and development banks could be flanked by initiatives promoting co-operative approaches and the exchange of knowledge and know-how between riparian countries.

The major user of water in the basin is irrigated agriculture in both Bulgaria and Turkey. Promoting joint riparian efforts to increase water use efficiency in irrigation could not only help to reduce pressure on available water resources, but also foster cooperation in water resource management. The same applies to efforts on reducing pollution by agricultural run-off.

Aspirations for EU accession require Turkey to adopt the WFD and UNECE Helsinki Convention (1992). Using the Meric river basin as a pilot area, Turkey could be assisted in implementing the relevant legislation, thus providing an agenda for transboundary cooperation. Such an initiative could finally aim at the establishment of a trilateral river basin commission, in the same way as they exist in other European transboundary basins.

Coruh river basin

Bilateral cooperation between Turkey and Georgia benefits from good political relations and extends to water-related fields such as nature protection. However, in order to tackle the 'sediment conflict' and to ensure sustainable management of the river, transboundary water cooperation is essential.

Referring to EU regulation, Turkey should be assisted in carrying out a state-of-the-art EIAs for the planned dam cascade on the Coruh River. In this context, the sediment question has to be addressed and carefully studied.

Despite good relations, water cooperation between the two countries still suffers from a weak legal foundation and a largely absent organisational back-up. A permanent bilateral cooperative structure should be established to strengthen bilateral monitoring of the planned infrastructure.

Since both countries already cooperate in ongoing activities to improve the ecological state of the Black Sea, the Coruh River dispute and the assumed effects of the infrastructure on the coastal zones could be addressed within this framework. In addition, the erosion issue on the Georgian Black Sea coast demands a comprehensive and long-term approach for which the multilateral Black Sea Cooperation might be an adequate platform.

The ecological state of the Coruh River and the related coastal ecosystems should be subject to supplementary scientific studies and analyses. International support may be helpful in developing sufficient capacity and in supplying adequate resources.

Based on a careful assessment of the sediment issue, various technical cooperation measures could be designed and implemented in order to reduce negative downstream effects. The transfer of international experience coupled with a technical approach to sediment management in the context of dam building should be analysed.

Current cooperation suffers, inter alia, from a lack of reliable data. Activity could be targeted on the exchange of water flow and water use data, water quality etc. Although there are already several projects ongoing, the set-up and/or the improvement of a working and well-adapted monitoring infrastructure is urgently needed.

Kura-Aras river basin

The agricultural development strategy in the Kura-Aras catchment area should be designed in a sustainable and participatory manner. In addition, diffuse pollution from agricultural sources is a matter of concern that provides room for cooperation.

Current programmes and initiatives largely focus on the three south Caucasian states of Georgia, Armenia and Azerbaijan, with Turkey as a marginally (if at all) involved partner. Clearly, water quality is a prime transboundary issue between Georgia and Azerbaijan, but is relevant along the Turkish-Armenian and the Turkish-Iranian border too. Here, bilateral measures to control emissions from point and non-point sources could serve as a basis from which to start. In addition, Turkey and Iran should be invited to join ongoing transboundary efforts in hazard prevention.

Because of the outstanding value of biodiversity in the basin, there are a couple of internationally relevant nature protection issues that have not been adequately addressed yet; for instance basin-wide approaches to wetland protection, maintaining of ecological in-stream uses and transboundary protection areas.

In the long-term, uncoordinated realisation of all existing water infrastructure development plans in the riparian states, in particular in Turkey and Iran, would negatively affect water availability downstream and harm water-dependent ecosystems. Therefore, enduring dialogue between the riparians on infrastructure development should be facilitated. In this context, the inclusion of all riparians is crucial in the long run.

In general, there is a need to work towards a joint platform which includes all riparian countries, and to establish consultation mechanisms to harmonise basin-wide development efforts. This would also mean taking an integrative perspective on water development, use and protection, including in-stream flows.

Euphrates and Tigris rivers system and Orontes River

At present, it seems very unlikely that Turkey would welcome direct involvement by third parties in negotiations on transboundary water issues. However, there are some areas of action such as financial, technological and scientific assistance which might positively impact on transboundary water management.

As mentioned earlier, there is a serious lack of reliable data. If wished, joint initiatives to create reliable data related to surface and groundwater resources, could be promoted.

Using the programme of the Joint Communiqué (2001) as a template, support could be provided, for example in developing and implementing participatory irrigation management approaches in agriculture, designing strategies for efficient water use and for controlling salinisation of agricultural land etc. Moreover, a broadened agenda of cooperation could be facilitated for the sustainable use of the region's land and water resources.

Joint projects could be encouraged in water-related development fields such as energy, agriculture, environment and health. This strategy may be able to gather the riparians around a basic objective to deal with water management within the larger picture of overall socioeconomic development and integration of the underdeveloped regions in Turkey, Syria and Iraq.

Institutional capacity of the riparians' administrations to implement the recently signed Memorandum of Understandings should be strengthened.

Track two (unofficial, professional networks) initiatives such as the Euphrates Tigris Initiative for Cooperation (ETIC) should be supported as it provides necessary scientific cooperation to operationalise the cooperative agreements related to water and socio-economic sectors.

Since all countries are pursuing water resource development by means of infrastructure (dams), support can be provided on how to adequately deal with environmental and social issues in the planning and implementation phases of dams.

Annex

Annex Bilateral water agreements

Annex	Basin	Parties	Year	Title
1	Meric	Greece -Turkey	1934	Agreement pertaining to the construction of hydraulic facilities on both banks of the Meric-Evros River [own translation]
2	Meric	Greece Turkey	1963	Protocol on the improvements of the River Meric watercourse that constitutes a significant portion of the Turkish-Greek Thracian Border [own translation]
3	Meric	Bulgaria Turkey	1968	Agreement between the Republic of Turkey and the People's Republic of Bulgaria on the Cooperation of the Utilization of the Waters of the Rivers Flowing in the Territories of the Two Countries [own translation]
4	Meric	Bulgaria Turkey	1975	Agreement on Long Term Economic, Technical, Industrial and Scientific Cooperation between the Government of the Republic of Turkey and the Government of the People's Republic of Bulgaria [own translation]
5	Meric	Bulgaria Turkey	1993	Agreement on Assistance and Cooperation in the Field of Water for Reducing the Negative Effects of the Drought of 1993 [own translation]

(continued)

A. Kibaroglu et al. (eds.), *Turkey's Water Policy*,
DOI 10.1007/978-3-642-19636-2, © Springer-Verlag Berlin Heidelberg 2011

Annex	Basin	Parties	Year	Title
6	Meric	Turkey Bulgaria	1998	Agreement on Cooperation in the Fields of Energy and Infrastructure Between the Government of the Republic of Turkey and the Government of the Republic of Bulgaria [original extract]
7	Meric	Turkey Bulgaria	2002	Protocol of the Fifteenth Session of the Turkish-Bulgarian Joint Committee For Economic and Technical Cooperation [original extract]
8	Kura-Aras	USSR Turkey	1964	Protocol and its attachments on the meeting of the Turkish-Soviet Joint Commission pertaining to the construction of a joint dam on Arpacay (Ahuryan) [own translation]
9	Kura-Aras	USSR Turkey	1975	Cooperation Agreement between the Government of the Republic of Turkey and the Government of the Union of the Soviet Socialist Republics on the construction of a dam on the boundary river Arpacay (Ahuryan) and the Constitution of a Dam Lake [own translation]
10	Kura-Aras Coruh	USSR Turkey	1990	Agreement on the Cooperation for the Construction of Hidrotechnical Facilities for the Prevention or Correction of the Riverbeds of Arpacay (Ahuryan) Coruh River, Posof and Caksu Streams extending between the border stone number 41 through border stone number 450 on the Turkish Soviet Union border [own translation]
11	Tigris and Euphrates	Turkey Iraq	1946	Treaty of Friendship and Good Neighbourly Relations signed between Turkey and Iraq [own translation]
12	Euphrates	Turkey Syria	1987	Protocol on Matters Pertaining to Economic Cooperation between the Republic of Turkey and the Syrian Arab Republic [original extract]

Annex 1

Agreement pertaining to the construction of hydraulic facilities on both banks of the Meric-Evros River [own translation]

The Republic of Turkey on one hand and

The Republic of Greece on the other

Desirous of regulating the hydraulic facilities on both banks of the river Meric within the friendly spirit that exists between the two nations and have decided to conclude this agreement to this effect:

The President of the Republic of Turkey has appointed:

His Excellency Menemenli Numan Bey, Secretary General of the Ministry of Foreign Affairs,

The President of the Republic of Greece has appointed:

His Excellency Sakellaropulos, Extraordinary Envoy and Minister Plenipotentiary to Ankara.

After presenting their credentials of authority have agreed on the following issues:

PART ONE

Provisions and conditions pertaining to the facilities currently in existence

1) Facilities to be removed

   Is currently existing on the river and the facilities mentioned below shall be removed as soon as possible.

   1) The foundation piles of the ancient Decauville bridge located immediately at the downstream of the Marassia-Maras village.
   2) Two dams at the mouth of the river and in the vicinity Gemicikoyu-Gematu towards Turkey and the spurs under construction (section to of the Kuckuncukadasi-border limitation protocol).
   3) The mill dam carried away by water and located at the upstream of Edekoyu-Pactii and the spur just a little above the said dam towards Greece.
   4) The brick wall at the point where the Cayade delta forks out and in the vicinity of the Turkish checkpoint.

2) Facilities to remain in place:

   All the other facilities that actually exist on the river such as the protective wall, dikes, spurs, small spurs, etc., shall be retained as they are until preparation of the project (B) sets out the rules in respect to these facilities. It should be well understood that this preservation of the facilities should not imply in any way, the development that will carry the facilities beyond their lines of operation, or

modification thereof. It is also well understood that the term retention shall not imply the repair of the devastated spur. The measures taken on these facilities in order to continue to deride their intended use shall not mean improvement or extension of the existing facility. Such maintenance works may only be performed after notifying the other party ten days in advance of the actual date of work.
As the Turkish delegation had requested the removal of the closing dike in the direction of the mouth where the rive meets at the Kuleburga (Pythion) downstream and which obstructs the natural flow of the river at the border marking number 23, marked by the border limitation commission during 1925-1926 on one hand, the Greek delegation had requested the removal of the spurs at the Nazimbey Ciftligi shores, which gave the appearance of encroachment, it was decided to incorporate this matter in the review to be conducted in accordance with section 11 of project (b) of this agreement, and the 20 kilometer portion of this project shall be included in the review.
Leaving or removing of the dikes on the shores of Turkey facing Kurtbucagi (south of Poros-Kaldirkos) in project B shall be stipulated after the review stated in part II.

PART TWO

Hydraulic facilities to be constructed

1) Provided that prior notification of one month is given, determination of the type of facility each party is to build at its liberty.
   These facilities shall be in three types:

   a) Frontal protection walls of the river shores,
   b) Dikes,
   c) Spurs

   a) Provided that the base is not broader than 4 meters in the river bed as of the revetment, and does not exceed the water level of 50 centimeters at most the frontal protection walls may be made in any shape or form and out of any material. The thickness of the revetment may not exceed one meter. However, the situation is different for sustaining walls. In this event the thickness of the wall shall be calculated according to the material to be used and the pressure it will be exposed to.
   If wooden piles are to be used for frontal protection, the width of the facility may not exceed 3 meters from the shores when the level of water is at its lowest.
   Generally frontal protection may be erected at the places where the river is eroded or where there is the possibility of erosion. They shall in no way be erected at the convex or the tips of the shores.

b) The dikes shall be submersible, and proportionately short in length, shall be parallel to the river or inclined to the river. The purpose of the dikes is to protect the inhabitants thereat from floods. These shall be erected at steep shores at points that are threatened with being destroyed by floods. Likewise, permission shall be given for the construction of these dikes around the villages and at desired heights to protect the villages from flood waters It is well understood that Turkey is authorized to construct any type of hydraulic facilities in the area between the Arda Bridge and border marking number 23, granted that, such works shall be restricted in accordance with the international law and other laws that Turkey is a part of, if the work performed adversely affects the Greek territory.
c) Permission shall only be granted to build small spurs at the concave places caused by erosion. The maximum length of these spurs shall be 7.50 m at the place of erosion and this length shall gradually be shortened as the erosion decreases and shall be drawn to it at the place where the erosion begins or ends. These spurs shall be perpendicular to the low flow and their heights shall not exceed the natural floor on the shore and its tip shall be equal to the level of the low water. It shall be sloped above the low flow and shall be 45 degrees at the base. Spurs shall not be built at places where the river separates and where islets are constituted.

2) Hydraulic facilities to be built after the joint review of the gradual stages for the reinforcement of the main bed of the river.
Both States shall begin the reinforcement of the main bed coasts in the joint section by taking into consideration the effects caused by the main and the secondary tributaries flowing into the river. If a correction needs to be made in the bed of the river in the study conducted as a result of this process, such correction shall be made with an agreement concluded between the parties in this regard.
The said process shall be conducted in sequences at the places where the river separates the common watercourse from the source to the downstream of the river. Each section must be 20km long. The said process shall be conducted by both parties together and shall be completed within the period specified in the study prepared in this regard.
The drawing of topographic maps shall begin from the direction of the source and obtaining of the first maps starting from the Bulgarian border shall begin in the summer of 1934. The competent authorities of both parties shall mutually determine the details that will go into making the topographic maps through the use of precision triangulation and measuring equipment. The points of triangulation determined by the border limitation commission shall be included in this triangulation study. As the topographic maps are completed piece by piece, the advance drawing for each section or the whole part and the advance drawing shall be submitted to a third party expert. The advance drawings to be prepared by the said party shall include the execution schedule.
After the approval of the advance project by both Governments, each government shall prepare the final project based on the advance drawing for the place

(s) where the advance drawing pertains. These final projects shall be submitted to the approval of both governments and shall only be executed after such approval. The disputes that may arise in this regard shall be solved by means of an arbitrator.

3) Urgent actions that could be taken, through agreement, during the time to elapse until the implementation of project B.
Without prejudice to the procedures stipulated in part III of this agreement, urgent individual works may be executed by each of the contracting parties provided it is submitted to the approval of the other party beforehand.
4) A preparatory hydraulic facility contained in (Plan A) for the improvement of the Meric-Evros valley and which in any case is beneficial.
The study and construction of a large hydraulic facility for the general improvement of the Meric-Evros River that is the works to be performed by both parties for the protection of the land from large floods. The construction of such a facility shall be postponed because it does not seem possible to build one for a period of ten years.
Along with this, as it is necessary to conduct observation for all types of studies, and not only the studies for the above, both parties have decided to establish observation stations (rainwater measuring stations and water measuring stations) as of now.
The competent authorities of both parties shall conclude an agreement on the quantity of such stations, their types and locations and the preparation of the observations to be made there, their operation, and the manner in which the parties shall inform each other of the observations made at these facilities.

PART THREE

Procedures to be followed in the determination of the studies and the execution of the works stipulated in paragraph 3 of Part 2

The party that wants to conduct such work at its own shore shall prepare a technical study related with the work, in duplicate copies, and submit it to the other party. The party receiving such a request must reply within 3 months approving such a request without any conditions, or accept it with certain changes, or that it is totally against such a request. If the party receiving such requests unconditionally approves the request within 3 months or remains silent during that period, the party which has submitted this request shall become entitled to perform the works provided that the works are in accordance with the study submitted. All modifications that need to be made prior to the start of such work or found necessary during the performance of the work shall be notified to the other party in the same manner. If the party receiving the request denies concurrence or sets forth conditions, which cannot be accepted by the requesting party, the dispute to arise in this regard shall be solved amicably and if this is not possible it shall be referred to arbitration.

If the party that wants to initiate any work finds it beneficial, it may submit an advance drawing that the other party may use for its approval. Nevertheless, this shall not bind the other party receiving the request on its discretion to approve or decline. That is, the party receiving the request may abstain, temporarily from giving the approval even after receipt of the final drawing.

Obtaining of the topographic maps and the performance of the hydraulic observations for the performance of the studies mentioned above shall be subject to the rules stipulated in the following paragraphs.

1) Topographic maps:
   If it becomes necessary to obtain topographic data of the other party to conduct the above studies or for the information of the other party on the measuring and observation to be made thereat, or to obtain intersections in length and width throughout the length of the river, the concerned party shall submit a letter to request such a map indicating the extent and the scale of the topographic map. The party receiving the request shall reply to the other party within one month as of the receipt of such a request that it concurs with this request, and to determine a date as close to the date of request, or if the expenses involved with such a study is to be borne the requestor or the party receiving the request, or if the requesting party shall use its own personnel and equipment in preparing such a map. If the disputes to arise from the performance of such work cannot be resolved amicably they shall be referred to arbitration.
2) Observation of Springs:
   Each of the parties, on their own territories and at the bridges whose sections end on the territories of that party, may construct observations stations without seeking the approval of the other party and conduct observations thereat giving the time of the observation and the place where the observation is to be made ten days prior to the date of the observation.
   If one of the contracting parties wants to establish an automatic reading water station, it must first of all submit the drawing of such a station and obtain the approval of the other party. The party receiving the request is obliged to reply within three months as of the date of the request. The contracting party may request that modifications be made in the project or may abstain from giving its approval. In this event, the dispute shall be resolved through arbitration. If the party receiving the request fails to reply within the prescribed period, the other party may construe this as approval and may begin construction at the end of the specified period.
3) Measuring the speed of water:
   Either of the parties shall only be able to measure the flow of the Meric-Evros River with measuring equipment provided that the party wanting to make the measuring gives, a ten days prior notice, informing the other party about the nature of the measuring involved, the duration of the measuring, and the equipment and materials to be used.
   If it is desired to conduct the measuring with fixed equipment and tools, the procedure to be followed shall be the same as the one prescribed for automatic water measuring stations in paragraph 2 of part 3.

If it becomes necessary to obtain the length and width of the intersections, the procedures prescribed in paragraph 2 of this section for topography shall be followed. That is, the prior notification that needs to be given is limited to only ten days.

4) Miscellaneous provisions and conditions:

   a) The competent authorities of both, by working hand in hand with each other, may prepare and implement regulation that will allow the production of fish in the main waters of the river. This regulation shall contain provisions, which will prohibit the damming of the river waters, and change the course of the river and direct the removal of such items.
   b) Because the border marking number 24 was washed away by the current it shall be re-constructed in the summer of 1934.
   c) Because there are numerous trees and brushes piled up in the bed of Meric-Evros, which hinder the flow of the water and cause damages in the course of the river, there will be tree cutting in accordance with plan B and this work shall constitute one of the elements of the said plan. Because there is a greater urgency for the trees to be cut near Kuleliburgaz-Pythion where the railroad crosses the river on the Turkish side, the tree cutting should be performed without waiting for the implementation of plan (B).
   d) The competent authorities of both parties shall reach an agreement to remove the trees and their stubs, which hinder the flow of the river at some sections, as soon as possible.
   e) New plantation either directly or through planting shall only be permitted in the concave and eroded sections of the shores.
   f) As set forth in the various paragraphs of this agreement, letters which convey the purpose of the work to be performed and which do not require the approval of the party shall be submitted to the governor of Edirne or to the Meric-Evros commission.

PART FOUR

Performance of the work and controls on the maintenance of the facility

In accordance with the provisions in the preceding sections, the party concerned shall notify the other party of the starting date of the construction and maintenance works for which permission has been obtained and for the other party to control the works during the performance of the work and until the completion of it. This control shall be performed by an expert and when necessary a military officer shall accompany him in his checks. Such experts shall check, on-site, the facility to see if it has been constructed in accordance with the conditions stipulated in the agreement and the procedures set forth above and if they comply with the approved drawings.

The concerned party shall render all the necessary assistance to the party performing the work from the beginning until its completion.

In the event, the above mentioned commission, after the on-site inspection, determines that the work performed does not comply with this agreement or the approved drawings or the modifications decided to be made later, the concerned party after the receipt of the notification to this effect, is obliged to suspend the works it is performing or to demolish the facility or to modify it in the manner recommended.

All disputes that may arise from the interpretation or the implementation of this agreement by the Supreme Contracting Parties shall be referred to arbitration if the efforts for a mutual solution fails, and if an agreement cannot be reached the matter shall be referred to the International Court of Justice in accordance with article 22 of the Agreement between Turkey and Greece on friendship, neutrality and reconciliation, dated 30 October 1930.

If the concerned party is found to be right with the decision of the court, the work that was suspended on grounds that it did not comply with the provisions of this agreement or the drawing or the modification thereto may be resumed.

The expenses incurred because of the suspension or the demolition of the works, and reconstruction of the facility shall be borne by the party that caused the delay or demanded the demolition. The amount of compensation shall be determined by the court.

This agreement shall remain valid for a period of ten years as of the date of exchange of the ratification. Provided that notification is served six months prior to the expiration of this period, this agreement may be terminated by one of the contracting parties. If neither of the parties exercises this right it shall be automatically extended for another ten years.

This agreement was concluded in Ankara on 20 June 1934.

Sakellaropoulos Numan (Menemenli)

Annex 2

Protocol on the improvements of the River Meric watercourse that constitutes a significant portion of the Turkish-Greek Thracian Border [own translation]

In order to resolve the dispute arising from the execution of the hydraulic works on both banks of the Meric-Evros riverbeds, the Turkish-Greek technical teams comprised of the individuals listed below shall:

The Turkish side:

1. Arif Onat: Vice Director General, General Directorate of State Hydraulic Works.
2. Sadettin Acar: Deputy Chief, Projects and Construction Department
3. Mufit Kulen: Deputy Chief, Projects and Construction Department
4. Ihsan Baltaoglu: Deputy Regional Director, XIth Regional Directorate of the General Directorate of State Hydraulic Works.

The Greek side:

1. Nicolas Chorafas: Professor at the University of Thessalonica and Inspector at Construction Affairs
2. Stavros Triantaphiyllidis: Director of Hydraulic Construction Works of the Greek Ministry of Construction.
3. Kimon Kyriacos: Technical Advisor
4. Constantin Koungoulos: Western Thrace Construction Works Chief.

This team, duly authorized, convened at the General Directorate of State Hydraulic Works and held negotiations during 8 January 1963 through 16 January 1963.

Through continued discussions and review in depth of the works performed and to be performed the parties have come to an agreement in the following manner on the land to be exchanged resulting from this.

SECTION I

Exchange of Land

ARTICLE 1

Because of the construction works to be performed in order to prevent floods from the rise in the water level of Meric-Evros River, some sectors of the borderline shall be modified. Therefore, it has become necessary to exchange some land in the said sectors. In order to accomplish this exchange, the borderline determined by the Joint Border Commission, established during 1925-1926 in accordance with the

Lausanne Treaty, shall be taken as the basis. The land to be exchanged between the two countries shall be equal in area and in case of an ultimate exchange the values of the lands exchanged shall be taken into consideration.

ARTICLE 2

It was found appropriate to perform the land exchange in two stages. In the first stage, a portion of the land in the Enez sector and the portion of the land in the Ferre sector shall be mutually exchanged. In this manner, the borderline existing in the form of a river in this sector shall be significantly converted to a land border. The area of the land to be exchanged in the first stage shall be equal.

The final determination of the borderline shall be in the second stage and shall be in the following manner.

The starting point of the land border line in the Enez region shall at a maximum of 50 meters to the west from the Ipsala barrier and the Meric-Evros riverbed intersection point, and shall terminate at 100 meters to the west of 0+000 of the HARZA Project barrier. There will be no change in the said border between the source and the Ferre intersection. The existing borderline between Ferre upstream intersection and the Peplos downstream intersection shall not be changed. The existing borderline up to the Saricaali (Tyhion) intersection that is not shown in the HARZA project but found appropriate by the parties during the negotiations shall remain unchanged. And finally, the Saricaali (Thyion) intersection, which will be dealt with in the near future, shall remain as a river border.

Both parties have agreed to have the borders drawn in the manner specified above and have agreed to exert all their efforts to accomplish this.

Without prejudice to the provisions in article 23, the exchange made in the first stage shall be final. Only the lands to be exchanged in the second stage, in order to insure equality in their areas, may be altered in the Enez sector. In any case, the borderline shall remain to the west of the barrier to be constructed in this sector.

ARTICLE 3

The area of the land to be exchanged between the parties in the Enez and Ferre sector shall be approximately 1,750 decares. This figure may increase or decrease by 10 percent in case a new route other than the one specified in HARZA project is used. A significant portion of the land that will be transferred by the Greek side in Enez for the land in Ferre shall be to the east of the barrier proposed in the HARZA study. The borderline to be determined according to the new route of the barrier may exceed the barrier route specified in the HARZA Project at most 50 meters to the west, in certain sectors. 9/10th of the land to be exchanged by Turkey at Ferre shall be to the west of the barrier route to be constructed according to the HARZA

Project and the remainder shall be to the east. This final unprotected section parallel to the barrier shall be equal in thickness throughout its length.

ARTICLE 4

Both governments shall exert their best efforts to complete all the necessary formalities for the realization of the land exchange following the entry into effect of this protocol. To the extent of enacting a law, to the constitution of a joint commission to draw the border line, if necessary and to erect barbwires throughout the border line, with the expenses split in half between the parties. Both governments agree to use all their means to accomplish the practical and legal aspects of the first stage by 15 July 1963.

ARTICLE 5

Along with the works to determine the borderline, a joint team, comprised of experts, shall begin to work on the assessment of the value of the lands to be exchanged. If there is a difference in value, the party owing the money shall deposit the amount in question in foreign currency to the account number of the recipient with a Swiss bank and this amount shall be blocked until the exchange in the second stage is completed. The said account shall be independent from the economic, trade or other relations between the two countries. If a dispute arises between the members of this team, upon the request of one of the parties, the matter shall be referred to a General Engineer to be appointed by the French Ministry of Agriculture.

ARTICLE 6

The Greek government undertakes to re-construct the Peplos barrier according to the HARZA project in the year 1963.

ARTICLE 7

The Turkish Government, at its discretion, shall be entitled to fortify the Peplos enclosing barrier in the dimensions specified in the HARZA project, with its own measures of protection.

ARTICLE 8

The Greek Government shall fully retain the right to establish a drainage network and a pumping station in the Peplos covered area to be created in this manner.

ARTICLE 9

The Greek Government, at its discretion, shall be fully entitled to construct a barrier according to the dimensions specified in the HARZA project or smaller, in the sector to be left to it in the Ferre sector.

The Turkish Government, at its discretion, may erect a barrier at Ferre and may establish a pumping station, within its own boundaries, to drain the water from the covered area to be created in this manner.

ARTICLE 10

The Greek Government fully retains the right to establish a protective barrier within the Greek soil and along the borderline in the Karaagac sector without having to pay any compensation to Turkey for any reason whatsoever.

ARTICLE 11

As of the entry into effect of this protocol, either of the parties, by submitting their applications in writing within 15 days in advance, may begin construction at Enez and Peplos for the Government of Turkey, and Ferre for the Government of Greece, at the site and in the size specified in the HARZA project. In such an event, it shall be assumed that permission has been automatically granted for one party to enter the land of the other and to begin the construction works at the expiration of the said period. If one of the parties does not comply with the provisions of this article the other party shall become entitled to wreck the construction made by the other.

ARTICLE 12

In the second stage, the exchange of land shall be accomplished within a reasonable period of time. The remaining portion of Ferre sector and the Saricaali (Tyhion) sector shall be exchanged against the land in Peplos. That is, the land in Ferre not exchanged up to the Ferre intersection and the Saricaali intersection in the west of Saricaali region shall be exchanged for the portion in the Peplos sector.

The route of the Saricaali (Tyhion) intersection to be dealt by Turkey shall be determined with an agreement to be concluded between the authorized administrations of both countries.

For the determination of the said route:

a. The area of land Turkey accepted to abandon to Greece in Saricaali may not be less than 1500 decares and may not exceed 2,500 decares.
b. The intersection shall be suitable for hydraulic conditions.
c. The intersection shall comply with the principles determined in the Harza study.

In keeping with article 1 of the protocol; in order the ensure the equality of the lands to be exchanged, it shall be possible to move the land border established at Enez to the east and to the west in order to reach the final decision on the border. The technical problem of the Tatarkoy (Vissa) intersection, one of the points discussed in the negotiations, shall be incorporated into the lands that will be exchanged during the second stage in the event it is settled by abandoning a portion of the Turkish soil to Greece or vice versa.

ARTICLE 13

The final marking of the land border and the assessment of the value of the land subject to exchange and compensation shall be subject to the procedures stipulated for the first stage. Along with the compensation to be determined, the values of the pumping stations and drainage network, and the difference from the costs stipulated in the (Harza project) for the barriers constructed in Peoplos and Ferre sectors up to date, shall be taken into consideration. Likewise, the blocked account mentioned in article 5 shall be taken into account in the calculation of the said amount.

ARTICLE 14

The finalization of the exchange of land in the second stage shall be made with the deposit in the Swiss bank mentioned in article 5, of the differences in value of the lands subject to exchange.

SECTION II

Technical Issues

ARTICLE 15

Tatarkoy (Vissa) intersection:

For now, construction of the Tatarkoy (Vissa) intersection shown on the Harza Project is not necessary. Because the riverbed has changed in such a way it will not require the construction of this intersection shown on the Harza project.

In spite of this, there is the possibility that the riverbed may damage the Greek costs and therefore, this matter needs to be solved. Therefore, a group of technicians from both sides need to conduct in-situ examination of the situation and need to find an appropriate solution and if need be prepare a project, and render a decision on how the construction should be done, and come up with results and these results need to be finalized after the exchange of Notes between the Ministers of Foreign Affairs of both parties. While a solution is being searched for this matter, action must be taken in accordance with the decision of the Joint Border Commission, which met during 28 June-7 July 1959. If the decision reached requires the exchange of land, this matter shall be taken into consideration in the second stage of the land exchange because of the implementation of the Harza Project specified in article 12.

The Joint Commission should have completed its work by September 1963.

ARTICLE 16

In order for Enez, Ferre and Peplos intersections to function as envisioned in the project, it has been agreed that the covering of Enez and Peplos shall be performed by the Turkish side and the Greek side shall do the covering of Ferre, at a time convenient for them after the entry into effect of this protocol. Provided that the locations and drawings of these coverings are in conformity with the Harza studies, there is no need to exchange the drawings or have them approved prior to the construction. Both sides may begin the construction works without requesting additional permission, but by notifying the local organization 15 days prior to the actual start of works. Both parties are at the liberty to construct the covering 50 meters from the tip of the counter-party and to work exclusively on the construction.

If repairs are required in the said covering they may exercise this right by notifying the local administration of the other party within 15 days prior to the start of work or perform the work at a later date.

ARTICLE 17

Control of Fuse-plugs

Since the construction works are coming to an end, it has been decided to mutually check the plans and altitudes of the barriers and the fuse plugs and to determine if they are in conformity with the guidelines contained in the Harza project. This mutual control shall be performed by a team of mixed technicians to be constituted within three months starting on 1 July 1963.

The details on the implementation of this article are found in enclosure 1.

ARTICLE 18

Summer barriers:

The parties have jointly decided to conduct studies on the protection of the lands in between the barriers, and places not protected by barriers and to improve the conditions of these arable lands. Likewise, it has been decided that the measures to be taken for these soils were to protect these soils from the possible floods that occur between the months of May and September and not against major floods. In that case, it is necessary to jointly decide on the extent of the floods for which measures need to be taken. The technical teams shall conduct the hydrological study of the river to this end and shall notify each other of their findings. Construction shall be made at the place found appropriate and work shall be performed on the advance drawing of the barriers.

A meeting at higher levels shall be conducted between the parties within a reasonable period of time for the control of the advance drawings and a protocol to this effect shall be signed. This protocol shall specify the priorities of the summer barriers and the start and end of the works on the barriers.

The details pertaining to the hydrological studies and the advance drawings are given in enclosure 2.

ARTICLE 19

Works performed without the consent of both parties:

Works performed by both parties without obtaining the prior consent of each other are listed below.

a) Turkish side:

1. Spurs constructed in three groups near Soufulu (Soufli) without notifying the other party and without obtaining their consent.
2. The existence of a poplar wood forest in the Ipsala bridge sector, on Greek soil that reduces the flow the river creates an undesired situation for the Turkish coast.
3. Allegations were made that barriers not specified in the HARZA project were built.

b) Greek side:

1. A drainage canal was built by the Turks from the Nazimbey islet to let the waters flow into the Meric river which is in violation of the border drawn by the Joint Team according to the Lausanne Treaty,
2. Constant loading of the excavation soil from the drainage canal specified in the HARZA project in the direction of Uzunkopru into the waters of the river, which prevent the free flow of the water.
3. Preservation of the cofferdams constructed at the start of the Enez intersection,

After the review of the issues enumerated above, it was observed,

a) Both parties should adhere more closely to the provisions on the water construction of the 1934 Turkish-Greek Agreement,

b) That any construction which obstructs the flow of water on the river bed should be avoided without exchanging of drawings,
c) It was found appropriate to review all the subjects and solve them so that they will not cause damages to both parties by a local team of experts from both parties,

ARTICLE 20

Organizing the riverbed:

Both parties have agreed that there is a need to regulate the bed of Meric remaining between the downstream of Ferre intersection and the upstream of Enez intersection.

This regulation shall be conducted according to the drawings prepared by a joint team to be constituted for this work, based on the flow rate of the river to be observed. The parties may perform these projects either jointly, or separately, or have one of the parties perform the work and the other contributes half of the expenses, after the ratification of the projects by the parties. During the works on the axis of the river, the condition that the area of the land which will pass to the other party shall be equal and that the area of these lands, for both parties, in any case may not exceed 1000 decares, because of the change in the axis of the river and the border. Because the said lands are equal in value, only their areas shall be taken into consideration in the exchange.

ARTICLE 21

Provision of irrigation water and its quantity and control:

It was observed that the irrigation problem of the plains on both sides of Meric-Evros River was not sufficiently studied by the HARZA project.

Therefore, the following issues have been found appropriate for the beneficial use of the waters from the river by both sides.

1. Measuring of the flow rate shall be continued by both parties both in the existing hydrometric stations and those that are to be installed.
2. Both sides shall continue the studies and preparation of projects on the irrigation facilities.
3. A senior level technical meeting shall be held in the month of September 1963 to render decisions on the said studies and the projects.
4. The water drawing facilities for irrigation purposes shall either be through a regulator to be built together by the parties, or in the form of water socket erected

by the concerned party on the coast. However, this unilateral construction of the socket shall be subject to an agreement concluded between the parties and shall not cause damages on the coast of the other party.

5. Until the completion of the issues mentioned above, both parties shall continue with the irrigation practices they have been applying up to now.

SECTION III

Recommendations

ARTICLE 22

Both parties have made the following recommendations on the matters that were not thoroughly discussed:

1. Both parties recommend that the technical delegates come together on certain days, regardless of the fact that there are agenda items or not, and to hold extraordinary meeting in case of emergency situations.
2. The Greek delegation has recommended that poles be erected to prevent the disputes to arise because of the change in the riverbed and to prevent border issues to the extent possible throughout the borderline determined by the Lausanne Treaty Mixed Commission and the corrections that may become necessary from the implementation of this protocol. The poles shall be erected in a manner to prevent danger from floods.
3. The Turkish delegation has recommended that land border in the Enez region be extended up to the river to prevent possible border issues. Because, when the HARZA project is implemented in full in this sector, the mouth of the river shall move to the Enez intersection.

Both parties decided to refer the above-mentioned recommendations to their governments to be reviewed in the future.

Final Provisions

ARTICLE 23

This protocol shall enter into effect upon the ratification by the concerned governments and through the exchange of notes between them.

However, the provisions pertaining to the exchange of land or abandonment thereof, shall enter into effect after their ratification in accordance with their respective constitutions and the pertinent laws.

The following individuals, whose signatures appear below, have prepared and signed the protocol in this regard.

This protocol was prepared in Ankara in duplicate copies on 19 January 1963.

Turkish Delegation	Greek Delegation
1. Arif Onat	1. Nicolas Chorafas
2. Sadettin Acar	2. Stavros Triantaphyllidis
3. Mufit Kulen	3. Kimon Kyriacos
4. Ihsan Baltaoglu	4. Constantin Koungoulos

Enclosure-1

The manner of checking the plans and heights of all the fuse plugs and barriers by both parties:

1. Taking into consideration the existing trigonometric symbols, one or several mixed technical teams shall perform the following measurements:
   a) The planning of the barriers according to the Harza study,
   b) The planning of the length of fuse-lugs, their locations and number as specified in the Harza study,
   c) Check the altitude of the barriers and fuse-plug sections,
2. The mixed technical teams mentioned above should be constituted by 1 June 1963 and should begin their studies. By this date, the parties will have mutually granted permission to work in the region of the other party and to roam there freely.
3. It has been determined beforehand that the in-situ studies shall last for three months.
4. The results of the measuring to be conducted by this mixed team shall be presented to the authorized administrations of both parties within two months as of the start of fieldwork.
5. In case of any disputes, the matter shall be referred to another mixed commission. The members of this commission shall be determined by the ministries of the two countries. The commission shall evaluate the dispute from both sides and shall determine its effect for proper functioning as stipulated in the Harza Project.
6. The said mixed commission shall present its findings and remarks in the form of a joint protocol and shall present this to the authorized administrations of both parties. In case of serious disputes, the concerned ministries of both parties shall exchange letters to make the necessary changes.
7. The party that caused the dispute must make its calculations according to the changed indicated in the letter of the other part's ministry and the remarks of the Joint Technical Commission.

Enclosure 2

About the determination of the locations where summer barriers could be erected (their heights and their distance from the shores of the Meric-Evros River), the purpose and the method of preparing the hydrological studies

A) Hydrology:

Purpose:

1. Every year, during the period of April through November the hydrometric measuring data of the size and frequency of the floods observed from the Edirne and Pythion intersection and Ipsala and other shores shall be compiled: As a result of this compilation, the largest flood observed during this period and its possible occurrence shall be determined. Obviously attention shall be paid in the compilation of data on the level of water at the above-mentioned sites. These figures should be available with the graphic and theoretic curves of water consumption.
2. Compilation of all the rain observations by Turkey and Greece in the Meric-Evros river basins, and if possible, efforts should be exerted to obtain the observations belonging to the Bulgarian side of the river basin.
3. Determination of the of the maximum flow rate of the river, through the application of one of the known methods (Analytic or American), based on the data compiled on rain and the correlation of the hydrometric measuring during the period mentioned. This method is applicable for the flow rates at the three points mentioned below:

   a) Karaagac borderline
   b) Pythion bridge
   c) Ipsala Bridge.
   If sufficient results cannot be obtained with these methods because of the lack of the necessary observations, it shall suffice to make an estimate of the possible consumption during the said period with the existing data.

4. After all of these, assessment shall be made of the economic expenditures for the protection of the summer barriers.
5. The estimated capacity of the current low riverbed shall be calculated based on the topographic sections obtained at intervals of 20 m starting from Ipsala bridge to the Karaagac border.
6. A five-month period, as of the entering into effect of this agreement, has been determined, for the compilation of the existing data and to reach the results mentioned above.
7. Both parties must pass the information available to them to the other party. Pursuant to this, the concerned administrations of both countries shall prepare appropriate studies at the points mentioned above. This study shall be prepared and submitted in English for the other party to review it.

8. A technical team shall be constituted two months later to meet at Edirne and Dedeagac to discuss the results obtained by both parties. The purpose of these meetings is to present the following issues to the authorized administration of both parties.

   a) The consumption involved with the summer barrier project,
   b) Their distance from the shores of the river
   c) Their altitudes
   d) The sections of the barriers and manner of construction, (Makta, fuse-plug).

Places where summer barriers shall be constructed:

1. Both parties are obliged to show the sites where they want protection throughout the summer season, as a result of the 5 month period hydrological study, on map of 1:50,000 scale.
2. The team which will review the hydrological studies shall indicate these points on the map and the protocol to be prepared in this regard shall be submitted to the respective authorized administrations.
3. It is desired that the respective administrations notify each other of their final decision within 3 months following the date of their decision.
4. Based on this concurrence, both parties shall prepare a joint program for the implementation of the project for the protection of the summer barriers. This program shall be ratified by the both governments.

Annex 3

Agreement between the Republic of Turkey and the People's Republic of Bulgaria on the Cooperation of the Utilization of the Waters of the Rivers Flowing in the Territories of the Two Countries [own translation]

The contracting parties have agreed on the following issues: by informing each other, in general terms, on the facilities built or envisioned to be built on the rivers crossing both countries, and on Meric, Tunca, Degirmendere (Veleka) and Rezve rivers which constitute the border and the quality of water in these rivers;

By acknowledging the necessity for close cooperation in the use of these waters flowing in their respective countries for irrigation and other needs which are of vital importance for the economic development of the respective countries and which require necessary measures for protection against adverse consequences from floods and icing;

By indicating that development of the water resources through the application of science and technology are important elements for the welfare of their people;

Based on the principles of international law and good neighborly relations.

ARTICLE 1

The Republic of Turkey and the People's Republic of Bulgaria shall cooperate in examination and studying the facilities to be built and operated on the rivers flowing through territories of two countries which will be beneficial to both parties.

ARTICLE 2

The contracting parties have agreed not to inflict serious damages to each other by constructing and operating facilities on the rivers flowing through their territories.

ARTICLE 3

The contracting parties have agreed to exchange information on floods and icing instantly.

Additionally, the contracting parties agree to exchange hydrological and meteorological data on the rivers flowing through their respective countries.

The procedure of communication and exchange of data in this regard shall be determined with a technical protocol to be signed between the two contracting parties.

ARTICLE 4

If one of the parties requests the compilation, preparation and delivery of the data and information mentioned in article 3, and required only by the party making the request, the party making the request shall pay for the expenses made by the other party complying with the request.

The balances of these expenses shall be cleared every year. Formula in this regard shall be prepared in the protocol mentioned in article 3.

ARTICLE 5

The agencies mentioned in article 6 shall act in accordance with the principles of this Agreement by signing separate implementation agreements for each individual undertaking that will provide new mutual benefits and conditions to the contracting parties.

ARTICLE 6

Following the entering into effect of this agreement, the contracting parties, within three months, shall notify each other of the addresses of the agencies authorized to implement this agreement.

The meetings of the agencies mentioned in this article shall be determined and arranged through diplomatic channels.

ARTICLE 7

Turkish and Bulgarian languages shall be used in the meetings. The documents to be prepared jointly (technical data, protocols, decisions, etc.) shall be prepared in both languages.

The information, data, documents, letters etc. to be exchanged between the contracting parties shall be prepared in the language of the party submitting such documents.

ARTICLE 8

The disputes that may arise from the implementation of this agreement shall be referred to the Joint Turkish-Bulgarian Commission, comprised of experts of both parties in equal numbers.

Through diplomatic channels, the contracting parties shall inform each other about the appointment of the experts to the Joint Commission as well as the place, date of the meetings and related issues on the works of the Joint Commission.

If the Joint Commission cannot reach an agreement on the matters they are reviewing, these matters shall be resolved through diplomatic channels through negotiations between the contracting parties.

ARTICLE 9

This agreement shall be subject to ratification and shall enter into effect thirty days after the date of exchange of the ratification documents in Sofia.

This agreement has been prepared and signed in duplicate original copies in Turkish and Bulgarian in Istanbul on twenty-three October of the year one thousand nine hundred and sixty-eight, and both texts are equally valid.

Annex 4

Agreement on Long Term Economic, Technical, Industrial and Scientific Cooperation between the Government of the Republic of Turkey and the Government of the People's Republic of Bulgaria [own translation]

With the desire to expand and deepen the benefits of mutual cooperation in economic, technical, industrial and scientific fields;

With the belief that the development of this cooperation will accelerate the economic and technical progress of both countries;

The Government of the Republic of Turkey and the Government of the People's Republic of Bulgaria have reached an agreement on the following issues:

ARTICLE 1

In order to intensify and diversify their economic relations, the contracting Parties have agreed to develop and strengthen the economic, technical, industrial and scientific cooperation to a maximum extent.

ARTICLE 2

The Parties shall facilitate the initiatives of the Turkish and Bulgarian agencies and enterprises to sign contracts and shall assist such organizations in ensuring that the said contracts are in accordance with this Agreement and the laws in effect.

ARTICLE 3

Both parties have observed that the cooperation covered under article 2 may notably be conducted in the following manner.

- Establishment of new industrial facilities and the extension and modernization of the facilities exiting in the two countries.
- Joint production and marketing of the items to be consumed in the markets of the two countries or in third country markets.
- Establishment of mixed production, commercial and transportation companies.
- Improvement of economic relations through the facilitation of transit of goods, materials and passengers travelling through the respective countries and cooperation in the fields of transportation and communication.

ARTICLE 4

The economic, technical, industrial and scientific cooperation covered under this agreement shall be accomplished through individual contracts to be signed between the concerned agencies and companies of the two countries and these contracts shall specify the conditions and methods of cooperation. When necessary, the said contracts shall be approved by the competent authorities of the two countries.

ARTICLE 5

The cooperation between the concerned Turkish and Bulgarian enterprises and organizations shall be facilitated in all the fields of economy and particularly in the following fields:

- Machinery production and metallurgy industry,
- Production, preparation and marketing of the agricultural and agricultural industry products produced in the two countries and cooperation in agricultural economy and food industry including the related methods and procedures,
- Energy production and irrigation, including the joint use of the waters flowing through both countries for energy production and irrigation purposes,
- Electronics, chemicals, petrochemical and medical industries,
- Transportation and communication,
- Tourism
- Commercial exchanges and other fields, which are of common interest to both countries.

ARTICLE 6

The Contracting Parties have decided to assign the Turkey-Bulgaria Joint Economic Commission, established with the letters exchanged in February 1972, to monitor the implementation of this Agreement.

This commission, henceforth to be referred to as the Turkish-Bulgarian Joint Committee for Economic and Technical Cooperation, is assigned with finding solutions for the problems that may arise during the implementation of this agreement, to make appropriate recommendations to their respective countries and to identify new possibilities of cooperation.

ARTICLE 7

The Contracting Parties shall realize the economic, technical, industrial and scientific cooperation, which constitute the subject of this agreement, within the framework of their respective laws in effect.

ARTICLE 8

This Agreement shall enter into effect on the date the Contracting Parties exchange Notes stating that the agreement has been ratified in accordance with the national laws of the respective countries, and shall remain valid for a period of 5 years.

Unless one of the parties notifies the other in writing three months prior to the expiration of this period that it wants to terminate the agreement, it shall be automatically extended on annual basis.

ARTICLE 9

If the contracts signed within the framework of this agreement have not been completed by the time this agreement expires, the provisions of this agreement shall remain valid for these contracts until they are completed.

This agreement has been prepared in two original copies in French in Sofia on 13 September 1975.

Government of the Republic of Turkey

Government of the People's Republic of Bulgaria

Annex 5

The Agreement between the Government of the Republic of Turkey and the Government of the Republic of Bulgaria on Assistance and Cooperation in the Field of Water for Reducing the Negative Effects of the Drought of 1993 [own translation]

The Government of the Republic of Bulgaria and the Government of Turkey mentioned as 'the parties' below have agreed on the following issues,

Based on Friendship, Good Neighborly Relations, Cooperation and Security Agreement between Bulgaria and Turkey,

On the acknowledgement of the need for a close cooperation in the use of the waters of the rivers crossing their countries as envisioned in the Agreement signed in 1968,

By taking into consideration the need for cooperation to alleviate the severe consequences of drought suffered by both parties due to the significant decline in the water level in the said rivers crossing their countries,

In response to the request by the Government of the Republic of Turkey due to the bad situation in the Meric valley on the Turkish side.

ARTICLE 1

The Republic of Bulgaria, on a one time basis and limited to the year 1993, shall provide additional water to the Republic of Turkey from river Tunca in accordance with the Volume and Graphic specified in the enclosure.

ARTICLE 2

1. The Republic of Turkey shall allocate US$ 0.12 per m3 of water provided in compensation for the measures taken by the Republic of Bulgaria to provide the water.
2. The manner and type of compensation and the graphic of allocation shall be determined between the parties by 5 September 1993 latest after the entry into effect of the Agreement.
3. If the parties cannot reach an agreement in accordance with paragraph 2 of this article, the amount determined in accordance with the conditions specified in paragraph 1 of article 2 shall be deposited by the Republic of Turkey in the bank account number designated by the Republic of Bulgaria.

This Agreement shall enter into effect on the date of the exchange of Diplomatic Notes containing the text of the Agreement.

Enclosure-1

ADDITIONAL VOLUME AND GRAPHIC OF WATER TO BE PROVIDED FROM RIVER TUNCA IN ACCORDANCE WITH ARTICLE 1 OF THE AGREEMENT.

1. The Republic of Bulgaria shall allocate additional water from River Tunca in the manner specified below.
   The allocation of water shall 25 m^3/sec for five days or total 10.8 x 10(6)m^3, for 21 data 12.3 m^3/sec or total 21.8 x 10(6) for a total of 32.6 x 10(6) m^3 additional water.
   With the proposal of the Republic of Turkey the said flow regime may be changed without exceeding the total volume.
   With the change in situation or meteorological conditions Turkey may request the suspension of the provision of additional water.
   In this event, the water flowing in within 48 hours shall be accepted as additional water provided by Bulgaria and shall be taken into consideration in the calculation of the compensation.
   Changes may not exceed 20 percent and new changes may not be proposed before an interval of 5 days.
2. In the determination of flow in river Meric and its tributaries the basic flow under natural conditions shall be "de facto" accepted in principle, and this amount shall be accepted outside of the calculation. This flow rate is accepted as 5 m3/sec for River Tunca during the month of August.
3. 20 percent of the water coming from JREBCEVO Dam shall be lost. This loss shall be shared as 10 percent each between the parties.
   After the calculation of the water provided at the border of the Republic of Turkey the water, exceeding 5 m^3/sec, shall be increased by 10 percent and this shall be compensated in accordance with article 2 of this agreement.
4. The amount of water that passes shall be determined by means of the flow observation stations located in both countries and at the vicinity of common border. The results obtained shall be audited at certain intervals or upon the request of one of the parties.

Annex 6

Agreement on Cooperation in the Fields of Energy and Infrastructure between the Government of the Republic of Turkey and the Government of the Republic of Bulgaria [original extract]

[...]

ARTICLE 3

The Parties agree that the cooperation envisaged in Article 2 [Article 2: continuation of the supply of electricity to increase gradually to the level of 4 billion kWh per year, reaching this level in the year 2002 and continuing with the same level until the year 2008] shall be effected with the understanding to allow Turkish companies to participate in joint companies in the realization of such infrastructure and energy projects as the hydro-energy station "Gorna-Arda" and the "Orizovo-Kapitan Andreevo" part of "Maritza" highway at the territory of Bulgaria, related to the construction of energy power stations, highways and motorways, in conformity with the Bulgarian laws and regulations in force.

ARTICLE 4

The Parties, within the framework of their competence, shall provide the necessary assistance to the respective institutions and entities for the conclusion and implementation of the contracts for the supply of electricity as well as for the infrastructure and energy projects at the territory of the Republic of Bulgaria. (...)

[...]

Done in Ankara on November 4, 1998 in two original copies in English.

Annex 7

Protocol of the Fifteenth Session of the Turkish-Bulgarian Joint Committee For Economic and Technical Cooperation [original extract]

The Fifteenth Session of the Turkish-Bulgarian Joint Committee for Economic and Technical Cooperation (hereinafter referred to as the Joint Committee) established in accordance with the exchange of letters dated February 9, 1972 was held in Sofia between January 22-23, 2002.

The Turkish Delegation was led by H.E. A. Mesut Yilmaz, Minister of State and Deputy Prime Minister of the Republic of Turkey and the Bulgarian Delegation was led by H.E. Kostadin Paskalev, Deputy Prime Minister and Minister of Regional Development and Public Works of the Republic of Bulgaria.

I. Trade and Economic Relations

[. . .]

5. Other Matters

[. . .]

Environment

The two Sides agreed to enhance environmental cooperation in the following fields:

- Water and water related ecosystems and underground water protection;
- Collection, transportation, recovery and disposal of municipal wastes and hazardous wastes;
- Integrated management of coastal zones and marine environment protection;
- Exchange of information of mutual interest regarding environmental protection in both countries;
- Ecological education and scientific cooperation;
- Air quality management in settlement areas,
- Reduction of emissions arising from industrial establishments;
- Environmental impact assessment and strategic environmental assessment implementations;
- Sustainable urban planning principles and implementations.

II. Energy and Infrastructure

[. . .]

The Turkish Side reiterated its request for the establishment of a Joint Technical Working Group to examine the possibilities of constructing the Suakacagi Dam on the Tunca/Tundja River. The Bulgarian Side undertook to respond to the Turkish Side at its earliest convenience.

Both Sides agreed to continue and to expand the exchange of hydraulic information that is highly important for flood protection (high water discharge from dams, water level of dams on Meric/Maritza, Arda and Tunca/Tundja rivers, snow deposits and rainfall data etc.). Both Sides decided to continue the periodical activities of the Technical Working Group that has been established under the Agreement signed in 1968 to exchange information related to flood protection and to establish the necessary measuring devices.

The two Sides agreed to hold the next session of the joint committee for Economic and Technical Cooperation in Ankara at a date to be determined through diplomatic channels.

Done and signed in Sofia on January 23, 2002, in two originals in the English language, both texts being equally authentic.

FOR THE GOVERNMENT OF THE REPUBLIC OF TURKEY	FOR THE GOVERNMENT OF THE REPUBLIC OF BULGARIA
A. Mesut YILMAZ	Kostadin PASKALEV
Minister of State and Deputy Prime Minister	Deputy Prime Minister and Minister of Regional Development and Public Works

Annex 8

Protocol on the meeting of the Turkish-Soviet Joint Commission pertaining to the construction of a joint dam on Arpacay (Ahuryan) [own translation]

Participants in this Turkish-Soviet Joint Commission:

Republic of Turkey	U.S.S.R.
Chairman: Neset Akmandor	Chairman: Bagramian G.A
Asst.Chairman: Arif Onat	Asst.Chairman: Melnikov V.M.
Member: Nazmi Karatekin	Member: Voronine A.L.
Member: Ayten Aydin	Member: Abramian A.A.
Member: Refik Akarun	Member: Avakian K.A.
Member: Turgut Dincer	Member: Grigorian S.O.
Member: Hidayet Turanli	Translator: Youri Savostianov
Member: Mehmet Palamutoglu	
Member: Saip Anadol	
Member: Ismail Hakki Demirel	
Member: Ibrahim Metiner	
Member: Ekrem Arikan	
Translator: Mehmet Kurbanoglu	

The Joint Commission comprised of the members whose signatures appear below has reached the following conclusions:

A- The height and type of the Arpacay (Ahuryan) Dam:

ARTICLE 1

The Soviet delegation, in consideration of the economic feasibility and the project prepared by them, has proposed to jointly construct a reservoir with a storage capacity of 440 million m3 volume on ARPACAY (AHURYAN).

The Turkish side, based on the estimate that 350 million m3 of water may be released to Arpacay (Ahuryan) from the Kars stream after the development of Kars and Cildir plateaus proposed that a reservoir with a capacity of (280-300) million m3 be constructed.

The Soviet delegation accepted the Turkish delegation's proposal in this regard.

The parties have agreed not to demand any claims from each other in the event they were wrong in their estimates and the water released from both countries to Arpacay (Ahuryan) was less in quantity.

The calculations of the dam was made based on the estimates that on an average 350 million m3 of water shall be released annually from the Turkish territory and on an average 150 million m3 of water shall be released annually from the Soviet territory and that an average of 50 million m3 of water shall be released from the mid-basin annually to Arpacay (Ahuryan).

The Turkish side shall provide the graphics of the water flowing from their territory to Arpacay to the Soviet side who will be preparing the project.

ARTICLE 2

The height, location and the volume of the reservoir shall be determined and decided in final form after the preliminary studies by taking into consideration the maximum amount of water that may flow into the dam from the Kars stream and the water contribution from the Soviet side with the capacity to regulate an amount of 500 million m3 of water, with the related studies, research and other data and the water inflow and outflow graphics.

ARTICLE 3

The Soviet side proposed that that the dam to be built on Arpacay (Ahuryan) should primarily be made of concrete based on the studies conducted by the Soviets in this regard in 1945 and the available data. At the end of the discussions the Joint Commission agreed to make a final decision on the type of the dam after obtaining additional data and conducting further research and during the preparation of the advance planning if the dam should be made of earth, rock fill, concrete or arch type by way of comparison.

B- Preparation of the Project:

ARTICLE 4

The drawings on the Arpacay (Ahuryan) dam shall be prepared in three stages first for the advance planning, second for the preparation of the final drawing and third for the implementation drawings.

ARTICLE 5

The advance drawing shall be completed within four months as of the entering into effect of this Protocol, in keeping with the schedule enclosed to the protocol and furthermore shall contain the instructions for the use of water at the upstream and downstream of the reservoir and shall be submitted for the governments' approval.

ARTICLE 6

Within two months as of the delivery of these advance drawings, the official agencies of both parties shall notify each other on whether they accept the drawings as they are or they want to make changes in them.

In order to make the changes the Turkish –Soviet Joint Commission shall convene at an appropriate place to discuss the changes and make the decisions on the necessary changes.

ARTICLE 7

If any government refuses the advance drawings (plans), the expenses related with the preparation of this drawing shall be shared equally between the parties. In this event this protocol shall become null and void. Except for the costs that went into the preparation of the advance drawing the parties shall not be entitled to make claims from each other as a result of this.

ARTICLE 8

The instructions for the joint operation of the Arpacay (Ahuryan) dam that will be approved with the advance drawing shall constitute the basis for the operation of this dam.

ARTICLE 9

The final drawing shall be prepared based on the advance drawing (planning) and in accordance with the schedule enclosed to this Protocol within four months as of the date of approval of the advance drawing (planning) and shall be approved by the authorized agencies of both parties within two months.

ARTICLE 10

The implementation drawings shall be prepared by the party performing the construction and the relevant costs shall be incorporated into the construction costs.

ARTICLE 11

The parties have agreed that the Soviet Union project representatives shall prepare the projects of the reservoir (Dam) on the Arpacay (Ahuryan) River.

The Turkish side retains the right to have six representatives at most, throughout the duration of the preparation of the project, by including the relevant expenses in the project costs.

During the preparation of the drawings, the Soviet side shall provide the necessary information and data to the Turkish representatives and provide them the necessary assistance.

Within one month following the ratification of this protocol, the Turkish side will also provide to the Soviet side all the information and data and the preparatory calculations on the reservoir. The Soviet side shall perform all the explorations and research and likewise prepare all the data for the project. The Turkish side shall assist the Soviet side in connection with these works.

ARTICLE 12

The parties have agree that the preparation of advance drawing shall cost TL 470,000 (47,000 Ruble) the breakdown of which is TL 125,000 (12,500 Ruble) for the preparation of the drawing, and the cost of drilling TL 380,000 or (Ruble 38,000). It was agreed during the preparation of the advance and final drawings that the initial estimates for the drilling would TL 250 or (Ruble 25) per meter and that there would be 1000m of drilling involved. If the amount of drilling to be performed changes, this would be deducted from or added to the preparation costs of the drawing and the cost of the project will become more realistic.

The expenses of having Turkish experts in the Soviet Union for the preparation of the drawings shall be shared equally between the parties.

ARTICLE 13

The following principles shall apply in the preparation of the advance and final drawings:

a. The parties retain the right to build and operate a hydroelectricity plant on the side of the reservoir that belongs to them and to make use of the water thereat. In order to facilitate the operation of the hydroelectricity plant it should be constructed on the territory of the respective party. Each party shall determine the capacity of the power to be obtained from the plant.
   The output graphic of the water to be released from the reservoir for the plant shall be provided to the project agency of the Soviet Union.
b. There shall be weirs at each side of the dam in order to obtain water for irrigation purposes. The controls for this weir shall be found in the respective sides of the dam and shall be operated individually.
c. The full weir shall not have a lid and therefore shall operate automatically.
d. Attention should be paid for the simple and safe operation of the dam during the preparation of the project.

ARTICLE 14

The following guidelines should be observed in the preparation of the instructions for the joint operation of the dam to be prepared together with the advance drawing and to be submitted for approval.

a. A permanent commission shall be established for the preparation of the annual operation schedule related with the operation of the dam and authorized to check the implementation of this schedule. This commission shall be constituted with equal participation of experts from both sides.
b. The maintenance, repair and renovation on the common parts of the dam and related facilities shall be performed with the decision of the above-mentioned commission and the related expenses shall be borne equally by the parties.
c. Each party shall operate the dam with its own materials, vehicles, equipment and personnel. Work on the additional facilities shall be performed individually.
   Water withdrawals from the reservoir for irrigation and energy production purposes shall be made in accordance with the annual operation schedule of the reservoir.
d. After the construction of the dam, the parties may draw water from the dam and from the Aras stream according to their shares either directly from the reservoir or from any border points of Arpacay (Ahuryan) and Aras Rivers.
   Both parties may withdraw water flowing in their territory from wherever they want, at any time and in any quantities they need extending up to the Iranian border and use it as they wish.
   Each party shall withdraw its own water according to the graphic approved by the commission. Changes in the graphic shall be made with the approval of the commission.
   Measuring devices shall be placed at the necessary places in order to determine that the total amount of water drawn from the reservoir by the parties is equal to each other at the end of the period of usage.
e. At the end of irrigation season every year, the operating commission working permanently, shall check whether the amount of water used by the parties is in conformity with the water usage schedule.
f. By taking into consideration the hydrological data, the water usage schedule shall be reviewed each year by the permanent commission and revised according to the requirements declared by the parties. This revised schedule shall contain the amount of water to be released from the reservoir, the places where water is to be taken from and the amount of water to be taken. The revision of water usage schedule may be performed monthly.
g. Water measuring weirs or facilities shall be established to measure the water taken from the dam lake either with pumps or through gravity and measuring shall be conducted at these points every month.
h. Either of the parties shall have the right to measure and control the amount of water taken by the other party when found necessary.

i. The location of the existing Guven (Talin) and Serdarabat (Oktonberian) regulators shall be modified to take water according to the drafted operating plan and water measuring facilities shall be established and new stations in additions to the existing flow observation stations shall be established.
j. Each party reserves the right to take the necessary measures to draw water at a time suitable for that party and to use half the water in the reservoir allotted to it without being subject to the other party,
The water drawing facilities of one side should not damage the water drawing facilities of the other party.

ARTICLE 15

During the preparation of the drawing both parties shall grant multiple entry rights to the personnel of the other party to cross the border and shall render the necessary assistance for the successful completion of the works. Therefore, the parties, through the border commissioners, shall provide information in timely manner on the personnel who need to cross the border, the type and the place of work to be performed by them, and the duration of the work each time.

ARTICLE 16

The disagreement mentioned in the Turkish-Soviet Joint Commission Protocol number 3 prepared in Erivan in 1962 (under the title of the sharing of the Arpacay (Ahuryan) and Aras waters) is now changed with the construction of the Arpacay (Ahuryan) dam and the waters to be shared is focused on the very small amount of water coming from the downstream of the dam and the Aras drainage basin.

However, the Turkish side's proposal to consider the natural flow within the instructions of the operation of the dam was not accepted by the Soviet side on ground that there was a dispute on this matter and that it should be left to the decision of the two governments to resolving the dispute. Hence, accordingly the parties agreed that the matter should be presented to the respective governments for their decision.

C- Usage of the reservoir by the parties:

ARTICLE 17

The parties have agreed to use the water to be regulated in the future reservoir on equal basis.

Likewise the parties agree that by the regulated water they mean the water drawn from the reservoir and all kind of water that flows through the dam.

ARTICLE 18

After the dam has been constructed the parties retain the right to use their share of the water from the reservoir and from the dam to the downstream through the boundary rivers of Arpacay (Ahuryan) and Aras up to the Iranian border, without violating the rights of the other party, at any point, at any time, in any quantities they desire.

D- Participation in the construction costs:

ARTICLE 19

The expenses mentioned in article 22 for the joint dikes, weirs, construction tunnels and other necessary auxiliary and secondary facilities and the losses mentioned in article 23 shall be all equally borne by the parties.

ARTICLE 20

The expenses for the hydroelectricity plant and the operation buildings, service ways, energy transport lines, telephone cables, and other facilities which are built with its own initiative and which are not stated in the approved project shall be borne by the party undertaking these constructions.

E- Compensation for inundated areas:

ARTICLE 21

The parties have agreed that the following values shall be considered in the determination of the damages for the compensation of areas which will be inundated.

These shall be; farmlands, settlements, highways, and railroads, and other facilities which will remain underwater in the regions.

The parties have taken into consideration that the damages to be incurred because the land will be inundated by the reservoir will be TL 60,000,000 million based on the declaration of the Turkish party and that the damages of the Soviet party will be 540,0000 Ruble based on the declaration of the Soviet party.

Because the losses of the Turkish party will be even greater owing to the expropriation of land, Turkey will contribute TL 3 million less to the construction of the dam and to the half the expenses of the joint facilities.

Both parties will conduct the evacuation of the land that will be inundated through their own construction organizations.

If the actual damages incurred because of the land inundated are found to be more, the parties shall not make claims to each other because of this.

F- Definition and evaluation of expenses and damages:

ARTICLE 22

With the word 'expense,' the values calculated according to the ratified projects and the unit prices used for the joint facilities mentioned in article 19 and the auxiliary facilities to be built during the construction of the joint facilities.

ARTICLE 23

The word 'damage' shall mean the following:

Remedying the damages incurred from floods, avalanches and wreckages, operation of material depots, preparation of the worksite area, building of storage areas for construction materials, service and operational roads.

ARTICLE 24

The unit prices will be used in the calculation of the advance and final projects as well as in the calculation of the construction works completed and the construction cost and damages.

The unit price used for final estimates shall be used in calculating the works performed by the parties.

ARTICLE 25

The parties have agreed that the financials issues such as the rate of exchange to be used in the calculation and how and over what period of time the payment will be made should be prepared by the authorized organs of the parties, and also added that the agreed version of the text upon the ratification by the respective governments shall be attached to this protocol.

G- Principles of performing the joint construction works:

ARTICLE 26

It has been agreed in principle that one of the parties shall construct the dam and the other shall check the construction.

ARTICLE 27

The parties have agreed that the governments will later decide as to which party will construct the dam according to the ratified final projects.

ARTICLE 28

The party that takes upon the construction work shall fulfil all the technical responsibilities. Any kind of changes to be made in the approved project may only be made with the concurrence of both parties.

ARTICLE 29

A joint permanent commission comprised of representatives of both parties in equal numbers shall be established to obtain concurrence on matters to arise during the construction works, and this commission shall convene upon the request of one of the parties. The decisions of the commission shall be ultimate.

ARTICLE 30

The construction site shall be fenced off with barbwires during the construction. People working on the construction and the control of the construction, together with the personnel, vehicles, construction machinery and equipment serving these people may move freely within the construction site. Entry into and exit from the fenced off area shall be made at certain points.

The personnel of both parties shall be able to conduct their business without any hindrance inside the fenced off area.

ARTICLE 31

The borders of the construction area shall be shown in the final project. Places of entry into and exit from the area, the procedures, protection, lighting and other issues shall be prepared and determined by the border commissioners. Entry into and exit from the construction site by certain personnel in connection with the construction works shall likewise be determined by the border commissioners.

The parties shall render all assistance to those who will be coming to inspect and check the construction site and the representatives of the other party.

ARTICLE 32

If an act in violation of the laws, regulations and general rules is committed in the temporarily fenced off area during the performance of the construction, the investigation and indictment shall be conducted by the laws of the party on whose territory the violation was committed.

ARTICLE 33

The parties undertake to fulfill all expropriation and evacuation works in the area which will be inundated before the dam begins to hold water and in this manner prepare the lake area ready to store water.

H- Preparing a topographic map:

ARTICLE 34

When the Turkish Government notified its approval in principle for the negotiation of the construction of a joint dam on Arpacay (Ahuryan), the Turkish Delegation had considered this joint dam project together with the development of the Cildir, Kars and Igdir plateaus including the smaller plateaus up to the border as a whole. The Turkish delegation believes that the projects to be prepared for the development of the land and water resources in these areas must be based on photogrametric maps in the scale of 1/25,000.

Therefore, Turkish delegation declares and proposes that with the initiation of the Arpacay (Ahuryan) Dam construction there is a need for the preparation of a photogrametric map in the scale of 1/25000 for the region covering from Lake Cildir up to the Iranian border of the Turkish-Soviet border and constituting a belt in the width of 30km from the Turkish side.

The Turkish Delegation declares and accepts that it will take all the necessary measures to eliminate all the concerns of the other side in the process of obtaining this map that is entirely of technical nature.

In order for this protocol to take effect, the Turkish delegation, based on the main reasons deliberated above, finds it necessary for appropriate circumstance to be created for the making of such a map, otherwise it does not see any urgency for the construction of Arpacay(Ahuryan) dam to begin immediately.

ARTICLE 35

The Soviet delegation is of the opinion that making of such a map covering the Turkish territory all along the Soviet-Turkish border is not relevant to the construction of the Arpacay (Ahuryan) reservoir. The Soviet delegation does not find it appropriate for the Turkish delegation to tie the effectiveness of this protocol with the taking of photograph of the border region from the air because this matter could only be resolved by the officials who will be ratifying this protocol.

I- Other Issues

ARTICLE 36

The existing State border line shall remain as it is, without any changes, after the construction of the dam and the making of the reservoir. The borderline shall be identified on the surface of the lake by means of buoys or the like items.

The parties have agreed to establish poles on each side before filling the reservoir and to tie buoys to the border poles.

ARTICLE 37

The parties reserve the right to fish or to hold benefits through other activities only at their side of the border of the dam lake. As regards to this matter, the parties have agreed that the authorized agencies of the parties shall prepare regulations in accordance with the Agreement on the Utilization of the Boundary Waters, signed on 8 January 1927.

ARTICLE 38

The parties have agreed to leave the issues discussed in the articles 16, 34 and 35 of this protocol to be resolved with the decisions of the governments in accordance with article 10 of the Turkish-Soviet Agreement on the Utilization of the Boundary Waters, dated 8 January 1927.

This protocol and its enclosures shall enter into effect after its ratification by the authorized agencies of the parties.

This protocol has been prepared in duplicate copies in Turkish and Russian in Ankara on 25 April 1963. Both texts are same, and shall be equally valid.

Turkish Delegation	USSR Delegation
Chairman: Neset Akmandor	Chairman: Bagramian G.A
Asst.Chairman: Arif Onat	Asst.Chairman: Melnikov V.M.
Member: Hidayet Turanli	Member: Voronine A.L.
Member: Saip Anadol	Member: Abramian A.A.
Member: Ismail Hakki Demirel	Member: Avakian K.A.
Member: Ekrem Arikan	Member: Grigorian S.O.

Supplement to the Protocol dated 25/4/1963 of the Turkish-Soviet Joint Commission

Program for research, study and project works of the Arpacay(Ahuryan) Dam and reservoir

A- PLANNING (Advance Project)

I. Review of the Existing Data:

1. Review of the meteorological, hydrological, topographic, geological and other existing documents.

II. Topographical studies:

1. Preparation of 1/5000 scale topographic and survey maps of the Arpacay (Ahuryan) Dam area (2-3km more to the downstream of the river and 1460 level as proposed by the Soviets).
2. Preparation of topographic maps in the scales of 1/5000 or 1/1000 of the dam sites.
3. Sufficient number of length and width sections in the river bed, weir and construction tunnel up to the downstream of the river at the dam site as proposed by the Soviets.

III. Hydrological and water requirement calculations:

1. Evaluation of the flow rates obtained from the Arpacay (Ahuryan) Kosavenk and Aras – Karakale flow observation stations.
2. Obtaining the repetition of flood over the last 5, 10, 25 and 50 years of the Arpacay (Ahuryan) River at the dam site and determination of a survey of flood hydrographies.

3. Calculation of the sedimentation amount for over a 50 years period (approximately) in the Arpacay (Ahuryan) river dam site.
4. Determination of the water level and drawing of a key curve corresponding to various consumption levels at a certain distance from the dam site downstream of the river for a number of sections.
5. Determination of the water requirements of both sides and preparation of graphics. (One party shall give its water requirement curve to the other party who is preparing the advance drawing. These curves shall be allocated to the parties according to the amount of water that could be regulated through the dam, by reserving half its rights.
6. Determining whether water quality is suitable for irrigation or not.

IV. Geological Engineering Works:

1. Preparation of geological maps of the reservoir area with a scale of 1/25000 and preparation of detailed geological maps of places found necessary and other geological engineering works with a scale of 1/5000.
2. Conducting of geological engineering studies at the sites of the dam and miscellaneous facilities.
3. Research of stone, sand-pebbles and territory for various types of dams, determination of sites, and determination of their quantities and physical characteristics, etc.

V. Designing of projects and preparation of reports:

1. Preparation of volume graphics with a scale of 1/5000 of the surface maps.
2. Calculation of the reservoir volume by taking into consideration the amount of water entering the dam on the one hand, and considering water leakage, evaporation losses and dead volume on the other.
3. Determination of the height of the dam by taking into consideration the load of the weir, waves and air factor.
4. By comparing various types of dams, selection of the most suitable dam in technical and economic terms.
5. Selection of the appropriate type and sizes by comparing various types of weirs.
6. Preparation of projects for the water conversion facilities necessary for the construction.
7. Preparation of work flow and organization charts.
8. Preparation of instructions for the joint operation of the dam.
9. Survey costs for the joint facilities to be built on both sides of the reservoir.
10. Preparation of the project drawings and report in five copies, in Turkish and Russian.

    The following issues should be contained in the report:

a) Introduction
b) Meteorological and hydrological data related to the dam site,
c) Geological engineering studies of the dam site and the lake area,
d) Facilities and their features,
e) Preparation of work flow and organizational drawings
f) Explanation of the alternatives related with the dam and miscellaneous facilities,
g) Preparation of the final project and results and recommendations on the construction

11. Preparation of the calculation and study files in 5 copies in Turkish and English.

B- Final Project

I. Topographic Studies:

1. Preparation of map of the construction site with a scale of 1/500.
2. Preparation of the width sector at the site selected for the construction of the dam.
3. Application works of the service ways, electricity and telephone cable routes necessary for the construction of the dam.

II. Geological Engineering studies:

1. Additional geological engineering works at the dam site and at various facilities.
2. Supplementary research on construction materials available for the type of dam selected.

III. Project Works:

1. Preparation of the projects for the selected type of dam and auxiliary facilities (To enable both sides to build a hydroelectricity plant individually).
2. Drawing of the construction sites, service roads and worksite facilities.
3. Preparation of the work flows and work organization charts.
4. Construction cost of all the facilities and annual expenses.
5. Preparation of the report on the final project.
6. Preparation in Turkish and Russian of the final project in five copies each (One copy of each picture shall be prepared an original copy).

REPUBLIC OF TURKEY
MINISTRY OF FOREIGN AFFAIRS
108.081-DI/3-36

MEMORANDUM

The Memoranda, dated 21 August, 14 December 1963 and 16 April 1964, submitted by the Embassy of the Union of Soviet Socialist Republics in Ankara to the Republic of Turkey, Ministry of Foreign Affairs, in connection with the Arpacay Dam have been carefully reviewed by the authorized Turkish authorities. The views and proposals of the concerned Turkish authorities in respect to the Soviet proposals addressed in the said memoranda are explained below:

1. The Turkish officials were very pleased to learn that the concerned Soviet officials have granted permission to take photographs in connection with the construction of Arpacay dam in Turkish territory from Lake Cildir up to the Iranian border along the Turkish –Soviet border within a belt of 30 km through photogrammetry at an altitude of 600 meters, via Turkish aircraft with Turkish and Soviet experts on board.
   As stated in the memorandum of the Embassy of the Union of the Soviet Socialist Republics, dated 16 April 1964, the Turkish and Soviet experts shall examine the aircraft and the cameras prior to each flight, will board the aircraft, they shall examine the films after they have been taken and shall immediately destroy any films that contain views of the Soviet territory.
2. The Turkish delegation which has expressed its opinion in article 4 of the Protocol number 3 prepared by the Turkish-Soviet Joint Commission in Erivan on 4 July 1962, on the provisions of article 1 of the "Agreement between the Republic of Turkey and the Union of Soviet Socialist Republics on the Utilization of the Rivers, Streams Constituting the Border" dated 8 January 1927 have concluded that since these two rivers, namely Aras and Arpacay should be considered as one river along the Turkish-Soviet border, the party which did not get is share of water from Arpacay should be able to get form any point on River Aras. The Turkish officials were very pleased to learn from the Soviet memorandum dated 21 August 1963 that the Soviet side could, in principle, accept the equal distribution of the water extending along the Turkish-Soviet border up to the Iranian border after the construction of the reservoir on Arpacay river. However, in order to prevent the arising of any disputes in the future, the Turkish officials request that the Soviet side officially accept and confirm the views expressed by the Turkish delegation in article 4 of the protocol number 3, dated 4 July 1962 on the sharing of the waters constituting the Turkish-Soviet border.
3. Article 32 of the Turkish-Soviet protocol, dated 25 April 1963 stipulates that the two sides reached a consensus on the matter concerning the punishment

of persons who commit crime, during the construction of the dam, jurisdiction of the country in whose side the crime is committed will be valid. Thus, the Turkish side does not see any need for a change to be made in connection with this matter.

The concerned Turkish officials have also stated that they are desirous to have the Turkish-Soviet Protocol, dated 25 April 1963 be ratified, as soon as possible, and have stated that the ratification by Turkey of the said Protocol will take place after the final concurrence between Turkey and the Soviet Union on the issues mentioned in articles 2 and 3 of the above protocol.

1 July 1964 Ankara, 6 May 1964

THE EMBASSY OF
THE UNION OF SOVIET SOCIALIST REPUBLICS
NO. 221/64

The Embassy of the Union of Soviet Socialist Republics expresses due respects to the Republic of Turkey, Ministry of Foreign Affairs and has the honour to explain the following points in response to the confirmation requested by the Ministry on the issues contained in the memoranda dated 16 April 1964 and 9 June 1964 in connection with the proposed construction of a reservoir on Ahuryan (Arpacay) river:

The Soviet side has carefully reviewed the proposals made by the Turkish side in the preparation of the protocol, dated 25 April 1963 and on which agreement was reached by the Soviet-Turkish Joint Commission which met in order to simplify the construction of a joint reservoir on River Ahuryan.

Based on the request made by the Turkish side, the Soviet authorities have agreed to have photographs taken from an altitude higher than 600m of the Turkish border line extending from Lake Cildir up to the Iranian border covering a belt in the width of 30 km from the border, via a Turkish aircraft jointly with Soviet and Turkish experts. However, the Soviet and Turkish experts shall examine the aircraft and the cameras prior to every flight, participate in the flights, and check the films after they are taken and shall immediately destroy any film that contains views from the Russian territory.

The Soviet authorities re-confirm their agreement on article 32 of the Protocol, dated 25 April 1963 which states that “If any violation of any law, regulation or rule should occur in the area fenced off during the construction period of the dam, the investigation and the prosecution of the perpetrator shall be done according to the laws of the party on whose side the violation was committed.”

The Soviet side also concurs with the Turkish proposal to decrease the volume of the Ahuryan reservoir from 440 million m3 to 280-300 million m3 as determined in the said protocol.

The Soviet side, which has taken into consideration Turkey’s desire to prevent any disputes on the equal distribution of the water, hereby officially gives its concurrence on the principle of using the water equally all along the Soviet-Turkish border up to the Iranian border after the construction of the reservoir on the River Ahuryan.

Concurrence is given on all the other articles of the protocol on which agreement has been reached by the Soviet-Turkish Joint Commission in connection with the joint construction of a dam on River Ahuryan (Arpacay).

Thus, the Soviet authorities having taken into consideration the benefits of developing good neighbourly relations and the cooperation between USSR and the Republic of Turkey have concurred with the Turkish side's requests. This action shall be ratified with the final approved version of the said Protocol and has lifted the obstacles for the initiation of the works on the joint construction of a reservoir on river Ahuryan.

On this occasion, the Embassy of the USSR confirms its deepest respect to the Ministry of Foreign Affairs of the Republic of Turkey.

Ankara, 22 June 1964

Translated from the original in Russian.

22 June 1964
Mehmet Kurbanoglu
Translator
Signature

The Ministry of Foreign Affairs of the Republic of Turkey presents its respects to the Embassy of the Union of Soviet Socialist Republics and has the honour to inform the Embassy that it has received the Note number 221/64, dated 22 June 1964:

The Soviet side has carefully reviewed the proposals made by the Turkish side in the preparation of the protocol, dated 25 April 1963 and on which agreement was reached by the Soviet-Turkish Joint Commission which met to negotiate and facilitate the construction of a joint reservoir on River Ahuryan.

Based on the request made by the Turkish side, the Soviet officials have agreed to have photographs taken from an altitude higher than 600 m of the Turkish border line extending from Lake Cildir up to the Iranian border covering a belt in the width of 30 km from the border, via a Turkish aircraft jointly with Soviet and Turkish experts. However, the Soviet and Turkish experts shall examine the aircraft and the cameras prior to every flight, participate in the flights, and check the films after they are taken and shall immediately destroy any film that contains views from the Russian territory.

The Soviet authorities re-confirm their agreement on article 32 of the Protocol, dated 25 April 1963 which states that "If any violation of any law, regulation or rule should occur in the area fenced off during the construction period of the dam, the investigation and the prosecution of the perpetrator shall be done according to the laws of the party on whose side the violation was committed.

The Soviet side also concurs with the Turkish proposal to decrease the volume of the Ahuryan reservoir from 440 million m3 to 280-300 million m3 as determined in the said protocol.

The Embassy of the Union of Soviet Socialist Republics
Ankara

"The Soviet side, which has taken into consideration Turkey's desire to prevent any disputes on the equal sharing of the water, hereby officially gives its concurrence on the principle of using the water equally all along the Soviet-Turkish border up to the Iranian border after the construction of the reservoir on the River Ahuryan."

"Concurrence is given on all the other articles of the protocol on which agreement has been reached by the Soviet-Turkish Joint Commission in connection with the joint construction of a dam on River Ahuryan (Arpacay)."

"Thus, the Soviet authorities having taken into consideration the benefits of developing good neighbourly relations and the cooperation between USSR and the Republic of Turkey have concurred with the Turkish side's requests. This action shall be ratified with the final approved version of the said Protocol and has lifted the obstacles for the initiation of the works on the joint construction of a reservoir on river Ahuryan."

"On this occasion the Embassy of the USSR confirms its deepest respects to the Ministry of Foreign Affairs of the Republic of Turkey."

This said Note conveys the concurrence of the Soviet officials on the issues of contention which were contained in articles 16,17, 18, 32, 34 and 35 of the Turkish-Soviet Protocol, dated 25 April 1963 in connection with the dam contemplated to be built on Arpacay river and which are listed below:

a) Equal sharing of all the waters of the rivers along the Turkish-Soviet border between the two countries,
b) The laws that shall be enforced in case of a violation of laws at the dam construction site,
c) Preparation of photogrametric maps of the Turkish territory from Lake Cildir up to the Iranian border within a 30km belt from the border,

And the Ministry of Foreign Affairs, with the memorandum, dated 6 May 1984, submitted to the Embassy of the USSR confirms that these are the view of the Turkish authorities.

The Turkish officials who have expressed their pleasure for the acceptance of all their proposals and views by the Soviet officials have initiated immediate action for the ratification of the said protocol instantly.

The Ministry of Foreign Affairs shall immediately inform the Embassy upon the completion of the approval of the said protocol.

On this occasion, the Ministry of Foreign Affairs of the Republic Turkey expresses its deepest respects to the Embassy of the USSR.

Ankara, 8 July 1964

Annex 9

Cooperation Agreement between the Government of the Republic of Turkey and the Government of the Union of the Soviet Socialist Republics on the Construction of a Dam on the Boundary River Arpacay (Ahuryan) and the Constitution of a Dam Lake [own translation]

The government of the Republic of Turkey and the government of the Union of Soviet Socialist Republics have agreed on the following issues by taking into consideration the good neighbourly relations existing between the countries and the provisions of the Agreement "On The Use of the Border Waters" signed in Kars on 8 January 1927, and based on the desire to further develop their economic and technical cooperation which will benefit the government of the Republic of Turkey and the government of the Union of Soviet Socialist Republics:

ARTICLE 1

The Government of the Republic of Turkey and the Government of the Union of Soviet Socialist Republics have decided to construct a dam on the Arpacay (Ahuryan) river which is at the common border of the two countries, and to constitute a dam lake with a volume of 525 million m3.

ARTICLE 2

For the realization of the cooperation envisioned in article 1 of this agreement, the Soviet organizations, shall prepare the drawings of the work in keeping with the technical drawing approved by the official agencies of both parties, to include the delivery of equipment and materials necessary for the construction, and shall construct a dam on the bordering Arpacay (Ahuryan) river and constitute a dam lake with a volume of 525 million m3.

ARTICLE 3

For the realization of the cooperation envisioned in article 1 of this agreement, the Turkish organizations, as half partners, shall fully participate in the technical control of the dam construction.

ARTICLE 4

The lump sum expenses of the dam according to the technical project prepared by the Soviet organization was found to be 16,6 million Ruble and this amount shall be equally financed by the parties, that is each party shall bear 50 percent of the expenses.

ARTICLE 5

Because the irrigation facilities on the Turkish side will not be completed on time, therefore, in order to preclude Turkey from incurring losses in the usage of half of the dam water in the initial stages, the Soviet side shall bear 4.4 million ruble of Turkey's share of the construction costs.

However, this implementation shall not prejudice the principles of providing half the expenses by Turkey and the Soviets of the dam construction and benefiting equally from the dam waters.

ARTICLE 6

With the balance remaining after the deduction of the 4.4 million Ruble from Turkey's 50 percent participation share in the construction of the dam, as stated in article 5 above, Turkey shall perform construction works on lump sum basis, preferably on the Turkish side, at the point to be mutually agreed.

The enumeration of this construction to be performed by the Turkey, preferably on the Turkish side, has been specified in Enclosure I which is an integral part of this agreement.

ARTICLE 7

All the necessary assistance, within the framework of the laws in effect, shall be rendered to the members of the Soviet agencies involved with the construction for their entry into and exit from the area determined and fenced of by Turkey and for them to lodge during the performance of the works.

Members of the Turkish agencies tasked with the construction and control of the dam shall be accorded the same kind of assistance from the Soviet side and shall be allowed to enter and exit.

Following the completion of the construction of the dam, the Turkish and Soviet personnel tasked with the operation of the facility shall be entitled to enter freely into the identified Soviet and the Turkish facilities.

All these issues shall be fulfilled in accordance with the guidelines outlined in Enclosure-II and Enclosure-III.

ARTICLE 8

The equipment and material to be shipped to the construction site in accordance with this agreement shall be mutually kept exempt from taxes, and customs duties.

The leftover equipment and material that will be taken into the Soviet and Turkish sides exempt from taxes and without any limitations because of the joint construction of the dam shall be removed in the same manner upon the completion of the construction as these material and equipment will no longer be required.

ARTICLE 9

The parties in an effort to provide services on the basis of equal shares, may jointly perform major and minor repair works on the dam and the dam lake through mutual consent.

ARTICLE 10

The joint operation of the dam on Arpacay (Ahuryan) river and the dam lake shall be performed in accordance with the Instructions given in Enclosure-III.

ARTICLE 11

For the cooperation related with the construction of a dam on the Arpacay (Ahuryan) boundary river and the constitution of the dam lake shall be prepared by the Turkish and Soviet agencies within 3 months pursuant to the signing of this Agreement and shall be signed after the Agreement enters into effect.

ARTICLE 12

No matter what the extent and the capacity of the dam lake which is created with the construction of the dam on Arpacay (Ahuryan) river will be, there shall be no change in the border existing between the two countries today.

When the dam lake is constituted, the existing border today shall be reflected on the surface of the water by means of buoys.

ARTICLE 13

This Agreement shall enter into effect on the date of exchange of the ratification document(s) after the completion of the approval process in accordance with the current laws in effect in both countries.

ARTICLE 14

This Agreement has been prepared in duplicate copies in Turkish, Russian and French.

In case of inconsistencies between the Turkish and Russian versions the French version shall prevail.

In witness hereof, the following fully authorized officials, whose names are written below, have signed this Agreement.

Signed in Ankara on 26 October 1973.

/S/ Oguz GOKMEN
Government of the
Republic of Turkey

/S/ V.A.SERGEYEV
Government of the
Union of Soviet Socialist Republics

ENCLOSURE-I

The enumeration of the construction works to be performed by the Turkish side in accordance with article 6 of this Agreement signed between the government of the Republic of Turkey and the government of the Union of Soviet Socialist Republics on 26 October 1973, for the joint construction of a dam on Arpacay (Ahuryan) River:

1. Make the dam foundation leak-proof
2. Upstream and downstream Cofferdams
3. Operation buildings
4. Conference buildings
5. Service road leading to the irrigation water output (left shore)
6. Right bank operating road
7. Development of the dam site

GUIDELINES PERTAINING TO THE PROVISION OF TEMPORARY ASSISTANCE IN CROSSING THE TURKISH-SOVIET BORDER AND FOR THE TURKISH AND SOVIET CITIZENS WORKING ON THE CONSTRUCTION AND CONTROL OF THE DAM ON ARPACAY (AHURYAN) DAM AND TEMPORARILY REMAINING AT THIS SITE.

The guidelines to which the pedestrians (or vehicles) will be subject to in temporarily crossing the Turkish-Soviet border at the Arpacay (Ahuryan) river:

1. In order to facilitate the crossing of the border for the Turkish and Soviet experts, workers, vehicles, equipment, technical equipment, materials, foodstuff, medicine and other cargos to be delivered to the construction site of water facilities in connection with the dam to be constructed on Arpacay dam and the related parts of this dam, a simplified process (for pedestrians or vehicles) for crossing the border is introduced with this agreement.

[. . .]

ENCLOSURE-II

PASSAGE NR: ————————————————————
For crossing the Turkish-Soviet Border by vehicle:

Driver: (Name, Surname) (Father's name)
Identification Card Nr:
Shall cross the border with the vehicle (Plate nr) ——————————
(Make) model (machine, engine number) and
follow ————————————————————road.
The crossing shall be valid until . . ./. . .. /197. . .

Individual in charge

Signature (Name, Surname)

Seal ——————— 197...

(Turkish text) (Russian Text)

Encl-3

LIST NR:.................................and the identification card nr belonging to the service letters sent from the Turkish Border to the Soviet border with the surnames, names and the names of fathers.

Sender:...................................
(Name of the organization)

Line Nr. Nr of the place where the package is sent to

Names of (Receivers)

-Individual in charge
(Signature) ————— (Name, Surname)

Seal....................... 197...

(Turkish text) (Russian (text)

ENCLSOURE III

INSTRUCTIONS ON THE JOINT OPERATION OF THE DAM AND RESERVOIR ON ARPACAY (AHURYAN) RIVER

After the construction of the dam on Arpacay (Ahuryan) river, the parties reserve the right to use the amount of water that is allotted to them, from the dam lake and the downstream of the dam up to the Iranian border via the border rivers of Arpacay (Ahuryan) and Aras, without violating the rights of the other party, from any point they want, at any time they want and in any quantities they want.

These instructions outline the rules of technical operation of all the hydrometric equipment used for observation and measuring and facilities of the dam and the reservoir.

A Permanent Working Commission, comprised of three representatives from each side, these being a head engineer, operation expert, and a hydraulic engineer, for the solution of matters related with the joint use of the water and the technical operation of the dam's facilities is established.

Both parties shall notify each other of the names, surnames and duties of their own representatives in the Permanent Working Commission through the border commissar.

The Permanent Commission shall operate within the framework of these regulations and act according to the principles of obtaining water not only from the regulated waters in the reservoir, but from the water flowing in Aras river, and from Arpacay (Ahuryan) corresponding to the half of the share of the party, and from any point from the Aras river on the border.

A sub-commission, with the participation of three representatives from the operation services of both parties, shall be established in order to execute the decisions of the Permanent Working Commission.

If a change needs to be made in composition of both commissions, the border commissars of the parties shall notify each other.

During the joint operation of the dam, all matters related with the maintenance of the state border regime and the place, time, and the manner of the meeting of Turkish and Soviet experts shall be arranged by the border commissars.

Both parties undertake to let the experts and members of the Permanent Working Commission and the sub-commission who carry the access documents showing the border crossing point, and the date (time), enter their soils in order to perform the works outlined in these regulations.

Duties of the Permanent Working Commission:

1. The primary duty of the commission is to prepare the annual operating schedule of the dam and to check the implementation of this schedule.
2. The commission shall decide on the following issues at the meetings held once a month:

   a) Matters related with the operation of the dam and the facilities,
   b) Use of the water by the parties according to the water usage schedule,
   c) Sanitation of the reservoir,
   d) Fish production in the dam lake.
   e) Disputes between the operating personnel of the two sides,

3. Upon the request of one of the parties, the Permanent Working commission shall make corrections on the water usage graphics and the maintenance and repair schedules of the facilities.
4. The Permanent Working Commission shall check once a month the conformity of the amount of water actually drawn by the parties with the water usage schedule in effect. A balancing shall be made in the waters from the Arpacay (Ahuryan) River and the reservoir waters according to Form 2 enclosed in the instructions.

   In order to do balancing between the water released from the dam and the water coming from the downstream of the reservoir and from other sources, new measuring devices equipped with automatic indicators shall be installed on the existing hydrometric equipment on the Arpacay and Aras rivers. The observation data evaluated by the operating services of both parties shall be submitted to the Commission in the form given in enclosure 3 of the Instructions.

5. Throughout the operation period, the dam and other facilities shall be systematically reviewed and any cracks, loose and deformed parts, and damages shall be entered in a report by the Permanent Working Commission.
   The Commission shall render decisions to correct the results of these deformations and take measures for them not to occur more frequently or in greater amounts, and shall check to see if these measures are being implemented.
6. The Permanent Commission retains the right to apply to the authorized agencies of the parties for the modification of any element of the facilities.
7. Commission's work will be performed on the basis of equality. Disputes that may arise during the operation of the Commission shall be referred to the authorized officials of the parties for solution.
8. The Commission shall hold meetings on rotational basis in Turkey and the Soviet Union.
9. Upon the request of either one of the parties, the meetings of the commission shall be held every month. The place and date of the meeting shall be determined through mutual agreement beforehand.
   One of the rooms shall be allocated for the meeting.

Maintenance and Repair:

11. The parties shall perform the maintenance of the reservoir and repair of the facilities on their part.
12. The dam shall systematically kept under surveillance by the operation services of both parties.
13. With the inspections conducted from time to time all the deformations and damages shall be noted.
    In order to determine the vertical and horizontal movements of the dam and other facilities shall be tied to the permanent ropers. The levels of the facilities and elements shall be checked by the operating personnel form time to time.
14. The expansion joints of the dam and the facilities shall constantly be checked and when necessary the joints shall be repaired to prevent any leaks.
15. Deformations and damages that occur for various reasons shall be eliminated as soon as possible.
16. The metal sections of the facilities (according to the rules of maintenance) shall be oiled, painted and shall be kept operational at all times.
17. Leaks from the dam and the reservoir shall be subject to special control.
18. During winter the mobility of lifting devices and other equipment shall be systematically checked and shall be protected from freezing.
19. The dam and its facilities shall be thoroughly inspected by a commission comprised of the responsible experts of both parties once every 3 to 5 years.
    The decisions taken by the said commission shall be implemented by the Permanent Working Commission.
20. The parties use their own personnel, materials and equipment during the operation of the facility.
    The parties shall mutually agree if additional operational facilities are to be built and each party shall build its own facility.

Water Distribution Arrangement:

21. The parties shall withdraw water from the reservoir according to the water usage schedule.
22. The parties may obtain their half share of water either from the regulated waters made in the reservoir or from the water flowing in the river Aras, at any point on the Arpacay (Ahuryan) river and Aras River that constitute the border.
23. Every year, at the end of the irrigation season, the Permanent Working Commission shall check to see if the amount of water used by the parties is in accordance with the water usage schedule.
24. In accordance with the water withdrawals graphic, the parties shall use their share of the water from the reservoir at any place and at any time they want. In order to determine the amount of water taken by both parties from the reservoir and from the rivers both parties shall install the necessary automatic recording apparatus.
25. The water usage programs prepared in advance shall be adjusted according to the actual weather conditions of the year and the changes in the hydrological regime of the rivers.
    The water distribution guidelines envisioned in the Protocol dated 25 April 1963 shall be observed in the adjustment of the water usage programs.
26. In the annual water usage program, the total volume of the water obtained by both parties should be equal. In the monthly meetings of the Permanent Working Commission, the amount of water they will take and the places where they will withdraw water are determined for the duration of the following month.
    Extraordinary meetings shall be held upon the request of one of the parties to solve the matters related with the water usage and to make changes in the monthly water usage graphics.
27. The parties shall be entitled to use their share (half each) of the water from the weirs at any place and at any time they want.
28. In case the parties construct a hydropower plant, water withdrawals for that purpose should not make any changes in the height of the dam and in the volume of the reservoir that are determined in the advance drawings. Additionally, this matter should not prejudice the other party's right to use its share of the water according to the water usage graphic as determined in the advance drawing.
29. The operation services of the parties have the mutual right to check each other's actual water drawing.
    Upon mutual consent, the parties shall conduct measuring to control the water and make corrections on the consumption curves, and the authority to control shall rest with all the water measuring facilities.
30. The party that has not used its share of water according to the water usage graphic by the end of the year shall not be entitled to claim the amount of water it has not used in the following year(s).

Operation Services:

31. Both parties shall perform operation services of the dam throughout the year.
32. Both parties shall have technicians and workers in sufficient numbers for the operation of the dam and the facilities.
33. Office buildings shall be constructed on both sides for the operation personnel to work and shall be maintained.
34. The operation service shall perform the following works:

    a) Maintenance, repair and renovation works of the dam and the facilities,
    b) The works envisioned in the water usage program
    c) Works related with establishment and maintenance of water measuring facilities,
    d) Calculation of the water inflowing and out flowing from the reservoir,
    e) Raising and lowering the lids to release water from the dam,
    f) Manipulation of the lifting device to eliminate to possibility of its freezing,
    g) To let the flood waters pass without causing any damage,
    h) Execute decisions of the Permanent Working Commission.

35. The chief engineer of the dam and the reservoir is a member of the Permanent Working Commission.
36. The operating service of each party shall have telephone connections to its own administrative centre and the dam facilities.

Miscellaneous Issues:

37. The parties shall withdraw water from the reservoir at the dam weirs envisioned by Turkey and the Soviet Union in accordance with the water usage program. In order to alleviate the load of the weirs in the transfer of large amounts of water, excess water shall be released from the weirs at the same time.
    The amount of released water shall not be considered in the water usage account.
38. In order to monitor the dynamics of the groundwater flows that may occur during the filtration of the reservoir, dam body and the nearby facilities, observation wells shall be opened during the construction. The cost of opening these wells shall be incorporated into the cost of the dam construction. The level of the groundwater shall be constantly monitored from these wells. This observation data should be exchanged between the parties every month.
39. The machinery and equipment shall be operated in accordance with the instructions given by the factory or specially written instruction.
40. In accordance with paragraph "I" of article 14 of the Protocol dated 25 April 1963, the regulators of Guven (Talin) and Serdarabat (Oktemberyan) may be modified according to the operating plan prepared in order to allow the intake of necessary water. These shall be equipped with water measuring devices.
    The modification shall be accomplished according to the project to be mutually agreed.
    The party that cannot use its share of water for any reason whatsoever, may not object to the other party's utilization of its share of water.

41. The dam and its facilities and their environs shall be lit up at night.
42. Each side will keep construction materials, equipment, etc. for possible emergency failures that may occur unexpectedly.
    The list of materials and equipment and their quantities shall be maintained by the Permanent Working Commission.
43. The Arpacay (Ahuryan) reservoir may be used for fishery. The conditions for fish production shall be determined between the parties with a special agreement.
44. Forms 1, 2 and 3 have been attached to these instructions to be used as guides during the operation of the dam. The Permanent Working Commission has the right to make additions and changes on these forms.

Annex 10

Agreement on the Cooperation for the Construction of Hydrotechnical Facilities for the Prevention or Correction of the Riverbeds of Arpacay (Ahuryan) Coruh River, Posof and Caksu Streams extending between the border stone number 41 through border stone number 450 on the Turkish Soviet Union border signed between the Government of the Republic of Turkey and the Government of the Union of Soviet Socialist Republics. [own translation]

Acting on the good neighbourly relations existing between the Government of the Republic of Turkey and the Government of the Union of Soviet Socialist Republics,

In accordance with the "Protocol On the Control, Maintenance, Repair and Improvement of the Border Markings between the Republic of Turkey and the Union of Soviet Socialist Republics and the Maintenance of the Border Forest Cutting Belts" dated 29 December 1973 and "the Final Protocol of the Joint Control Commission on the Results of the Joint Control conducted during 1984-1988 on the passing of the rivers, and streams between the border of the Republic of Turkey and the Union of Soviet Socialist Republics, dated 6 December 1989,

The parties have agreed on the following issues for the prevention or correction in the changes in the beds of Arpacay (Ahuryan) river, Posof stream, Coruh river and Caksu stream extending between the border stone number 41 and border stone number 451 of the border region.

ARTICLE 1

The contracting parties shall enter into cooperation in the fulfillment of the measures envisioned in the Final Protocol for the correction of the existing changes and for the prevention of possible changes in the beds of the:

a) Arpacay (Ahuryan) river around the border stone number 50/2
b) Caksu stream between the border stone number 229 and 230,
c) Posof stream between the border stone number 230 and 232,
d) Coruh river between the border stone number 418 and 423.

ARTICLE 2

The cooperation envisioned in article 1 of this agreement shall be executed in the following stages:

i. Preparation of maps (plans)
ii. Research should be conducted based on the preliminary and final projects should be prepared thereafter,
iii. Construction of Hydrotechnical Facilities

Each party shall conduct necessary activities individually. The details of the works shall be specified by a Work Protocol to be jointly prepared by the concerned agencies of the parties within the framework of this Agreement.

ARTICLE 3

The cooperation envisioned in article 1 of this Agreement shall be executed by the General Directorate of the State Hydraulic Works of the Government of the Republic of Turkey and the Ministry of Water Construction Works of the Government of the Union of Soviet Socialist Republics.

ARTICLE 4

In order for the persons, vehicles, cargo, equipment and other materials to cross the border for the performance of the activities envisioned in article 2 of this agreement, the crossing of the border shall be at the crossing points determined in accordance with "the Guidelines for the Simplified Crossing Arrangements and Temporary Stay in the Territories of the Other Party, between the Turkish- Soviet State Border of the Persons, Vehicles, Cargo, Equipment and Other Materials for the Works related with the Hydrotechnical Facilities related with prevention or correction of the changes in the beds of Arpacay (Ahuryan) River, Posof Stream, Coruh River, and Caksu Stream extending between the border stone number 41 through the border stone number 450 on the Turkish-Soviet border" and which is an integral part of this agreement and given in the enclosure.

ARTICLE 5

Each party shall construct the units of the facilities on their territories, including the provision of the necessary equipment, materials and labor, following the approval of the Final Project, through its own means. The necessary construction works shall be performed in accordance with the joint work schedule to be specified in the Work Protocol and in parallel with both sides and in a manner not to harm the river banks of the other party. The parties shall reserve the right to supervise the performance of these works.

ARTICLE 6

The equipment and materials to be sent by each party for the works to be performed on their respective soils in accordance with this Agreement and which need to cross the border in order to be transported according to the local requirements shall be exempt from all taxes and duties. The leftovers of the said equipment and materials shall be taken back in the same manner after the completion of the construction works.

ARTICLE 7

The parties shall operate and maintain the facilities to be constructed on their soils through their own means.

ARTICLE 8

The parties shall jointly check that the beds of the rivers and streams to be recreated during the performance of the hydrotechnical facilities are in conformity with the conditions stipulated in the border marking documents of 1973.

After the completion of the construction the parties shall prepare and approve the joint operation guidelines of the hydrotechnical facilities.

ARTICLE 9

This Agreement shall be subject to ratification in accordance with the laws of both parties and shall enter into effect as of the date of the Note notifying that the internal formalities have been completed.

This Agreement has been signed in Ankara on 7 March 1990 in duplicate copies in Turkish and Russian and both texts are equally valid.

YURI A.SOLMOV	ERKAN GEZER
Government of the Union of	Government of the
Soviet Socialist Republics	Republic of Turkey

Annex 11

The Treaty on Friendship and Good Neighbourly Relations between Turkey and Iraq [own translation]

His Excellency Ismet Inonu, the President of Turkey, and His Excellency the Prince Regent Abdullah of the Kingdom of Iraq have acted together in order to improve the friendship and the very good neighbour relations existing between Turkey and Iraq and to fortify the coalescence and the brotherly relations between the two nations over centuries.

Both countries have accepted the ideals of peace and security as the unchanging foundation of their foreign policy in the search of peace and security in the world and especially in the Middle Eastern countries, and are pleased to see that the provisions of the United Nations Charter, signed in San Francisco, with the intent developing international solidarity, solidifies their own purpose and will accelerate them in pursuing their goals,

With the consideration that mutual understanding and assistance in the area of economy will facilitate closer relations in its true sense,

With the rightful bliss of taking the first step in the implementation of the principles upheld in the United Nations Charter and with the intent of remaining loyal to the commitments arising from the Charter,

Have come to the conclusion that an agreement has to be reached in order to accomplish all of the thoughts above and have, therefore, appointed the individuals whose names are written below for this purpose:

His Excellency Ismet Inonu, the President of Turkey:
His Excellency Hasan Saka,
Deputy of Trabzon, Minister of Foreign Affairs
His Excellency Feridun Cemal Erkin
Ambassador, Under Secretary, Ministry of Foreign Affairs

His Excellency Regent Prince Abdullah of the Kingdom of Iraq:

His Excellency Al Farik Nuri Essaid
Speaker of the Senate, holder of the 1st Degree Rafidain Order

His Excellency Abdullah Hafidh
Member of the Saylavlar Assembly

After these representatives have presented their certificates of authority in the prescribed manner, they have decided on the following provisions:

ARTICLE 1

The parties to the Treaty undertake to comply with the territorial integrity of each other's countries and with the border delimited and demarcated between the two countries with the Treaty dated 1926.

ARTICLE 2

The parties to the Treaty undertake to definitely avoid intervening with each other's internal affairs.

ARTICLE 3

The parties to the Treaty undertake to cooperate on international matters of mutual interest in general, and to consult each other on regional matters and to provide full support and collaboration to each other in pursuing policies related to these matters within the framework of the United Nations Charter.

ARTICLE 4

The parties to the Agreement undertake to immediately notify the authorized organs of the United Nations Organization in case of a threat of attack on the territorial integrity of the respective countries or to the inviolability of borders or in case of an attack to either one of the parties.

ARTICLE 5

The parties to the Agreement undertake to resolve all disputes that may arise between the parties in peaceful manner in accordance with the provisions of article 33 of the United Nations Charter and if they cannot resolve the dispute in this manner they agree to take the matter to the Security Council in accordance with article 37 of the said Charter.

Likewise, the parties to the Treaty undertake to do best of their efforts to resolve the disputes that may arise between one of the parties to the Treaty and a third neighbouring country or between two neighbouring countries in accordance with the same provisions.

ARTICLE 6

The parties to the Treaty, in keeping with the spirit of the Treaty, have accepted the protocols listed below as integral parts of this agreement in order to promote their cooperation, and develop their relations in every field:

1. Supplementary Protocol Nr. 1: Protocol to regulate the waters of the Tigris and the Euphrates and their tributaries.
2. Supplementary Protocol Nr. 2: Protocol on Bilateral Cooperation on Security Matters
3. Supplementary Protocol Nr. 3: Protocol on Cooperation in Education, Training and Culture.
4. Supplementary Protocol Nr. 4: Protocol on Post, Telegraph and Telephone.
5. Supplementary Protocol Nr. 5: Protocol on Economic Affairs.
6. Supplementary Protocol Nr. 6: Protocol on the Border Affairs.

Parties to the Treaty have also signed the below agreements:
Agreement on the extradition of criminals.
Agreement on Legal Assistance Pertaining to Law, Punishment and Trade.

ARTICLE 7

This Treaty has been concluded for an indefinite period of time however; this Agreement may be reviewed upon the request of either of the parties at the end of every five years.

This Treaty shall be ratified by the parties and the certificates of ratification shall be exchanged at Baghdad as soon as possible.

In case of disputes on this Treaty concluded in Ankara in Arabic, Turkish and French, on twenty-nine March of the year one thousand and forty-six, the French version of the Agreement shall prevail.

Hasan Saka	*Nuri Essaid*
Feridun Cemal Erkin	*Abdulilah*

Supplementary Protocol Number 1

Protocol on Flow Regulation of the Tigris and Euphrates Rivers and of Their Tributaries

Turkey and Iraq

In order to ensure the maintenance of a regular water supply, to regulate the water flow and to avoid the danger of floods during the annual periods of high water, have found it necessary to construct dams and permanent observation stations in Turkish territory,

They believe that after studies the most appropriate places for the dikes and other constructions all the expenses of which will be met by Iraq, will be found on the Turkish territory,

They have agreed to establish permanent observation stations on Turkish territory to record the flow of the said rivers and also agreed that the records to be obtained in this manner need to be delivered to Iraq regularly,

Since they have agreed, that the water control facilities to be built on these rivers should in principle also be used for irrigation and hydropower production serving the interest of both countries,

They have agreed on the following issues:

ARTICLE 1

Iraq may send to Turkey groups of technical experts to make investigations and surveys, collect hydraulic and geological information needed for the selection sites for the construction of the dams and observation stations to be constructed on the Tigris, Euphrates and their tributaries.

The maps to be drawn from the result of the land measurements shall be prepared by the authorized Turkish agencies.

All the expenses for the works mentioned in this article shall be met by Iraq.

ARTICLE 2

The technical experts from Iraq shall collaborate with Turkish technical experts; Turkey shall authorise them to proceed to the places to be visited and shall provide them with the information, assistance and facilities necessary for the accomplishment of their task.

ARTICLE 3

Turkey shall install and operate permanent flow measurement facilities, and transmit periodically the readings and the recorded data to Iraq. Iraq and Turkey shall meet the operating expenses of these stations, on equal basis, as of the date of entry into effect of this protocol.

Turkish and Iraqi technicians shall inspect the permanent observation stations at specific intervals.

During times of overflow, the river level observations made every day at 8 o'clock in the morning shall be wired to the officials to be designated by Iraq via the communication stations in Diyarbakir, Cizre for the Tigris, and via the communication stations in Keban, etc. for the Euphrates.

The results of the level observations during normal times when there is no overflow shall be sent to the same agencies as bi-monthly bulletins.

Iraq shall pay for the notification expenses mentioned above.

ARTICLE 4

Turkey in principle accepts to construct flow regulation works needed in the interest of Iraq in Turkish territory.

Each of the facilities except the permanent observation stations shall be a subject of separate agreements with respect to their location, expense, operation and maintenance and in meeting Turkey's irrigation and electricity requirements.

ARTICLE 5

Turkey shall inform Iraq of projects for waterworks on any of the Protocol watercourses, and shall consult with Iraq with a view to accommodating the interests of both countries.

ARTICLE 6

Pursuant to the signing of this Protocol, the parties to the treaty shall appoint a representative respectively, as soon as possible. The representatives shall discuss all the matters related with the fulfilment of the provisions of this protocol and shall act as go-between for keeping communication between parties.

Hasan Saka	*Nuri Essaid*
Feridun Cemal Erkin	*Abdulilah*

[...]

Supplementary Protocol Number 5

Protocol on Economic Affairs

The parties to the agreement have agreed to the following issues for the two countries to realize the conditions to work more closely and in cooperation in the field of economy.

Section I

Joint Economic Commission

ARTICLE 1

A Joint Economic Commission shall be established and this commission shall hold meetings at specific intervals to discuss how the provisions contained in sections 1, 2, 3, 4 and 5 of this protocol and the provisions contained in the Supplementary Protocol numbers 1 and 4 shall be implemented and shall jointly prepare drafts as to what will be necessary to clearly define these provisions, and submit these drafts for the review and approval of both Governments.

ARTICLE 2

The committees in the Joint Economic Commission shall be the following:

1. Trade, Customs and Finance Committee
2. Tourism Committee
3. Transportation and Ports Committee
4. Committee for regulating the waters of the Tigris and Euphrates rivers.

[. . .]

Annex 12

Protocol on Matters Pertaining to Economic Cooperation Between the Republic of Turkey and the Syrian Arab Republic [original extract]

The Republic of Turkey and the Syrian Arab Republic,
Recalling historic and cultural ties existing between the two countries,
Desirous to add new dimensions to already existing good neighbourly relations,
Bearing in mind numerous complementarities in the economies of their countries,
agreed to sign the present Protocol.

[. . .]

WATER

6. During the filling up period of the Atatürk Dam reservoir and until the final allocation of the waters of the Euphrates among the three riparian countries, the Turkish Side undertakes to release a yearly average of more than 500 m3/sec. at the Turkish-Syrian border and in cases where the monthly flow falls below the level of 500 m3/sec., the Turkish Side agrees to make up the difference during the following month.
7. The two Sides shall work together with the Iraqi Side to allocate the waters of the rivers Euphrates and Tigris in the shortest possible time.
8. The two Sides agreed to expedite the work of the Joint Technical Committee on Regional Waters.
9. The two Sides agreed in principle to construct and operate jointly projects in the lands of both countries on the Euphrates and Tigris rivers for irrigation and power generation provided that the technical and economic feasibility studies of these projects are carried out in cooperation by the experts of the two countries.
10. The Turkish Side explained the details of the "Peace Pipe Line" planned to carry a portion of the waters of the Seyhan and Ceyhan rivers in Turkey through Syria by two pipe-lines, one going to countries of the Gulf and the other to the Hashemite Kingdom of Jordan and Kingdom of Saudi Arabia to supply water for household purposes and limited irrigation in the region.<v>The Syrian Side agreed in principle to the project and showed interest provided that the Turkish Side carries out its technical and economic feasibility study by an international consultancy firm. The Syrian Side undertakes to facilitate the feasibility studies pertaining to the Syrian portion of the project. In case of its

positive conclusion, the Syrian Side will enter into negotiations for the final realization of the project.

[...]

Done and signed in in Damascus on July 17, 1987, in two original copies in English language.

Turgut Özal	Dr. Abdul Raouf El-Kassem
Turkish Prime Minister	Syrian Prime Minister

Index

DOI 10.1007/978-3-642-19636-2, © Springer-Verlag Berlin Heidelberg 2011

FSC
www.fsc.org
MIX
Papier aus verantwortungsvollen Quellen
Paper from responsible sources
FSC® C105338

If you have any concerns about our products,
you can contact us on
ProductSafety@springernature.com

In case Publisher is established outside the EU,
the EU authorized representative is:
Springer Nature Customer Service Center GmbH
Europaplatz 3, 69115 Heidelberg, Germany

Printed by Libri Plureos GmbH
in Hamburg, Germany